Second Edition

Problems and Solutions on SOLID STATE PHYSICS, RELATIVITY AND MISCELLANEOUS TOPICS

Major American Universities Ph.D. Qualifying Questions and Solutions - Physics

ISSN: 1793-1487

Published

Problems and Solutions on Solid State Physics, Relativity and Miscellaneous Topics
Second Edition
edited by Swee Cheng Lim, Choy Heng Lai and Leong Chuan Kwek (NUS, Singapore)

Problems and Solutions on Quantum Mechanics
Second Edition
edited by Swee Cheng Lim, Choy Heng Lai and Leong Chuan Kwek (NUS, Singapore)

Problems and Solutions on Thermodynamics and Statistical Mechanics
Second Edition
edited by Swee Cheng Lim, Choy Heng Lai and Leong Chuan Kwek (NUS, Singapore)

Problems and Solutions on Mechanics (Second Edition)
edited by Swee Cheng Lim, Choy Heng Lai and Leong Chuan Kwek (NUS, Singapore)

Problems and Solutions on Thermodynamics and Statistical Mechanics
edited by Yung-Kuo Lim (NUS, Singapore)

Problems and Solutions on Optics
edited by Yung-Kuo Lim (NUS, Singapore)

Problems and Solutions on Electromagnetism
edited by Yung-Kuo Lim (NUS, Singapore)

Problems and Solutions on Mechanics
edited by Yung-Kuo Lim (NUS, Singapore)

Problems and Solutions on Solid State Physics, Relativity and Miscellaneous Topics
edited by Yung-Kuo Lim (NUS, Singapore)

More information on this series can also be found at https://www.worldscientific.com/series/mauqqsp

Major American Universities Ph.D. Qualifying Questions and Solutions - Physics

Second Edition

Problems and Solutions on SOLID STATE PHYSICS, RELATIVITY AND MISCELLANEOUS TOPICS

Editors

Swee Cheng Lim

Choy Heng Lai

National University of Singapore, Singapore

Leong Chuan Kwek

Nanyang Technological University, Singapore &

National University of Singapore, Singapore

World Scientific

NEW JERSEY • LONDON • SINGAPORE • BEIJING • SHANGHAI • HONG KONG • TAIPEI • CHENNAI • TOKYO

Published by

World Scientific Publishing Co. Pte. Ltd.
5 Toh Tuck Link, Singapore 596224
USA office: 27 Warren Street, Suite 401-402, Hackensack, NJ 07601
UK office: 57 Shelton Street, Covent Garden, London WC2H 9HE

Library of Congress Control Number: 2024932139

British Library Cataloguing-in-Publication Data
A catalogue record for this book is available from the British Library.

Additional problems also created by Dr. R. Nimma Elizabeth, G. Priscilla and Dr. S. Sinthika, Lady Doak College, India.

Major American Universities Ph.D. Qualifying Questions and Solutions - Physics
PROBLEMS AND SOLUTIONS ON SOLID STATE PHYSICS, RELATIVITY AND MISCELLANEOUS TOPICS
Second Edition

ISBN 978-981-12-9149-4 (hardcover)
ISBN 978-981-12-9163-0 (paperback)
ISBN 978-981-12-9150-0 (ebook for institutions)
ISBN 978-981-12-9151-7 (ebook for individuals)

For any available supplementary material, please visit
https://www.worldscientific.com/worldscibooks/10.1142/13794#t=suppl

Desk Editor: Joseph Ang

Typeset by Diacritech Technologies Pvt. Ltd.
Chennai - 600106, India

PREFACE TO THE FIRST EDITION

This series of physics problems and solutions, which consists of seven volumes — Mechanics, Electromagnetism, Optics, Atomic, Nuclear and Particle Physics, Thermodynamics and Statistical Physics, Quantum Mechanics, Solid State Physics and Relativity, contains a selection of **2550** problems from the graduate school entrance and qualifying examination papers of seven US. universities — California University Berkeley Campus, Columbia University, Chicago University, Massachusetts Institute of Technology, New York State University Buffalo Campus, Princeton University, Wisconsin University — as well as the CUSPEA and C. C. Ting's papers for selection of Chinese students for further studies in U.S.A. and their solutions which represent the effort of more than 70 Chinese physicists plus some **20** more who checked the solutions.

The series is remarkable for its comprehensive coverage. In each area the problems span a wide spectrum of topics while many problems overlap several areas. The problems themselves are remarkable for their versatility in applying the physical laws and principles, their uptodate realistic situations, and their scanty demand on mathematical skills. Many of the problems involve order of magnitude calculations which one often requires in an experimental situation for estimating a quantity from a simple model. In short, the exercises blend together the objectives of enhancement of one's understanding of the physical principles and ability of practical application.

The solutions as presented generally just provide a guidance to solving the problems, rather than step by step manipulation, and leave much to the students to work out for themselves, of whom much is demanded of the basic knowledge in physics. Thus the series would provide an invaluable complement to the textbooks.

The present volume, under the title "Solid State Physics, Relativity and Miscellaneous Topics" consists of 165 problems. Both special and general relativity problems are included, although some other special relativity problems have been solved elsewhere, particularly in the volumes on Mechanics and on Electromagnetism. Problems that are not appropriate for the other six volumes, such as mathematical techniques, are here grouped in the section Miscellaneous Topics.

In editing, no attempt has been made to unify the physical terms, units and symbols. Rather, they are left to the setters' and solvers' own preference so as to reflect the realistic situation of the usage today. Great pains has been taken to trace the logical steps from the first principles to the final solution, frequently even to the extent of rewriting the entire solution. In addition, a subject index to problems has been included to facilitate the location of topics. These editorial efforts hopefully will enhance the value of the volume to the students and teachers alike.

Yung-Kuo Lim
Editor

PREFACE TO THE SECOND EDITION

This is the second edition of a former popular series on Problems and Solutions in various topics of physics ranging from Mechanics, Electromagnetism, Optics, Atomic, Nuclear and Particle Physics, Thermodynamics and Statistical Mechanics, and Quantum Mechanics. However, we have not addressed several interesting topics at undergraduate and graduate level. In this volume, we attempt to collect problems in Solid State Physics, Relativity and various miscellaneous interesting topics.

As usual, the volume is divided into several subtopics, and there are new interesting problems and solutions. Altogether, we have compiled several thousand problems and solutions, spanning an entire undergraduate physics course. These questions are carefully chosen at the level of the PhD Qualifying Examinations at American universities. Moreover, these questions can also serve as an excellent resource for Physics Competitions, like the International Physics Olympiad or the regional Physics Olympiads.

We believe that a good grounding in problem-solving is essential for the study of physics, and we believe that this compendium serves as an invaluable resource for the preparation of graduate study in physics, or competitions. We suggest that the student should first attempt the problems themselves before consulting the solutions. We have tried to elucidate solutions with appropriate figures and diagrams.

Finally, we hope that the reader (student or teacher) will continue to enjoy this compendium.

11 January 2024
Lim Swee Cheng, Lai Choy Heng, Kwek Leong Chuan
Editors

CONTENTS

Part I

Solid State Physics

1.1

CRYSTAL STRUCTURES AND PROPERTIES (1001–1027)

1001

Figure 1.1 shows a hypothetical two-dimensional crystal consisting of atoms arranged on a square grid.

- **a.** Show an example of a primitive unit cell.
- **b.** Define "the reciprocal lattice" and explain its relation to Bragg reflection.
- **c.** Show the reciprocal lattice and the first Brillouin zone. How is this zone related to Bragg reflection?
- **d.** State and explain the theorem due to Bloch that says an electron moving in the potential of this lattice has traveling-wave functions. What boundary conditions must be used with this theorem?

(*SUNY, Buffalo*)

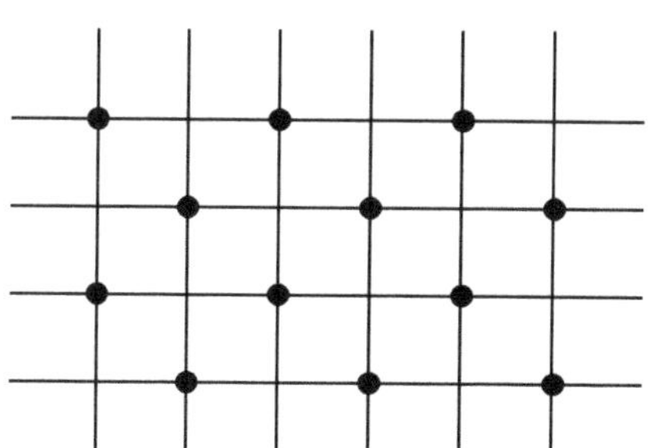

Fig. 1.1

Sol:

a. A primitive unit cell is a unit cell that contains lattice points at corners only, such as shown in Fig. 1.2. The basis vectors of the unit cell are

$$\mathbf{a}_1 = a(\mathbf{i} - \mathbf{j}),$$
$$\mathbf{a}_2 = a(\mathbf{i} + \mathbf{j}),$$

where a is the edge of the square lattice.

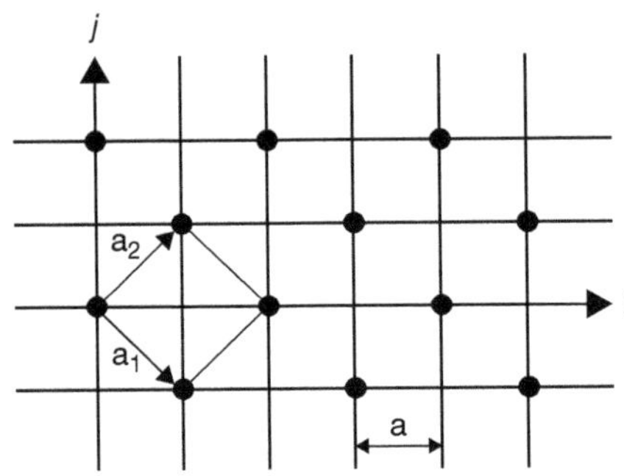

Fig. 1.2

b. If $\mathbf{a}_i$ $(i = 1, 2)$ are the basis vectors of the direct lattice, vectors $\mathbf{b}_j$ $(j = 1, 2)$ satisfying the relation

$$\mathbf{a}_i \cdot \mathbf{b}_j = 2\pi\,\delta_{ij} = \begin{cases} 2\pi, & i = j, \\ 0, & i \neq j \end{cases}$$

are the basis vectors of the reciprocal lattice. In the reciprocal space the condition for Bragg reflection is that the difference between the reflected wave vector $\mathbf{k}$ and the incident wave vector $\mathbf{k}_0$ is an integer multiple n of a reciprocal lattice vector $\mathbf{k}^*$

$$\mathbf{k} - \mathbf{k}_0 = n\mathbf{k}^*.$$

c. From the direct basis vectors

$$\mathbf{a}_1 = a(\mathbf{i} - \mathbf{j}),$$
$$\mathbf{a}_2 = a(\mathbf{i} + \mathbf{j})$$

the reciprocal basis vectors are obtained as

$$\mathbf{b}_1 = \frac{\pi}{a}(\mathbf{i} - \mathbf{j}),$$
$$\mathbf{b}_2 = \frac{\pi}{a}(\mathbf{i} + \mathbf{j}).$$

The reciprocal lattice and the first Brillouin zone are shown in Fig. 1.3. Bragg reflection takes place at the boundaries of the Brillouin zone.

d. The wave representing an electron moving in the periodic potential field $V(\mathbf{r}+\mathbf{R}) = V(\mathbf{r})$, $\mathbf{R}$ being a lattice vector, of the lattice has the form of a Bloch function

$$\psi_{\mathbf{k}}(\mathbf{r}) = e^{i\mathbf{k}\cdot\mathbf{r}} u_{\mathbf{k}}(\mathbf{r}),$$

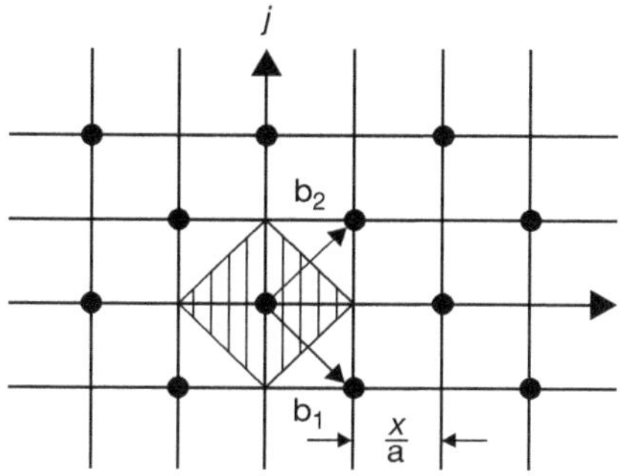

Fig. 1.3

where the function $u_{\mathbf{k}}(\mathbf{r})$ has the same translational symmetry as the lattice:

$$u_{\mathbf{k}}(\mathbf{r}) = u_{\mathbf{k}}(\mathbf{r}+\mathbf{R}).$$

It is a plane wave modulated by the periodic potential field. This is Bloch's theorem. The exponential part of the Bloch wave is a plane wave which describes the global behavior of electrons in a crystal lattice, while the periodic function describes the local motion of those electrons around the nuclei. Therefore, Bloch waves characterize the motion of the electrons in a crystal.

The Born–von Kármán periodic boundary condition must be employed with Bloch's theorem.

1002

A 100 g of lithium crystallizes in FCC structure at 295 K and 8 GPa, which undergoes phase transitions to BCC crystal structure upon cooling to room temperature and removal of pressure.

a. If its density at room temperature is $\rho_{\mathrm{BCC}} = 0.53\ \mathrm{g/cm^3}$, what is its density at 295 K? Based on your answer, answer whether lithium expands or contracts after the transition.

b. Determine the highest linear density of lithium at room temperature.

c. Calculate the number of unit cells and lithium atoms before and after transition, and justify your answer.

Sol:

a. The density of Li is given by the formula $\rho = \dfrac{nM}{N_A a^3}$

where,

$$n\text{—number of atoms per unit cell}$$
$$M\text{—Atomic mass of Li} = 6.941\text{ g/mol}$$
$$N_A\text{—Avogadro's number} = 6.022 \times 10^{23}\text{mol}^{-1}$$
$$a\text{—edge length of the unit cell}$$

At 295 K, Li crystallizes in FCC structure, so the number of atoms per unit cell is 4 and the edge length of the unit cell can be determined based on the assumption that the phase transition does not alter the radius of the atom.

At room temperature, Li transitions to BCC structure so the number of atoms per unit cell is 2.

$$0.53\text{ g/cm}^3 = \frac{2 \times 6.941\text{ g/mol}}{6.022 \times 10^{23}\text{ mol}^{-1} \times a^3}$$

$$a^3 = \frac{2 \times 6.941\text{ g/mol}}{6.022 \times 10^{23}\text{ mol}^{-1} \times 0.53\text{ g/cm}^3}$$

$$a^3 = 4.349\text{ cm}^3 \to a = 3.51 \times 10^{-8}\text{ cm}$$

For a BCC structure, the edge length "a" and radius "r" are related through,

$$r = \frac{\sqrt{3}\,a}{4} = \frac{\sqrt{3}(3.51 \times 10^{-8}\text{ cm})}{4} = 1.52 \times 10^{-8}\text{ cm}$$

For an FCC structure, the edge length "a" and radius "r" are related through $r = \dfrac{a}{\sqrt{8}}$ so,

$$a = \sqrt{8}\,(1.52 \times 10^{-8}\text{ cm}) = 4.3 \times 10^{-8}\text{ cm}$$

Now, the density of Li at 890 K can be calculated:

$$\rho_{\text{FCC}} = \frac{4 \times 6.941\text{g/mol}}{6.022 \times 10^{23}\text{mol}^{-1} \times (4.3 \times 10^{-8}\text{ cm})^3} = 0.58\text{ g/cm}^3$$

Since the density has decreased after the phase transition from FCC to BCC structure, it is evident that Li expands during its transition from FCC to BCC structure which is in accordance with the packing fraction calculations where $\dfrac{FCC}{BCC} = \dfrac{\rho_{FCC}}{\rho_{BCC}} = \dfrac{0.74}{0.68} = \dfrac{0.58}{0.53} = 1.09$

b. Highest linear density for a BCC structure is along the [1 1 1] plane because it is the closest packed direction with two atoms per $a\sqrt{3}$. So,

$$\text{Highest linear density} = \frac{2}{a\sqrt{3}} = \frac{2}{3.51 \times 10^{-8} \times \sqrt{3}} = 3.28 \times 10^7\text{ atoms/cm}$$

c. We know that the number of unit cells $= \frac{\text{Total mass}}{\text{Mass of a unit cell}}$, so we have to determine the mass of a unit cell in BCC and FCC structures.

Before transition (FCC),

$$\text{Mass of a unit cell} = \frac{nM}{N_A} = \rho a^3 = 0.58 \times (4.3 \times 10^{-8})^3 = 4.61 \times 10^{-23}\,\text{g}$$

$$\text{Number of unit cells} = \frac{100}{4.61 \times 10^{-23}} = 2.17 \times 10^{24} \text{ unit cells}$$

No of atoms in the sample = No of atoms in a unit cell × No of unit cells

$$\text{No of atoms in the sample} = 4 \times 2.17 \times 10^{24} = 8.7 \times 10^{24} \text{ atoms}$$

After transition (BCC),

$$\text{Mass of a unit cell} = \frac{nM}{N_A} = \rho a^3 = 0.53 \times (3.51 \times 10^{-8})^3 = 2.29 \times 10^{-23}\,\text{g}$$

$$\text{Number of unit cells} = \frac{100}{2.29 \times 10^{-23}} = 4.36 \times 10^{24}$$

No of atoms in the sample = No of atoms in a unit cell × No of unit cells

$$\text{No of atoms in the sample} = 2 \times 4.36 \times 10^{24} = 8.7 \times 10^{24} \text{ atoms}$$

During the structural phase transition, the number of atoms will not change so to accommodate the atoms, the number of unit cell increases in a BCC structure since each unit cell can have only 2 atoms.

1003

A plane rectangular lattice with lattice parameters $a = 3$Å, $b = 2$Å becomes a plane oblique lattice with $\gamma = 75°$. If a translation vector in both lattices is given by $\vec{T} = 3\vec{a} + 5\vec{b}$, calculate the magnitude and direction of the translation vectors. Draw the two lattice types and comment on your findings.

Sol:

The translation vector is given as $\vec{T} = 3\vec{a} + 5\vec{b}$

We know, $\vec{T} = n_1\vec{a} + n_2\vec{b}$

So, $n_1 = 3$ $n_2 = 5$, and $a = 3$Å, $b = 2$Å

The magnitude of the translation vector is given by the formula

$$T = \sqrt{(n_1 a)^2 + (n_2 b)^2 + 2(n_1 a)(n_2 b)\cos\gamma}$$

For a plane rectangular lattice, $\gamma = 90°$; $\cos 90° = 0$, the equation simplifies to

$$T = \sqrt{(n_1 a)^2 + (n_2 b)^2}$$
$$T = \sqrt{(3 \times 3)^2 + (5 \times 5)^2} = \sqrt{81 + 625} = \sqrt{706} = 26.57Å$$

The direction of the translation vector is $[n_1 \;\; n_2] = [3\ 5]$

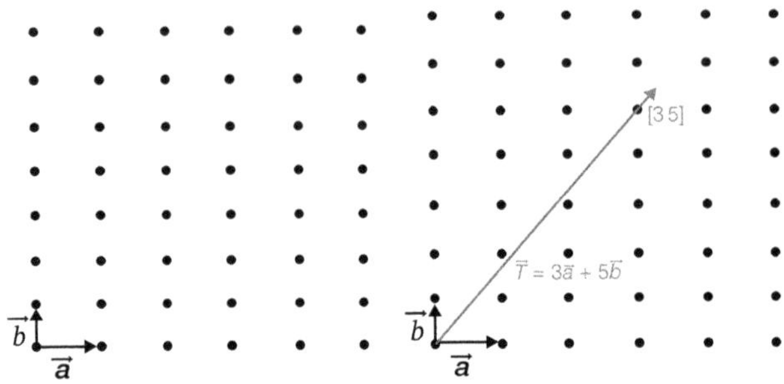

For a plane oblique lattice, $\gamma = 75°$, the magnitude is

$$T = \sqrt{(3 \times 3)^2 + (5 \times 5)^2 + 2(3 \times 3)(5 \times 5)\cos 120}$$
$$T = \sqrt{81 + 625 - 225} = \sqrt{481} = 21.93Å$$

The direction of the translation vector is $[n_1 \;\; n_2] = [3\ 5]$

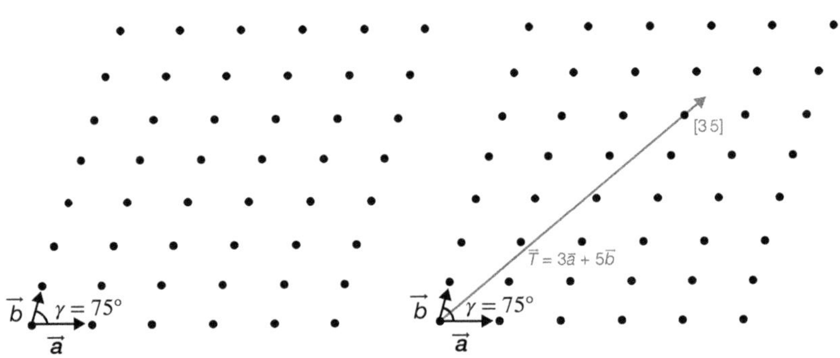

Thus, it can be concluded that the magnitude of the translation vector decreases when the lattice type changes from rectangular to oblique for the same lattice parameter.

1004

Show that the Fermi circle of lithium lies within the First Brillouin Zone obtained from a square lattice of side "*a*" and goes well beyond the First Brillouin Zone for magnesium.

Sol: Let us first calculate the area of the First Brillouin Zone and Fermi Circle. We know that the reciprocal lattice of a square lattice with lattice parameter "a" is also a square lattice of side $\frac{2\pi}{a}$.

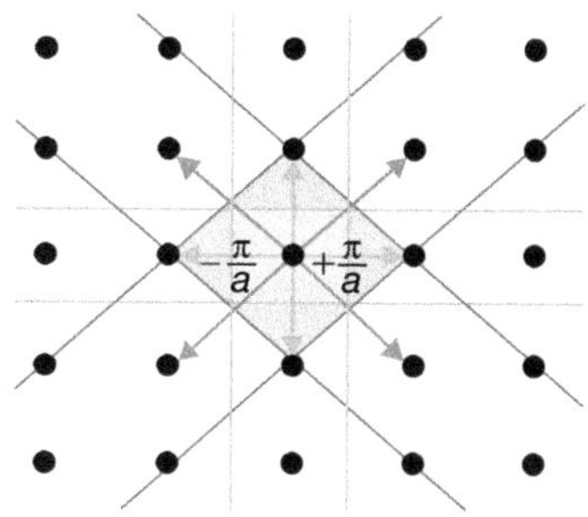

We know the first Brillouin zone is a square for a square lattice, so its area is given by

$$A = \left(\frac{2\pi}{a}\right)^2 = \frac{4\pi^2}{a^2}$$

Then the area of the Fermi circle is $A_{FC} = \pi K_F^2$, where K_F is the radius of the Fermi circle.

Lithium is a monovalent metal, so the electron occupies only half the area of the first Brillouin Zone.

$$\pi K_F^2 = \frac{1}{2}\frac{4\pi^2}{a^2}$$

$$K_F^2 = \frac{1}{2\pi}\frac{4\pi^2}{a^2} = \frac{2}{\pi}\frac{\pi^2}{a^2}$$

$$K_F = \left(\sqrt{\frac{2}{\pi}}\right)\left(\frac{\pi}{a}\right) = 0.798\left(\frac{\pi}{a}\right)$$

The radius of the Fermi circle is within $\left(\frac{\pi}{a}\right)$, which is the zone boundary of the first Brillouin Zone that lies between $\pm\left(\frac{\pi}{a}\right)$ as shown in the following figure.

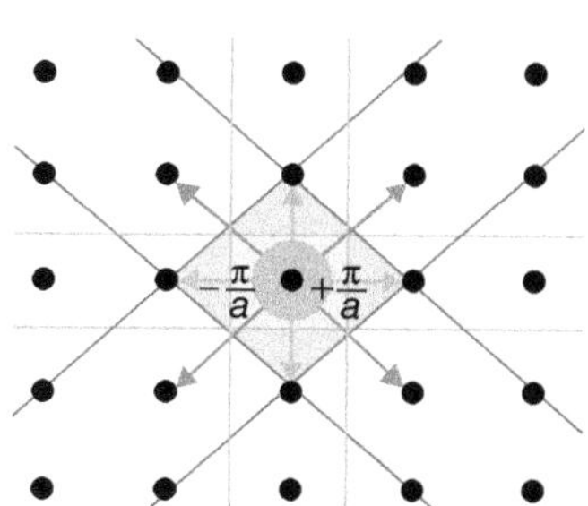

Magnesium is a divalent metal, so the electron occupies double the area of the first Brillouin Zone occupied by an electron in the monovalent metal.

$$\pi K_F^2 = 2\left(\frac{1}{2}\frac{4\pi^2}{a^2}\right)$$

$$K_F^2 = 2\left(\frac{1}{2\pi}\frac{4\pi^2}{a^2}\right) = \frac{4}{\pi}\frac{\pi^2}{a^2}$$

$$K_F = \left(\frac{2}{\sqrt{\pi}}\right)\left(\frac{\pi}{a}\right) = 1.228\left(\frac{\pi}{a}\right)$$

The radius of the Fermi circle is beyond $\left(\frac{\pi}{a}\right)$, which is the zone boundary of the first Brillouin Zone as shown in the figure.

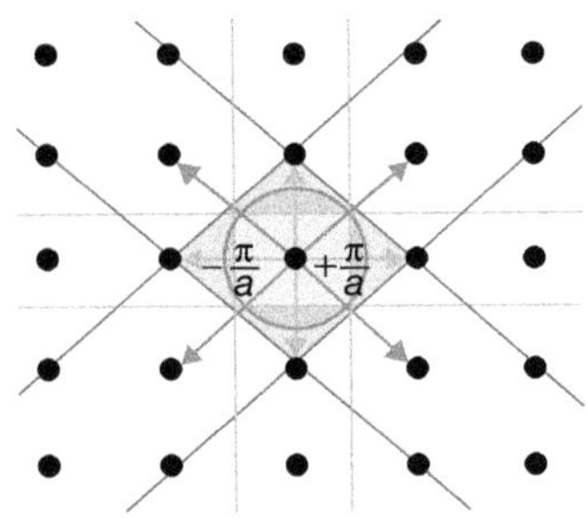

1005

NaCl crystallizes in a face-centered cubic lattice with a basis of Na and Cl ions separated by half the body diagonal of the cube. The atomic numbers of Na and Cl are 11 and 17 respectively.

a. Determine which X-ray reflections will be observed (indexed for the conventional cubic unit cell).

b. Of these which group will be strong and which group weak?

(Wisconsin)

Sol:

a. The unit cell of NaCl includes eight atoms occupying the following positions: Na^+ at (000), $\left(\frac{1}{2}\frac{1}{2}0\right)$, $\left(\frac{1}{2}0\frac{1}{2}\right)$, $\left(0\frac{1}{2}\frac{1}{2}\right)$, as indicated by solid circles in Fig. 1.4; Cl^- at $\left(\frac{1}{2}00\right)$, $\left(0\frac{1}{2}0\right)$, $\left(00\frac{1}{2}\right)$, $\left(\frac{1}{2}\frac{1}{2}\frac{1}{2}\right)$, as indicated by open circles in the figures.

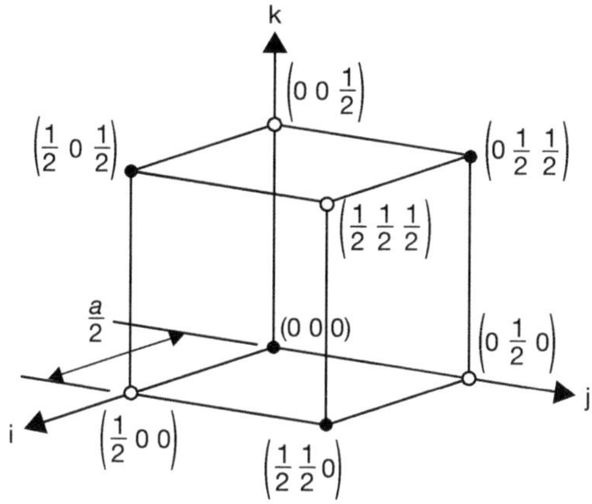

Fig. 1.4

The diffraction intensities are given by

$$\begin{aligned}I_{hkl} &\propto |F_{hkl}|^2 = F_{hkl} \cdot F^*_{hkl} \\ &= \left[\sum_j f_j \cos 2\pi\left(h u_j + k v_j + l w_j\right)\right]^2 \\ &\quad + \left[\sum_j f_j \sin 2\pi\left(h u_j + k v_j + l w_j\right)\right]^2,\end{aligned}$$

where h, k, l are integers. Substitution of the ions' coordinates leads to

$$\begin{aligned}I_{hkl} = f^2_{\mathrm{Na}^+}\{&[1 + \cos\pi(h + k) + \cos\pi(k + l) + \cos\pi(l + h)] \\ &+\alpha[\cos\pi h + \cos\pi k + \cos\pi l + \cos\pi(h + k + l)]\}^2 \\ &+f^2_{\mathrm{Na}^+}\{[\sin(h + k) + \sin\pi(k + l) + \sin\pi(l + h)] \\ &+\alpha[\sin\pi h + \sin\pi k + \sin\pi l + \sin\pi(h + k + l)]\}^2,\end{aligned}$$

where $\alpha = f_{\mathrm{Cl}^-}/f_{\mathrm{Na}^+} = 17/11$.

It is noted that the intensities $I_{hkl} \neq 0$ only when h, k and l are all odd numbers or all even numbers. Thus two different groups of diffracted beams can be observed.

b. When h, k and l are all odd numbers,

$$I \propto 16(1 - \alpha)^2,$$

giving rise to weak reflection. When h, k and l are all even numbers,

$$I \propto 16(1 + \alpha)^2,$$

giving rise to strong reflection.

1006

Titanium exhibits BCC structure at high temperatures and undergoes an allotropic phase transition upon cooling to room temperature. Identify its most stable crystal structure at room temperature based on the following information:

Density of Ti at room temperature, $\rho = 4.5$ g/cm^3

Atomic mass of Ti, $M = 47.867$ g/mol

Edge length of Ti, $a = 473$ pm $= 473 \times 10^{-10}$ cm

Sol: The crystal structure of Ti at room temperature can be determined by estimating the number of atoms in a unit cell using the density formula.

The density of Ti is given by the formula $\rho = \dfrac{nM}{N_A a^3}$

where,

$$n \text{ –number of atoms per unit cell}$$
$$M \text{ –atomic mass}$$
$$N_A \text{ – Avogadro's number} = 6.022 \times 10^{23} \text{mol}^{-1}$$
$$a \text{ –edge length of the unit cell}$$

Rearranging the formula to calculate n gives,

$$n = \frac{\rho N_A a^3}{M}$$

So,

$$n = \frac{4.5 \text{ g/cm}^3 \times 6.022 \times 10^{23} \text{ mol}^{-1} \times (473 \times 10^{-10} \text{ cm})^3}{47.867 \text{ g/mol}}$$

$$n = 5.99 \text{ atoms per unit cell} \approx 6 \text{ atoms per unit cell}$$

We know that the hexagonal close-packed (HCP) crystal structure has 6 atoms per unit cell, so the most stable crystal structure of titanium at room temperature is HCP structure.

1007

Diffraction studies involving X-rays, electrons or neutrons give information about the crystallographic properties of solids. Compare these three techniques with reference to particle energies and types of information that can be obtained. Which technique is most appropriate for studying surface crystallography? Which technique is used to determine magnetic structure?

(*Wisconsin*)

Sol: The typical energy of X-rays is several thousand eV, corresponding to a wavelength of about 10^{-10} m, which is of the same order of magnitude as the interplanar distance in a crystal. Hence X-rays are suitable for determining crystal lattice structures. For low energy diffraction studies, energies 20–50 eV are usually employed. Because of the large cross section of crystal atoms for scattering of low energy electrons, the incident electrons cannot penetrate deeply into a crystal. Thus low energy electron diffraction is an important technique for studying the surface structures of solids. A neutron sees two aspects of a crystal: distribution of nuclei and distribution of electronic magnetization. Hence the diffraction of neutrons by a magnetic crystal allows the determination of the distribution, orientation and order of the magnetic moments.

In brief, low energy electron diffraction is most suitable for studying surface crystallography, and neutron diffraction for determining magnetic structures of crystals.

1008

X-rays are reflected from a crystal by Bragg reflection. If the density of the crystal which is of an accurately known structure is measured with an rms error of 3 parts in 10^4, and if the angle the incident and reflected rays make with the crystal plane is 6° and is measured with an rms error of 3.4 minutes of arc, then what is the rms error in the determination of the X-ray wavelength?

(*Wisconsin*)

Sol: For simplicity consider a crystal whose primitive cell is simple cubic with edge d (to be multiplied by a factor of about one for the primitive cells of other crystal structures). For first order reflection, $n = 1$ and Bragg's law gives

$$2d\sin\theta = \lambda.$$

Differentiating, we have

$$\left|\frac{\Delta\lambda}{\lambda}\right| = \left|\frac{\Delta d}{d}\right| + \cot\theta|\Delta\theta|.$$

The volume of a unit cell is

$$d^3 = \frac{M}{\rho N_0},$$

where $\boldsymbol{M}$ is the molar weight and ρ the mass density of the crystal, and $\boldsymbol{N}_0$ is Avogadro's number. This differentiates to give

$$\left|\frac{\Delta d}{d}\right| = \frac{1}{3}\left|\frac{\Delta\rho}{\rho}\right|.$$

Thus

$$\left|\frac{\Delta\lambda}{\lambda}\right| = \frac{1}{3}\left|\frac{\Delta\rho}{\rho}\right| + \cot\theta \cdot |\Delta\theta|,$$

and, in terms of rms errors,

$$\frac{\sigma_\lambda}{\lambda} = \left[\left(\frac{1}{3}\frac{\sigma_\rho}{\rho}\right)^2 + (\sigma_\theta \cot\theta)^2\right]^{1/2}.$$

As $\frac{\sigma_\rho}{\rho} = 3\times10^{-4}$, $\sigma_\theta\cot\theta = \frac{\sigma_\theta\cos\theta}{\sin\theta} \approx \frac{\sigma_\theta}{\theta} = \frac{3.4}{60\times6} = 9.4\times10^{-3}$,

$$\frac{\sigma_\lambda}{\lambda} = \sqrt{10^{-8} + (9.4\times10^{-3})^2} = 9.4\times10^{-3}.$$

1009

Estimate the pressure needed to compress a solid to several times its normal density.

(Columbia)

Sol: To compress isothermally an elastic solid of volume $\boldsymbol{v}$ by $\boldsymbol{dv}$, the pressure increment dp required is given by

$$K = -v\left(\frac{\partial p}{\partial v}\right)_T,$$

where K is a constant, the compressibility of the material of the solid. Thus

$$p = -\int_{v_0}^{v}\frac{K dv}{v} = K\ln\left(\frac{v_0}{v}\right).$$

To compress the solid to α times its normal density, i.e.,

$$\frac{v_0}{v} = \frac{\rho}{\rho_0} = \alpha,$$

the pressure required is $p = K\ln\alpha$.

For solids, $K \sim 10$ $Pa = 10^8$ mb. For $\alpha = 10$ say, we require $p = 10^8 \ln 10 = 2.3\times10^8$ mb $= 2.3\times10^5$ atmospheres.

1010

Consider a line of $2N$ ions of alternating charges $\pm q$ with a repulsive potential $\boldsymbol{A/R^n}$ between nearest neighbors in addition to the usual Coulomb potential.

a. Find the equilibrium separation R_0 for such a system and evaluate the equilibrium energy $U(R_0)$.

b. Let the crystal be compressed so that $R_0 \to R_0(1-\delta)$. Calculate the work done in compressing a unit length of the crystal to order $\boldsymbol{\delta^2}$.

(*Princeton*)

Sol:

a. Neglecting surface effects, the lattice energy of the system is

$$U(R) = N\left(-\frac{aq^2}{R} + \frac{A}{R^n}\right),$$

where a is the Madelung constant. $U(R)$ is a minimum at equilibrium. So the equilibrium separation R_0 is given by

$$\left(\frac{dU}{dR}\right)_{R=R_0} = 0,$$

whence the equilibrium separation

$$R_0 = \left(\frac{nA}{aq^2}\right)^{\frac{1}{n-1}}$$

and hence the equilibrium energy

$$\begin{aligned} U(R_0) &= -\frac{N}{R_0}\left(aq^2 - \frac{A}{R_0^{n-1}}\right) \\ &= -\frac{Naq^2}{R_0}\left(1 - \frac{1}{n}\right), \end{aligned}$$

with $a = 2\ln 2$ for a one-dimensional chain.

b. When the crystal is compressed so that R_0 becomes R, the increase in the lattice energy is

$$\begin{aligned} U(R) - U(R_0) &= N\left[-aq^2\left(\frac{1}{R} - \frac{1}{R_0}\right) + A\left(\frac{1}{R^2} - \frac{1}{R_0^n}\right)\right] \\ &= -\frac{Naq^2}{R_0}\left[\frac{R_0}{R} - 1 - \frac{1}{n}\left(\frac{R_0^n}{R^n} - 1\right)\right] \end{aligned}$$

and is equal to the work done W by the applied forces. As $R = R_0(1-\delta)$, retaining terms up to δ^2 we have

$$\frac{R_0}{R} - 1 = (1-\delta)^{-1} - 1 \approx \delta + \delta^2,$$

$$\left(\frac{R_0}{R}\right)^n - 1 = (1-\delta)^{-n} - 1 \approx n\delta + \frac{n(n+1)}{2}\delta^2,$$

and thus

$$W \approx -\frac{Naq^2}{R_0}\left(1 - \frac{n+1}{2}\right)\delta^2 = \frac{Naq^2}{2R_0}(n-1)\delta^2.$$

The total length of the crystal is approximately $2N R_0$. Hence the work done in compressing a unit length of the crystal is

$$\frac{W}{2NR_0} = \frac{q^2}{R_0^2}\frac{\ln 2}{2}(n-1)\delta^2.$$

1011

For a polyethylene single crystal undergoing phase transition from orthorhombic to hexagonal structure, find the equivalent hexagonal planes for its orthorhombic planes (110) and (200).

Sol: Let's determine the transformation matrix for the hexagonal to orthorhombic transition.

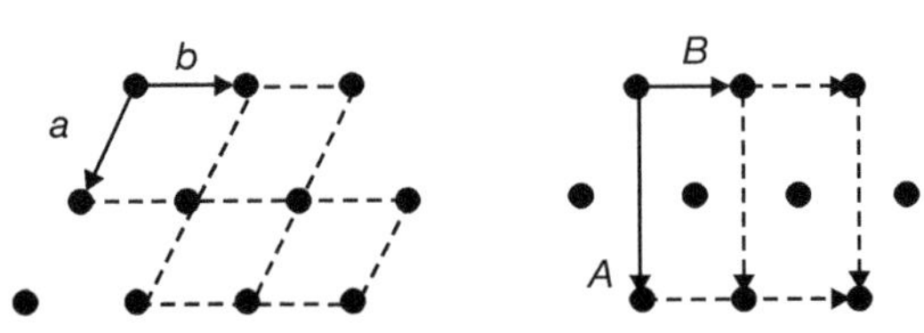

2D representation of the hexagonal to orthorhombic transition

$$A = 2a + b \rightarrow H = 2h + 1k + 0l$$

$$B = b \rightarrow K = 0h + 1k + 0l$$

$$C = c \rightarrow L = 0h + 0k + 1l$$

where, HKL and hkl represent the indices of hexagonal and orthorhombic structures, respectively.

Then it can be represented as,

$$\begin{pmatrix} H \\ K \\ L \end{pmatrix} = \begin{pmatrix} 2 & 1 & 0 \\ 0 & 1 & 0 \\ 0 & 0 & 1 \end{pmatrix} \begin{pmatrix} h \\ k \\ l \end{pmatrix}$$

For the orthorhombic to hexagonal transition, the inverse of the transformation matrix must be taken

$$\begin{pmatrix} 2 & 1 & 0 \\ 0 & 1 & 0 \\ 0 & 0 & 1 \end{pmatrix}^{-1} = \frac{1}{2} \begin{pmatrix} 1 & -1 & 0 \\ 0 & 2 & 0 \\ 0 & 0 & 2 \end{pmatrix}$$

Its corresponding matrix representation is given by

$$\begin{pmatrix} h \\ k \\ l \end{pmatrix} = \frac{1}{2} \begin{pmatrix} 1 & -1 & 0 \\ 0 & 2 & 0 \\ 0 & 0 & 2 \end{pmatrix} \begin{pmatrix} H \\ K \\ L \end{pmatrix}$$

For the orthorhombic plane (HKL) (110),

$$\begin{pmatrix} h \\ k \\ l \end{pmatrix} = \frac{1}{2} \begin{pmatrix} 1 & -1 & 0 \\ 0 & 2 & 0 \\ 0 & 0 & 2 \end{pmatrix} \begin{pmatrix} 1 \\ 1 \\ 0 \end{pmatrix} = \begin{pmatrix} 0 \\ 1 \\ 0 \end{pmatrix}$$

the equivalent hexagonal plane (hkl) is (010).

For the orthorhombic plane (HKL) (200),

$$\begin{pmatrix} h \\ k \\ l \end{pmatrix} = \frac{1}{2} \begin{pmatrix} 1 & -1 & 0 \\ 0 & 2 & 0 \\ 0 & 0 & 2 \end{pmatrix} \begin{pmatrix} 2 \\ 0 \\ 0 \end{pmatrix} = \begin{pmatrix} 1 \\ 0 \\ 0 \end{pmatrix}$$

the equivalent hexagonal plane (hkl) is (100).

1012

An ideal two-dimensional crystal consists of only one kind of atom (of mass m), and each atom has an equilibrium location at a point of a square lattice $\mathbf{R} = (ra, sa)$, where $r, s = 1, 2, \ldots, N$. The displacements from equilibrium are denoted by (x_{rs}, y_{rs}), i.e.,

$$\mathbf{R}_{ra} = (ra + x_{rs}, sa + y_{rs}),$$

and in the harmonic approximation the potential is given by

$$V(x_{rs}, y_{rs}) = \sum_{r,s} \left\{ k_1 \left[(x_{(r+1)s} - x_{rs})^2 (y_{r(s+1)} - y_{rs})^2 \right] \right.$$
$$\left. + k_2 \left[(x_{r(s+1)} - x_{rs})^2 (y_{(r+1)s} - y_{rs})^2 \right] \right\}$$

For the case $k_2 = 0.1k_1$,

a. determine the general phonon dispersion relation $w_{q\lambda}$ throughout the Brillouin zone,

b. sketch $w_{q\lambda}$ as a function of q for

$$q = (\xi, 0), \quad 0 \le \xi \le \frac{\pi}{a}.$$

(Princeton)

Sol:

a. The vibration of an atom at $\mathbf{R}_{rs} = (ra, sa)$is given by Newton's second law

$$m\ddot{\mathbf{R}}_{rs} = -\nabla_{rs} V,$$

where

$$\nabla_{rs} \equiv \mathbf{e}_x \frac{\partial}{\partial x_{rs}} + \mathbf{e}_y \frac{\partial}{\partial y_{rs}}.$$

Consider a solution representing a wave traveling in the crystal:

$$\mathbf{R}_{rs} = \mathbf{A}e^{i(\mathbf{q}\cdot\mathbf{R}-\omega t)},$$

where $\mathbf{A}$ is a constant vector. As $\ddot{\mathbf{R}}_{rs} = -\omega^2\mathbf{R}_{rs}$ and x_{rs} appears in the sum for V only in the terms

$$k_1\left[(x_{(r+1)s} - x_{rs})^2 + (x_{rs} - x_{(r-1)s)})^2\right],$$
$$k_2\left[(x_{r(s+1)} - x_{rs})^2 + (x_{rs} - x_{r(s-1))})^2\right],$$

substitution of $\mathbf{R}_{rs}$, in the equation of motion gives

$$-m\omega^2 x_{rs} = 2k_1(x_{(r+1)s} + x_{(r-1)s} - 2x_{rs}) + 2k_2(x_{r(s+1)} + x_{r(s-1)} - 2x_{rs})$$

and, similarly,

$$-m\omega^2 y_{rs} = 2k_1(y_{r(s+1)} + y_{r(s-1)} - 2y_{rs}) + 2k_2(y_{(r+1)s} + y_{(r-1)s} - 2y_{rs}).$$

With $\mathbf{q} = (q_x, q_y)$, $\mathbf{R} = (ra, sa)$, we have

$$x_{rs} = A_x e^{i[(q_x r + q_y s)a - \omega t]},$$
$$y_{rs} = A_y e^{i[(q_x r + q_y s)a - \omega t]}.$$

Substitution gives

$$-m\omega^2 x_{rs} = 2k_1(e^{iq_x a} + e^{-iq_x a} - 2)x_{rs} + 2k_2\left(e^{iq_y a} + e^{-iq_y a} - 2\right)x_{rs},$$

or

$$\begin{aligned}\omega^2 &= \frac{4k_1}{m}[1 - \cos(q_x a)] + \frac{4k_2}{m}[1 - \cos(q_y a)] \\ &= \omega_0^2\left[\sin^2\left(\frac{q_x a}{2}\right) + \left(\frac{k_2}{k_1}\right)\sin^2\left(\frac{q_y a}{2}\right)\right],\end{aligned}$$

where $\omega_0 = \sqrt{\frac{8k_1}{m}}$. Actually q_x, q_y in the solution x_{rs}, y_{rs} can be interchanged. Hence there are two phono dispersion relations for $k_2 = 0.1\, k_1$:

$$\omega_{q_1} = \omega_0\left[\sin^2\left(\frac{q_x a}{2}\right) + 0.1\sin^2\left(\frac{q_y a}{2}\right)\right]^{1/2}$$

$$\omega_{q_2} = \omega_0\left[\sin^2\left(\frac{q_y a}{2}\right) + 0.1\sin^2\left(\frac{q_x a}{2}\right)\right]^{1/2}.$$

b. Since $q_x = \xi$, $q_y = 0$,

$$\omega_{q_1} = \omega_0\left|\sin\left(\frac{\xi a}{2}\right)\right|,$$

$$\omega_{q_2} = \omega_0\sqrt{0.1}\left|\sin\left(\frac{\xi a}{2}\right)\right| = 0.316\,\omega_0\left|\sin\left(\frac{\xi a}{2}\right)\right|.$$

As ξ has values between 0 and $\frac{\pi}{a}$ inclusively, ωq varies with ξ as shown in Fig. 1.5.

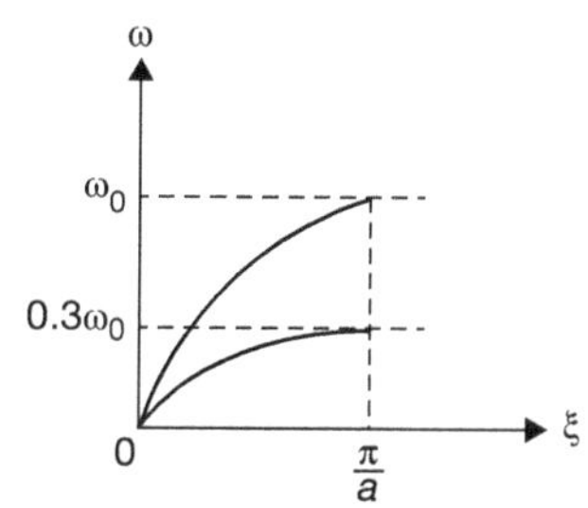

Fig. 1.5

1013

Consider a two-dimensional square array of atoms with lattice constant a. The atoms interact so that when the atom at (x_0, y_0) is displaced to $(x_0 + \Delta x, y_0)$, it is subject to a restoring force $-c_1\Delta x$ due to its two nearest neighbors at $(x_0 \pm a, y_0)$ and a restoring force $-c_2\,\Delta x$ due to its two nearest neighbors at $(x_0, y_0 \pm a)$. Assume $c_2 < c_1$.

a. Find the dispersion relation for sound waves propagating in the x direction, and draw the dispersion diagram.

b. What is the speed of sound in this two-dimensional crystal?

If a few atoms in the lattice are removed from their sites, creating vacancies, the average restoring force for a plane wave of sound is reduced by an amount proportional to the concentration of vacancies. Furthermore, each vacancy will scatter any incident sound wave, and the amplitude of the scattered wave is A/λ for $\lambda \gg a$, where λ is the wavelength and A is a constant.

c. For a concentration of vacancies, find the attenuation length for sound as a function of wavelength (to within a multiplicative constant). Assume the vacancies scatter independently.

d. Assume that an energy ε is needed to remove an atom from the crystal and create a vacancy. What is the temperature dependence of the sound attenuation length?

e. Under the same assumption as in (d), what is the temperature dependence of the sound velocity?

(Princeton)

Sol:

a. Take the origin at (x_0, y_0), then $(x_n, y_m) = (na, ma)$. Let the x displacement of the atom at (x_n, y_m) be $u_{n,m}$. The atom's x displacement relative to the nearest neighbors situated parallel to the x-axis is

$$-(u_{n+1,m} - u_{n,m}) + (u_{n,m} - u_{n-1,m}) = 2u_{n,m} - u_{n+1,m} - u_{n-1,m},$$

and that relative to the nearest neighbors situated parallel to the y-axis is

$$2u_{n,m} - u_{n,m+1} - u_{n,m-1}.$$

Hence the equation of motion of the atom at (x_n, y_m) is

$$\begin{aligned} m\frac{d^2u_{n,m}}{dt^2} &= -C_1(2u_{m,n} - u_{n+1,m} - u_{n+1,m}) \\ &\quad -C_2(2u_{m,n} - u_{n,m+1} - u_{n,m-1}). \end{aligned}$$

Try a solution of the form

$$u_{n,m} = Ae^{i(q_1na+q_2ma-\omega t)}.$$

Substitution gives

$$\begin{aligned} m\omega^2 &= C_1(2 - e^{iq_1a} - e^{-iq_1a}) + C_2(2 - e^{iq_2a} - e^{-iq_2a}) \\ &= 2C_1[1 - \cos(q_1a)] + 2C_2[1 - \cos(q_2a)], \end{aligned}$$

and hence

$$\omega = \left\{\frac{4C_1}{m}\sin^2\left(\frac{q_1a}{2}\right) + \frac{4C_2}{m}\sin^2\left(\frac{q_2a}{2}\right)\right\}^{1/2}.$$

For sound waves propagating in the x direction, $q_2 = 0$ and the dispersion relation is as plotted in Fig. 1.6, where $\omega_m = \sqrt{\frac{4C_1}{m}}$.

b. The velocity of sound in the crystal is

$$\mathbf{v} = \nabla_q \omega = \frac{\partial \omega}{\partial q_1}\mathbf{e}_x + \frac{\partial \omega}{\partial q_2}\mathbf{e}_y$$
$$= \frac{a}{m\omega}\left[C_1 \sin(q_1 a)\,\mathbf{e}_x + C_2 \sin(q_2 a)\,\mathbf{e}_y\right],$$

and has magnitude

$$v = \frac{a}{m\omega}\left[C_1^2 \sin^2(q_1 a) + C_2^2 \sin^2(q_2 a)\right]^{1/2}.$$

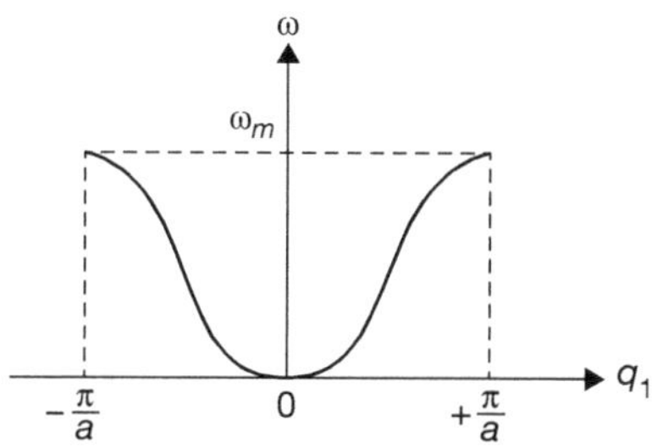

Fig. 1.6

c. Consider sound waves of intensity I and cross sectional area S traveling in the lattice. In a small distance dx it will encounter $nSdx$ vacancies, where n is the concentration of vacancies. Each scattering by a vacancy will cause an energy loss proportional to $\left(\frac{A}{\lambda}\right)^2$, and, since each vacancy will scatter any incident sound wave, the total energy loss by scattering will also be proportional to I. Thus

$$-SdI = CI\left(\frac{A}{\lambda}\right)^2 nSdx\ ,$$

where C is the proportionality constant. Hence

$$I = I_0 \exp\left(-\frac{CnA^2}{\lambda^2}\right).$$

The attenuation length l, defined as the distance over which the intensity decreases by a factor e^{-1}, is then

$$l = \frac{\lambda^2}{CnA^2}.$$

d. The vacancy density is related to the absolute temperature T by

$$n \propto e^{-\varepsilon/k_B T},$$

where k_B is Boltzmann's constant. Hence

$$l \propto \frac{1}{n} \propto e^{\frac{\varepsilon}{k_B T}}.$$

e. As the reduction in the restoring force is proportional to the vacancy concentration, the force constants are reduced from their original values C_1^0, C_2^0 to

$$C_1 = C_1^0(1 - Bn),$$
$$C_2 = C_2^0(1 - Bn),$$

where B is the proportionality constant. The speed of sound is therefore

$$v \propto \left[1 - B\exp\left(-\frac{\varepsilon}{k_B T}\right)\right]^{1/2}.$$

1014

Consider a d-dimensional vibrating lattice at zero temperature. The number density is η and the atoms have mass m. Using the Debye approximation and assuming all sound modes to have the same velocity v.

a. Evaluate the mean square displacement $\langle R^2 \rangle$ for $d = 3$.

b. Evaluate $\langle R^2 \rangle$ for $d = 1$, and discuss its relevance to experiment.

c. Evaluate the mean square strain $\left\langle \left(\frac{\partial R}{\partial x}\right)^2 \right\rangle$ for $d = 1$.

(Princeton)

Sol: Consider an atom oscillating with angular frequency ω_j and amplitude y_{0j}. Its displacement from the equilibrium position is

$$y_j = y_{0j}\cos(q_j x - \omega_j t)$$

and its kinetic energy is

$$\frac{1}{2}m\dot{y}_j^2 = \frac{1}{2}m\,\omega_j^2 y_j^2.$$

Its kinetic energy averaged over time is

$$\left\langle \frac{1}{2}m\dot{y}_j^2 \right\rangle = \frac{1}{2}m\,\omega_j^2 \langle y_j^2 \rangle = \frac{1}{4}m\,\omega_j^2 y_{0j}^2.$$

The oscillating atom is equivalent to a quantum harmonic oscillator of the same frequency whose total energy is $\left(n + \frac{1}{2}\right)\hbar\,\omega_j$. As the average kinetic energy of an oscillator is equal to half its total energy,

$$\frac{1}{4}m\,\omega_j^2 y_{0j}^2 = \frac{1}{2}\left(n + \frac{1}{2}\right)\hbar\,\omega_j,$$

we have

$$y_{0j}^2 = \frac{(2n+1)\hbar}{m\,\omega_j}.$$

At zero absolute temperature, all oscillators are in the ground state for which the quantum number $n = 0$. The mean square displacement is then

$$\langle y_j^2 \rangle = \frac{1}{2} y_{0j}^2 = \frac{\hbar}{2m\,\omega_j}.$$

Hence, averaged over all the atoms in the lattice, the mean square displacement at zero temperature is

$$\langle R^2 \rangle = \frac{1}{N} \sum_j^N \langle y_j^2 \rangle = \frac{\hbar}{2Nm} \sum \omega_j^{-1} = \frac{\hbar}{2\rho V} \sum_j \omega_j^{-1},$$

where $\boldsymbol{\rho}$, $\boldsymbol{V}$ axe respectively the mass density and volume, and $\boldsymbol{N}$ is the total number of atoms of the lattice.

a. $d = 3$. In the Debye approximation, the number of modes of oscillation with wave numbers less than q is given by the volume, measured in units of $\left(\frac{2\pi}{L}\right)^3$, $\boldsymbol{L}$ being the length of the lattice, assumed cubic, of a sphere of radius q with center at the origin of the $\mathbf{q}$-space. As there are three possible polarizations, the number of modes of vibration with wave numbers less than q is

$$\mathcal{N} = 3 \cdot \frac{4}{3} \pi q^3 \left(\frac{2\pi}{L}\right)^{-3} = \frac{L^3 q^3}{2\pi^2}.$$

Assuming all the modes have the same sound velocity v, we have $q = \frac{\omega}{v}$ and the density of state

$$D(\omega) = \frac{d\mathcal{N}}{d\omega} = \frac{3L^3 q^2}{2\pi^2} \frac{dq}{d\omega} = \frac{3V\omega^2}{2\pi^2 v^3}.$$

Furthermore as $\mathcal{N} = 3N$, the Debye cutoff angular frequency is

$$\omega_{\mathrm{D}} = \left(\frac{6\pi^2 N}{V}\right)^{\frac{1}{3}} v = (6\pi^2 \eta)^{\frac{1}{3}} v.$$

Hence

$$\begin{aligned} \langle R^2 \rangle &= \frac{\hbar}{2\rho v} \int_0^{\omega_{\mathrm{D}}} \omega^{-1} D(\omega) d\omega \\ &= \frac{3\hbar\,\omega_{\mathrm{D}}^2}{8\pi^2 m\eta v^3}. \end{aligned}$$

b. $d = 1$. The Debye approximation gives

$$\mathcal{N} = N = 2q\left(\frac{2\pi}{L}\right)^{-1} = \frac{Lq}{\pi} = \frac{L}{\pi}\frac{\omega}{v},$$

and hence

$$D(\omega) = \frac{L}{\pi v}$$

and

$$\omega_{\mathrm{D}} = \frac{\pi N v}{L} = \pi\eta v,$$

where η is the number of atoms per unit length of the one-dimensional lattice. Hence

$$\begin{aligned}\langle R^2\rangle &= \frac{\hbar}{2\rho L}\sum_j \omega_j^{-1} = \frac{\hbar}{2\pi\rho v}\int_0^{\omega_D}\frac{d\omega}{\omega} \\ &= \frac{\hbar}{2\pi m\eta v}\ln\omega\Big|_0^{\omega_D} = \infty\,.\end{aligned}$$

$\langle R^2\rangle$ is divergent for the one-dimensional case. It is seen that the divergence arises from the lower integration limit being zero. Physically, $\omega \approx 0$ corresponds to the lattice atoms move together as a rigid body, for which $\langle y^2\rangle$ is zero, not $\sim \omega^{-1}$. Thus an experimental determination of $\langle R^2\rangle$ would not yield infinity but same finite value.

c. For the jth atom,

$$\left(\frac{\partial y_j}{\partial x}\right)^2 = q_j^2 y_{0j}^2 \sin^2(q_j x - \omega_j t),$$

which, when averaged over time, becomes

$$\left\langle\left(\frac{\partial y_i}{\partial x}\right)^2\right\rangle = \frac{1}{2}q_j^2 y_{0j}^2 = \frac{\hbar\,\omega_j}{2mv^2}.$$

Thus for the one-dimensional lattice,

$$\begin{aligned}\left\langle\left(\frac{\partial R}{\partial x}\right)^2\right\rangle &= \frac{1}{2N}\sum_j^N \frac{\hbar\,\omega_j}{2mv^2} \\ &= \frac{\hbar}{4Nmv^2}\int_0^{\omega_D}\frac{L\omega}{\pi v}d\omega \\ &= \frac{\hbar\,\omega_D^2}{4\pi m\eta v^3},\end{aligned}$$

where $\omega_D = \pi\eta v$, as $\frac{N}{L} = \eta$.

1015

For an orthorhombic crystal system,

a. Represent the three twofold rotations about the axes in matrix form.

b. Show that two mutually perpendicular mirror planes produce another mirror plane perpendicular to the initial two planes, whereas three mutually perpendicular mirror planes produce inversion.

Sol:

a. In a proper twofold rotation, the angle of rotation will be $\alpha = 180°$. When the rotation is carried out about any axis, then the new axes will be as follows:

$$x_1 \rightarrow x'_1\,;\, x_2 \rightarrow x'_2;\, x_3 \rightarrow x'_3$$

The transformation matrix is given by

$$T = \begin{pmatrix} \cos(x_1x'_1) & \cos(x_1x'_2) & \cos(x_1x'_3) \\ \cos(x_2x'_1) & \cos(x_2x'_2) & \cos(x_2x'_3) \\ \cos(x_3x'_1) & \cos(x_3x'_2) & \cos(x_3x'_3) \end{pmatrix}$$

For a twofold rotation about x_3 axis [001], $x_1 \rightarrow x'_1\,;\, x_2 \rightarrow x'_2;\, x_3 \rightarrow x_3$, its **T** will be

$$2[001] = \begin{pmatrix} \cos(\alpha) & \cos(90+\alpha) & \cos(90) \\ \cos(90-\alpha) & \cos(\alpha) & \cos(90) \\ \cos(90) & \cos(90) & \cos(0) \end{pmatrix}$$
$$= \begin{pmatrix} \cos\alpha & -\sin\alpha & 0 \\ \sin\alpha & \cos\alpha & 0 \\ 0 & 0 & 1 \end{pmatrix}$$

We know, $\alpha = 180°$

$$2[001] = \begin{pmatrix} -1 & 0 & 0 \\ 0 & -1 & 0 \\ 0 & 0 & 1 \end{pmatrix}$$

For a twofold rotation about x_2 axis [010], $x_1 \rightarrow x'_1\,;\, x_2 \rightarrow x_2;\, x_3 \rightarrow x'_3$, its **T** will be

$$2[010] = \begin{pmatrix} \cos(\alpha) & \cos(90) & \cos(90+\alpha) \\ \cos(90) & \cos(0) & \cos(90) \\ \cos(90-\alpha) & \cos(90) & \cos(\alpha) \end{pmatrix}$$
$$= \begin{pmatrix} \cos\alpha & 0 & -\sin\alpha \\ 0 & 1 & 0 \\ \sin\alpha & 0 & \cos\alpha \end{pmatrix}$$

We know, $\alpha = 180°$

$$2[010] = \begin{pmatrix} -1 & 0 & 0 \\ 0 & 1 & 0 \\ 0 & 0 & -1 \end{pmatrix}$$

For a twofold rotation about x_1 axis [100], $x_1 \rightarrow x_1\,;\, x_2 \rightarrow x'_2;\, x_3 \rightarrow x'_3$, its **T** will be

$$2[100] = \begin{pmatrix} \cos(0) & \cos(90) & \cos(90) \\ \cos(90) & \cos(\alpha) & \cos(90+\alpha) \\ \cos(90) & \cos(90-\alpha) & \cos(\alpha) \end{pmatrix}$$
$$= \begin{pmatrix} 1 & 0 & 0 \\ 0 & \cos\alpha & -\sin\alpha \\ 0 & \sin\alpha & \cos\alpha \end{pmatrix}$$

We know, $\alpha = 180°$

$$2[100] = \begin{pmatrix} 1 & 0 & 0 \\ 0 & -1 & 0 \\ 0 & 0 & -1 \end{pmatrix}$$

b. We know that a proper twofold rotation can also be considered as a reflection. So let's take $m[100]$ and $m[010]$ as our two perpendicular mirror planes

$$m[100] \times m[010] = \begin{pmatrix} 1 & 0 & 0 \\ 0 & -1 & 0 \\ 0 & 0 & -1 \end{pmatrix} \begin{pmatrix} -1 & 0 & 0 \\ 0 & 1 & 0 \\ 0 & 0 & -1 \end{pmatrix}$$

$$m[100] \times m[010] = \begin{pmatrix} -1 & 0 & 0 \\ 0 & -1 & 0 \\ 0 & 0 & 1 \end{pmatrix} = m[001]$$

It results in a twofold axis, which represents a reflection along the x_3 axis.

Now let's take $m[100]$, $m[010]$, and $m[001]$ as our three mutually perpendicular mirror planes

$$m[100] \times m[010] \times m[001] = \begin{pmatrix} 1 & 0 & 0 \\ 0 & -1 & 0 \\ 0 & 0 & -1 \end{pmatrix} \begin{pmatrix} -1 & 0 & 0 \\ 0 & 1 & 0 \\ 0 & 0 & -1 \end{pmatrix} \begin{pmatrix} -1 & 0 & 0 \\ 0 & -1 & 0 \\ 0 & 0 & 1 \end{pmatrix}$$

$$m[100] \times m[010] \times m[001] = \begin{pmatrix} -1 & 0 & 0 \\ 0 & -1 & 0 \\ 0 & 0 & -1 \end{pmatrix}$$

It results in an inversion.

1016

a. What is the specific heat (per mole) of a monatomic gas at constant volume?

b. What is its specific heat (per mole) at constant volume for a diatomic gas? Explain.

c. What is the specific heat of a monatomic crystalline solid?

(Wisconsin)

Sol: Considered as ideal, a gas has internal energy per mole at absolute temperature T of

$$E = \frac{f}{2} N k_B T = \frac{f}{2} RT,$$

where f is the number of degrees of freedom of a molecule and $R = Nk_B$ is the gas constant. The molar specific heat at constant volume is thus

$$C_v = \left(\frac{\partial E}{\partial T}\right)_v = \frac{f}{2} R.$$

a. For a monatomic gas, $f = 3$ and

$$C_v = \frac{3}{2}R.$$

b. For a diatomic gas, $f = 5$ and

$$C_v = \frac{5}{2}R.$$

It is noted that the specific heat per mole is independent of temperature for both monatomic and diatomic gases if they can be approximated as ideal gases.

c. Consider one mole of a monatomic crystalline solid. It has a volume V and constains N (the Avogadro number) atoms. As there are $3N$ normal modes of vibration, the number of modes in the angular frequency range ω to $\omega + d\omega$ is (Problem **1014**)

$$D(\omega)d\omega = \frac{3V\omega^2 d\omega}{2\pi^2 v_0^3}.$$

where v_0 is the speed of sound in the crystal, subject to a maximum frequency

$$\omega_{\mathrm{m}} = \left(\frac{6\pi^2 N}{V}\right)^{1/3} v_0.$$

In a solid the average energy of an oscillator of frequency ω is

$$\frac{\hbar\omega}{e^{\hbar\omega/k_{\mathrm{B}}T} - 1}.$$

Hence the total internal energy of the mole of crystal is

$$U = \int_0^{\omega_{\mathrm{m}}} \frac{\hbar\omega}{e^{\hbar\omega/k_{\mathrm{B}}T} - 1} \cdot D(\omega)d\omega$$

$$= \frac{3V\hbar}{2\pi^2 v_0^3} \int_0^{\omega_{\mathrm{m}}} \frac{\omega^3}{e^{\hbar\omega/k_{\mathrm{B}}T} - 1},$$

and the specific heat per mole at constant volume is

$$C_v = \left(\frac{\partial U}{\partial T}\right)_v = \frac{3Vk_{\mathrm{B}}}{2\pi^2}\left(\frac{Tk_{\mathrm{B}}}{\hbar v_0}\right)^3 \int_0^{x_{\mathrm{m}}} \frac{e^x x^4 dx}{(e^x - 1)^2},$$

where $x_m = \dfrac{\hbar\omega_m}{k_{\mathrm{B}}T}$. Introducing the Debye temperature

$$\Theta = \frac{\hbar\omega_{\mathrm{m}}}{k_{\mathrm{B}}},$$

we can write this as

$$C_v = 9Nk_B\left(\frac{T}{\Theta}\right)^3 \int_0^{\frac{\Theta}{T}} \frac{e^x x^4 dx}{(e^x - 1)^2} = 9Nk_B F\left(\frac{T}{\Theta}\right).$$

when $T \gg \Theta$, $x \ll \frac{\Theta}{T}$ is small so that

$$\frac{e^x x^4}{(e^x - 1)^2} \approx \frac{x^4}{x^2} = x^2$$

and

$$F\left(\frac{T}{\Theta}\right) \simeq \frac{1}{3}\left(\frac{T}{\Theta}\right)^3\left(\frac{\Theta}{T}\right)^3 = \frac{1}{3},$$

giving

$$C_v \approx 3Nk_B = 3R.$$

Thus C_v is independent of temperature at high temperatures.

When $T \ll \Theta$, $\frac{\Theta}{T}$ is large and can be taken as ∞. Then

$$C_v \approx 9Nk_B\left(\frac{T}{\Theta}\right)^3 \int_0^{\infty} \frac{e^x x^4 dx}{(e^x - 1)^2},$$

showing that $C_v \propto T^3$ at low temperatures. For a quantitative estimate, we note that

$$\int_0^{\infty} \frac{x^3 dx}{e^x - 1} = 6\sum_1^{\infty}\frac{1}{n^4} = \frac{\pi^4}{15}.$$

This gives

$$U = \frac{3}{5}\pi^4 Nk_B \frac{T^4}{\Theta^3}$$

and hence

$$C_v = \left(\frac{\partial U}{\partial T}\right)_v = \frac{12}{5}\pi^4 Nk_B\left(\frac{T}{\Theta}\right)^3 = 234Nk_B\left(\frac{T}{\Theta}\right)^3.$$

1017

Use the Debye model to calculate the heat capacity of a monatomic lattice in one dimension at temperatures small compared with the Debye temperature $\Theta_D = \frac{\hbar\pi v}{k_B a}$, where v is the sound velocity, a is the lattice spacing and k_B is

Boltzmann's constant. Numerical constants in the form of integrals need not be evaluated.

(*Columbia*)

Sol: In the Debye model, the density of states for a one-dimensional monatomic lattice is (Problem **1014**)

$$\rho(\omega) = \frac{L}{\pi v}.$$

As a mode of frequency ω has average energy

$$\frac{\hbar w}{e^{\hbar\omega/k_B T} - 1},$$

the internal energy is

$$U = \int_0^{\omega_D} \frac{L}{\pi v} \frac{\hbar\omega}{e^{\hbar\omega/k_B T} - 1} d\omega,$$

where ω_D is given by

$$\int_0^{\omega_D} \frac{L}{\pi v} dw = N,$$

i.e.,

$$\omega_D = \frac{\pi N v}{L} = \frac{\pi v}{a} = \frac{k_B \Theta_D}{\hbar}.$$

Hence

$$C_v = \frac{\partial U}{\partial T} = N k_B \left(\frac{T}{\Theta_D}\right) \int_0^{\Theta_D} \frac{x^2 e^x}{(e^x - 1)^2} dx.$$

When $T \ll \Theta_D$, we can take $\frac{\Theta_D}{T} \approx \infty$, so that

$$\int_0^{\Theta_D} \frac{x^2 e^2}{(e^x - 1)^2} dx \approx \int_0^{\infty} \frac{x^2 e^2}{(e^x - 1)^2} dx = \frac{\pi^2}{3},$$

giving

$$C_v = \frac{\pi^2}{3} N k_B \left(\frac{T}{\Theta_D}\right) = \frac{\pi k_B^2 N a}{3\hbar v} T.$$

1018

A potassium chloride crystal is mounted on an X-ray spectrometer that uses monochromatic X-rays of wavelength 0.586 Å. If the glancing angles of first-order reflections from the planes (100), (110), and (111) are 5.38°, 7.61°, and 9.38°, respectively, then determine the nature of the crystal.

Sol: The nature of the crystal can be determined by calculating the interplanar distance ratios for these planes $d_{100}:d_{110}:d_{111}$.

We know, $n\lambda = 2d\sin\theta \rightarrow d = \frac{\lambda}{2\sin\theta}$, since $n = 1$ for first-order reflections

$$d_{100}:d_{110}:d_{111} = \frac{\lambda}{2\sin 5.38}:\frac{\lambda}{2\sin 7.61}:\frac{\lambda}{2\sin 9.38}$$

$$d_{100}:d_{110}:d_{111} = \frac{1}{\sin 5.38}:\frac{1}{\sin 7.61}:\frac{1}{\sin 9.38}$$

$$d_{100}:d_{110}:d_{111} = \frac{1}{0.0938}:\frac{1}{0.1326}:\frac{1}{0.1620} = 1:0.704:0.575$$

This ratio corresponds to simple cubic as shown below

$$d = \frac{a^2}{\sqrt{h^2+k^2+l^2}}$$

$$d_{100}:d_{110}:d_{111} = \frac{1}{\sqrt{1^2+0^2+0^2}}:\frac{1}{\sqrt{1^2+1^2+0^2}}:\frac{1}{\sqrt{1^2+1^2+1^2}}$$

$$d_{100}:d_{110}:d_{111} = \frac{1}{\sqrt{1}}:\frac{1}{\sqrt{2}}:\frac{1}{\sqrt{3}} = 1:0.704:0.575$$

So the nature of KCl is simple cubic.

1019

Acoustic properties of dielectric solids dominate their thermodynamic behavior and other properties such as photoconducting resistance. Diamond is a monatomic dielectric solid of carbon having 10^{21} atoms cm^{-3}.

a. In Fig. 1.7, sketch, roughly, its specific heat (per atom) as a function of absolute temperature.

Fig. 1.7

b. How is T_{Debye} related to the Debye frequency ω_D?

c. If the acoustic velocity at low frequencies is 5×10^5 cm/sec, what is approximately the value of ω_D?

(*Wisconsin*)

Sol:

a. The specific heat curve (Problem **1016**) is sketched in Fig. 1.8.

b. The Debye temperature T_D is defined in terms of the Debye frequency ω_D by

$$k_B T_D = \hbar \omega_D .$$

c. For a 3-dimensional lattice, the density of states is (Problem **1014**)

$$\rho(\omega) = \frac{3V\omega^2}{2\pi^3 v^3} d\omega.$$

As

$$\int_0^{\omega_D} \rho(\omega) d\omega = 3N,$$

$$\omega_D = \left(6\pi^2 \frac{N}{V}\right)^{1/3} v = (6\pi^2 n)^{1/3} v.$$

With $n = 10^{27}$ m^{-3}, $v = 5 \times 10^3$ m/s, we find

$$\omega_D = 1.95 \times 10^{13} \text{s}^{-1}.$$

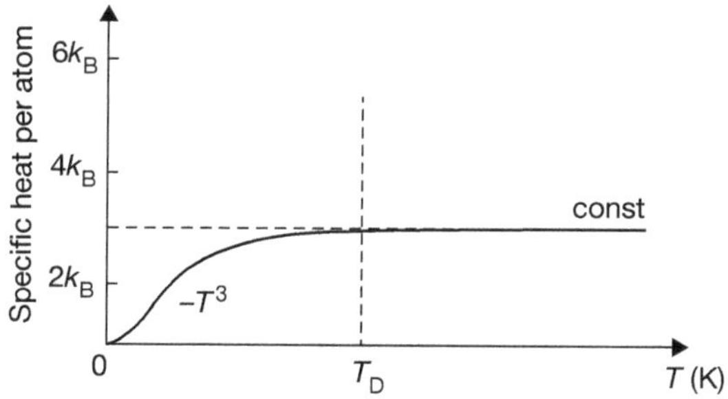

Fig. 1.8

1020

The conduction of heat by a non-metallic, crystalline solid requires a mechanism by which the phonon distribution may be brought into thermal equilibrium.

a. Show that a three-phonon collision process of the type $\mathbf{q}_1 + \mathbf{q}_2 = \mathbf{q}_3$ will not establish equilibrium.

b. Describe the form of the processes which establish equilibrium.

(*Wisconsin*)

Sol: Suppose that two phonons of wave vectors $\mathbf{q}_1$ and $\mathbf{q}_2$ collide and produce a third phonon of wave vector $\mathbf{q}_3$. Conservation of quasimomentum allows the following two processes:

$$\mathbf{q}_1 + \mathbf{q}_2 = \begin{cases} \mathbf{q}_3, \\ \mathbf{q}_3 + \mathbf{k}, \quad \mathbf{k} \text{ being an appropriate reciprocal lattice vector.} \end{cases}$$

These two types of three-phonon collision process are illustrated in Fig. 1.9 (a) and (b). The first is the normal process. The second, known as an umklapp process, can take place in a discrete lattice. In this case, $\mathbf{q}_1$, $\mathbf{q}_2$ produce a $\mathbf{q}_3$ that goes outside the limits $\pm\mathbf{q}_m$ which define the boundaries of the first Brillouin zone. It must be brought back to the latter by adding to it a reciprocal lattice vector $\mathbf{k}$, since all wave vectors are physically meaningful only in the first Brillouin zone, according to our convention.

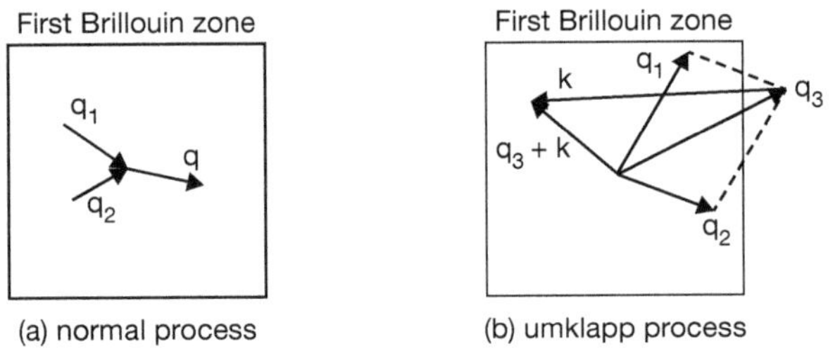

Fig. 1.9

a. In the process $\mathbf{q}_1 + \mathbf{q}_2 = \mathbf{q}_3$, the total momentum of the phonon system remains the same before and after collision, which means that the heat current density is not affected by the collision. Hence a thermal equilibrium cannot be established, and there is no thermal resistivity, or, in other words, the thermal conductivity is infinite.

b. In the umklapp process $\mathbf{q}_1 + \mathbf{q}_2 = \mathbf{q}_3 + \mathbf{k}$, the total momentum of the phonon system will change because of the inclusion of $\mathbf{k}$. The effective phonon $\mathbf{q}_4 = \mathbf{q}_3 + \mathbf{k}$ produced by the collision travels in a direction almost opposite to either of the original phonons $\mathbf{q}_1$, $\mathbf{q}_2$. Then, within a certain relaxation time the system will reach a state of thermal equilibrium. It is this process that gives rise to a finite thermal resistivity.

1021

Briefly describe the mechanism for thermal conductivity of crystalline insulating solids. The thermal conductivity of a solid electrical insulator (and gases, etc.) can be written as $k = \frac{1}{3}cvl$ where c is a heat capacity per unit volume, v a velocity, and l a mean free path. Provide estimates of these parameters with physical justifications for a typical crystal at room temperature.

(*Wisconsin*)

Sol: If there is a temperature gradient in a crystalline, insulating solid, the lattice vibrations excited at the hot end will have more vibrational modes and larger amplitudes, or, in other words, more phonons. As these lattice waves propagate toward the cold end, the latter's lattice vibrations will be enhanced and approach the same number of vibrational modes and the same larger amplitudes. This means that the phonons propagating in the lattice transfer from the hot to the cold end. On account of the nonlinearity of lattice vibrations, there are interactions among the phonons as they propagate. They collide with one another and bump against the imperfections in the crystal. The phonon collision processes may be the normal process or the umklapp process. The latter process plays the dominant role in thermal conduction and causes the distribution of phonons to approach an equilibrium. Hence the mean free path in the expression for thermal conductivity should be that of the umklapp process.

Take as example the NaCl crystal at room temperature. The typical values are molar specific heat $C_v \approx 25$ J/mol·K, speed of sound $v = 6 \times 10^4$ m/s, phonon mean free path $l = 2.3 \times 10^{-9}$ m, molar volume $V = 27 \times 10^{-6}$ m^3. These data give the thermal conductivity as

$$k = \frac{1}{3}\left(\frac{C_v}{V}\right)vl = 43\,\text{W}\,\text{m}^{-1}\text{K}^{-1}.$$

1022

a. Write down definitions and formulas which describe the thermal conductivity of a solid.

b. Identify the excitations which carry the heat current.

c. Sketch a graph of the thermal conductivity of a metal as a function of temperature.

d. Identify the characteristic temperature dependence at high and low temperatures and describe the dominant physical effects in these two regions.

(*Wisconsin*)

Sol:

a. The thermal conductivity of a solid is defined as the heat current density Q per unit temperature gradient, i.e., the constant coefficient k in the equation $Q = -k\frac{\partial J}{\partial x}$. For a nonmetallic solid, heat is conducted by phonons and the conductivity is given by

$$k = \frac{1}{3}C_v\bar{v}l,$$

where C_v is the specific heat per unit volume at constant volume of the solid, $\bar{v}$ and l are respectively the average speed and mean free path of the phonons.

b. For nonmetallic solids the excitations which carry the heat current are phonons.

c. The thermal conductivity of a metal consists of two parts, the lattice thermal conductivity k_a contributed by phonons and the electronic thermal conductivity k_e contributed by the free electrons:

$$k = k_a + k_e$$

with

$$k_a = \frac{1}{3}C_a\bar{v}_a l_a,$$
$$k_e = \frac{1}{3}C_e\bar{v}_e l_e,$$

where C is the specific heat at constant volume per unit volume, $\bar{v}$ and l are the average speed and mean free path of the particles (phonons or free electrons as the case may be) respectively. As for normal pure metals

$$\frac{k_a}{k_e} = \frac{C_a\bar{v}_a l_a}{C_e\bar{v}_e l_e} \approx 10^{-2},$$

the conductivity of a typical metal is principally that due to free electrons. Thus

$$k \approx k_e = \frac{1}{3}C_e\bar{v}_e l_e = \frac{\pi^2 N k_B^2}{3m\bar{v}_e} l_e T,$$

where we have used the result from Fermi–Dirac statistics

$$C_e = \frac{\pi^2 N k_B^2 T}{m\bar{v}_e^2},$$

N being the number of free electrons per unit volume. As l_e is due mainly to electron–phonon scattering,

$$l_e \propto \frac{1}{n_{\text{ph}}},$$

where n_{ph} is the mean number of phonons per unit volume at temperature T.

1. At high temperatures,

$$n_{\text{ph}} \propto T,$$

or $l_e \propto T^{-1}$, giving

$$k = \text{constant},$$

independent of temperature.

2. At low temperatures,

$$n_{\text{ph}} \propto T^3,$$

or $l_e \propto T^{-3}$, giving

$$k = T^{-2}.$$

3. At very low temperatures, the number of phonons is very small and scattering of the free electrons is due mainly to impurities. Thus $l_e \propto \frac{1}{N_I}$, N_I, being the number density of the impurity atoms, which is independent of T. Thus

$$k \propto T.$$

The above may be summarized in a plot of the metallic solid thermal conductivity against temperature as shown in Fig. 1.10.

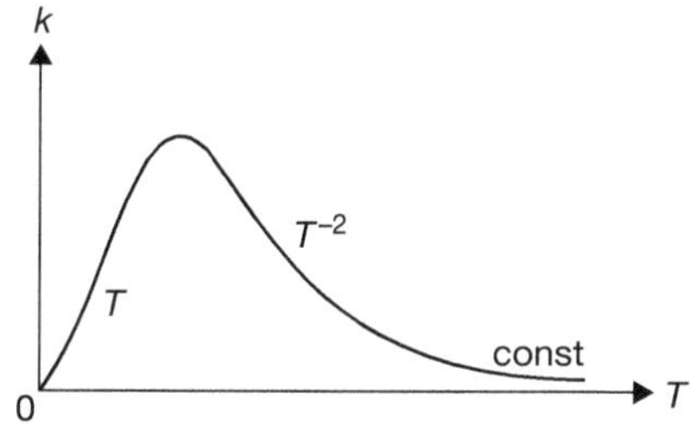

Fig. 1.10

1023

A material of density ρ has a face-centered cubic lattice with cube edge length a_0 and an Einstein temperature θ_E. For the high temperature limit, i.e., $T > \theta_E$, express the atomic mean square displacement in the x direction, $\overline{M_x^2}$, in terms of ρ, a_0 and θ_E plus the necessary physical constants.

(*Wisconsin*)

Sol: For a face-centered cubic lattice, the volume of a primitive cell per atom is $a_0^3/4$ so that

$$p = \frac{4m}{a_0^3},$$

where m is the mass of an atom.

In the Einstein model a lattice of N atoms is considered as a set of $3N$ independent harmonic oscillators in one dimension, each oscillator having an independent angular frequency ω_E related to the Einstein temperature θ_E by

$$\hbar\,\omega_E = k_B\,\theta_E\,.$$

For a harmonic oscillator the average potential energy $\frac{k}{2}\overline{\mu_x^2}$ and the average kinetic energy $\frac{m}{2}\overline{\dot{x}^2} = \frac{m}{2}\omega_E^2\overline{\mu_x^2}$ are equal, each being equal to half of the total energy $\left(n+\frac{1}{2}\right)\hbar\,\omega_E$ given by quantum mechanics. Hence the mean square displacement is

$$\overline{\mu_x^2} = \left(n+\frac{1}{2}\right)\hbar\,\omega_E/k = \left(n+\frac{1}{2}\right)\hbar/m\,\omega_E.$$

At temperature T, the oscillators are distributed over the eigenstates according to the Boltzmann distribution law $\exp\left[-\left(n+\frac{1}{2}\right)\frac{\hbar\,\omega_E}{k_B\,T}\right]$. Thus

$$\overline{\mu_x^2} = \frac{\sum\limits_{n=0}^{\infty}\left(n+\frac{1}{2}\right)\frac{\hbar}{m\,\omega_E}e^{-(n+\frac{1}{2})\hbar\omega_E/k_BT}}{\sum\limits_{n=0}^{\infty}e^{-(n+\frac{1}{2})\hbar\omega_E/k_BT}}$$

$$= \frac{\hbar}{m\,\omega_E}\left(\frac{\sum\limits_{n=0}^{\infty}ne^{-n\hbar\omega/k_BT}}{\sum\limits_{n=0}^{\infty}e^{-n\hbar\omega/k_BT}} + \frac{1}{2}\right).$$

The first term in the brackets can be written as

$$\frac{\sum ne^{nx}}{\sum e^{nx}} = \frac{\sum\frac{d}{dx}e^{nx}}{\sum e^{nx}} = \frac{\frac{d}{dx}\sum e^{nx}}{\sum e^{nx}} = \frac{d}{dx}\ln\sum e^{nx}$$

$$= \frac{d}{dx}\ln\left(\frac{1}{1-e^x}\right) = \frac{e^x}{1-e^x} = \frac{1}{e^{-x}-1}$$

by putting $x = -\frac{\hbar\omega_E}{k_B T}$. Hence

$$\overline{\mu_x^2} = \frac{\hbar}{m\omega_E}\left(\frac{1}{e^{\hbar\omega E/k_B T} - 1} + \frac{1}{2}\right).$$

With $m = \frac{a_0^3 \rho}{4}$, $\omega_E = \frac{k_B \theta_E}{\hbar}$, we have for the high temperature limit

$$\begin{aligned}\overline{\mu_x^2} &= \frac{4\hbar^2}{k_B\theta_E a_0^3 \rho}\left(\frac{1}{e^{\theta_E/T} - 1} + \frac{1}{2}\right)\\ &\approx \frac{4\hbar^2}{k_B\theta_E a_0^3\rho}\cdot\frac{T}{\theta_E}\\ &= \frac{4\hbar^2 T}{k_B \theta_E^2 a_0^3 \rho}.\end{aligned}$$

1024

Consider a vibrating solid.

a. Evaluate the Helmholtz free energy *F* of a phonon mode of frequency ω at temperature *T*.

b. Assume that the solid is harmonic with a bulk modulus *B* and that Δ is the fractional volume change. Ignoring any dispersion of the phonon modes, i.e. taking $\omega_k = \omega$, write down the free energy of the crystal.

c. If the volume dependence of ω is $\delta\omega/\omega = -\gamma\Delta$, where γ is known as the Grüneisen constant, how much contraction exists at temperature *T*?

d. Discuss the physical significance of the Grüneisen constant.

(*Princeton*)

Sol:

a. For a phonon mode of frequency ω, the partition function is

$$Z = \sum_{n=0}^{\infty} e^{-(n+\frac{1}{2})\hbar\omega/k_B T} = \frac{e^{-\hbar\omega/2k_B T}}{1 - e^{-\hbar\omega/k_B T}}.$$

The Helmholtz free energy of a crystal is therefore

$$\begin{aligned}F &= -k_B T \ln Z\\ &= k_B T\left[\frac{\hbar\omega}{2k_B T} + \ln\left(1 - e^{-\hbar\omega/k_B T}\right)\right].\end{aligned}$$

b. The free energy of a crystal is

$$F(V, T) = U(V) + k_B T \sum_k \left[\frac{\hbar\omega}{2k_B T} + \text{In}\left(1 - e^{-\hbar\omega/k_B T}\right)\right],$$

where V is the volume of the crystal. If $\omega_k = \omega$ and N is the total number of degrees of freedom of the crystal, we have

$$F(V,\ T) = U(V) + Nk_B T\left[\frac{\hbar\omega}{2k_B T} + \ln\left(1 - e^{-\hbar\omega/k_B T}\right)\right],$$

where $U(V)$ is the internal energy of the crystal at $T = 0$ K.

c. Because of the nonlinearity of the vibration, ω changes with volume. The equation of state of the crystal is therefore

$$p = -\left(\frac{\partial F}{\partial V}\right)_T = -\left(\frac{\partial U}{\partial V}\right)_T - N\left(\frac{1}{2}\hbar + \frac{\hbar e^{-\hbar\omega/k_B T}}{1 - e^{-\hbar\omega/k_B T}}\right)\frac{\partial\omega}{\partial V}.$$

As $\frac{\delta\omega}{\omega} = -\gamma\Delta = -\gamma\frac{\delta V}{V}$, we have

$$\frac{\partial\omega}{\partial V} = -\gamma\frac{\omega}{V},$$

and

$$p = -\left(\frac{\partial U}{\partial V}\right)_T + \gamma\frac{\bar{E}}{V},$$

where

$$\bar{E} = N\left(\frac{1}{2}\hbar\omega + \frac{\hbar\omega}{e^{\hbar\omega/k_B T} - 1}\right)$$

is the vibrational energy of the lattice at angular frequency ω.

Since the thermal expansion occurs as a result of the vibration, in the absence of applied pressure, we have $p = 0$ and so

$$\frac{\partial U}{\partial V} = \gamma\frac{\bar{E}}{V}.$$

Expanding $\frac{\partial U}{\partial V}$ around the static lattice volume V_0 by Taylor's expansion, we have

$$\frac{\partial U}{\partial V} = \left(\frac{\partial U}{\partial V}\right)_{V_0} + \left(\frac{\partial^2 U}{\partial V^2}\right)_{V_0}\delta V + \cdots$$
$$\approx \left(\frac{\partial^2 U}{\partial V^2}\right)_{V_0}\delta V.$$

Using the definition of the volume elastic modulus of a static lattice,

$$B = V_0\left(\frac{\partial^2 U}{\partial V^2}\right)_{V_0},$$

we have

$$\frac{\partial U}{\partial V} = B\frac{\delta V}{V_0} = \gamma\frac{\bar{E}}{V},$$

or

$$\frac{\delta V}{V_0} = \gamma \frac{\bar{E}}{BV}$$

at temperature T.

d. Differentiating the two sides of the last equation with respect to T, we have as $\frac{\partial \bar{E}}{\partial T} \gg \frac{\partial V}{\partial T}$,

$$\frac{1}{V_0}\frac{\partial \delta V}{\partial T} = \frac{\gamma}{BV}\frac{\partial \bar{E}}{\partial T} = \frac{\gamma}{BV}C_v.$$

Then from the definition of the coefficient of volume thermal expansion

$$\alpha = \frac{1}{V_0}\left(\frac{\partial \delta V}{\partial T}\right)_p,$$

we find the Grüneisen relation

$$\alpha = \gamma \frac{C_v}{BV}.$$

It can be seen that if the vibration of a crystal were strictly linear, $\gamma = 0$ and hence $\alpha = 0$, i.e., there would be no thermal expansion. Since thermal expansion does occur, the coefficient of thermal expansion gives a measure of the nonlinearity of crystal lattice vibration.

1025

Neutron diffraction may be used to measure ω vs. k for an excitation in a crystalline solid. To describe this, assume the crystal symmetry is known, write down the energy and momentum conservation laws for the diffraction, and then indicate what parameters must be measured in order to obtain ω vs. k.

(Wisconsin)

Sol: Let M be the mass of a neutron, $\mathbf{p}$ and $\mathbf{p}'$ be the momenta and $\mathbf{k}$ and $\mathbf{k}'$ be the wave vectors of the incident and scattered neutrons respectively. Then $\mathbf{p} = \hbar\mathbf{k}$, $\mathbf{p}' = \hbar\mathbf{k}'$ and the corresponding energies are $\frac{p^2}{2M}, \frac{p'^2}{2M}$. Energy conservation gives

$$\frac{\hbar^2 k^2}{2M} = \frac{\hbar^2 k'^2}{2M} \pm \hbar\omega(\mathbf{q})$$

and momentum conservation gives

$$\hbar\mathbf{k} = \hbar\mathbf{k}' \pm \hbar\mathbf{q} - \hbar\mathbf{G},$$

where $\mathbf{G}$ is an appropriate reciprocal lattice vector, and $\mathbf{q}$ is the phonon wave vector generated (+) or absorbed (–) in the process.

To obtain ω vs. k from the above equation it is necessary to measure the energy difference between the incident and scattered neutrons as a function of the scattering direction $\mathbf{k} - \mathbf{k}'$.

1026

Consider the presence of the point defects in the form of vacancies in thermal equilibrium in a monatomic simple cubic crystal of N sites.

a. Write down or evaluate the number n of such defects in a crystal held at temperature T if it costs an energy E to create one vacancy. (Assume $n \ll N$.)

We now consider the effect of lattice vibrations (i.e., phonons) on the result. Mark a simple Einstein theory of normal modes of the crystal with vacancies, i.e., treat each ion as an independent oscillator, choose two different frequencies, say ω and ω', depending on whether a given ion has one of its six nearest-neighbor sites vacant.

b. Which of the two frequencies do you expect to be larger?

c. How is the number n modified in the presence of the phonons?

(Princeton)

Sol:

a. Neglecting the effect of phonons, the number of vacancies at thermal equilibrium at temperature T is

$$n = Ne^{-E/k_B T},$$

where E is the energy required to create one vacancy.

b. When there are vacancies surrounding an atom, the restoring force it suffers when displaced from the equilibrium position becomes smaller. Consequently ω' is smaller than ω.

c. Consider now the effect of phonons. As $n \ll N$, surrounding each vacancy there are six nearest-neighbor atoms whose vibrational frequency has changed from ω to ω'. Since each site represents three modes of oscillation, there are $3(N - 6n)$ phonons of frequency ω and $3 \times 6n$ phonons of frequency ω'. The contribution of a phonon of frequency ω to the free energy of the crystal is (Problem **1024**)

$$k_B T\left[\frac{\hbar\omega}{2 k_B T} + \ln\left(1 - e^{\hbar\omega/k_B T}\right)\right] \approx k_B T \ln\frac{\hbar\omega}{k_B T}$$

in the classical limit $\hbar\omega \ll k_B T$. Hence the free energy of the crystal is

$$F = U_0 + nE + 3(N - 6n)\,k_B\,T\ln\frac{\hbar\omega}{k_B\,T} + 18n\,k_B\,T\ln\frac{\hbar\omega'}{k_B\,T} - k_B\,T\ln\frac{N!}{(N-n)!n!},$$

where U_0 is the internal energy of the crystal at 0 K, and the last term is contributed by the mixing entropy.

At equilibrium,

$$\left(\frac{\partial F}{\partial n}\right)_T = 0,$$

or

$$\frac{\partial}{\partial n}\left[\frac{nE}{k_B\,T} + 18n\ln\frac{\omega'}{\omega} + \ln(N-n)!+\ln n!\right] = 0.$$

Using Stirling's formula

$$\ln x! \approx x\ln x$$

for $x \to \infty$, as N, n are both large even though $n \ll N$ we have

$$\frac{E}{k_B\,T} + 18\ln\frac{\omega'}{\omega} + \ln\frac{n}{N} = 0,$$

or

$$n = N\left(\frac{\omega}{\omega'}\right)^{18} e^{-\frac{E}{k_B T}},$$

in the presence of phonons.

1027

A sodium chloride crystal plane is mounted on an X-ray spectrometer that uses monochromatic X-rays of wavelength 0.586 Å. The glancing angles recorded for three reflections are 5.97°, 12.01°, and 18.18°.

a. Show that it corresponds to reflections of successive order from the same plane.

b. Determine the interplanar spacing and specify whether it changes for different orders of reflections on the same plane.

Sol:

a. To find out whether these angles correspond to successive orders of reflections, we have to check the ratio between the $\sin\theta$ of the glancing angles.

$$\begin{aligned}\sin\theta_1 : \sin\theta_2 : \sin\theta_3 &= \sin 5.91 : \sin 12.01 : \sin 18.18 \\ &= 0.1040 : 0.2080 : 0.\,3120 = 1:2:3\end{aligned}$$

From this, it can be concluded that these reflections correspond to successive orders.

b. The interplanar distance can be calculated from the formula, $n\lambda = 2d\sin\theta \rightarrow d = \frac{n\lambda}{2\sin\theta}$

For first order, sin5.91 = 0.1040

$$d = \frac{n\lambda}{2\sin\theta} = \frac{0.586 \times 10^{-10}}{2 \times 0.1040} = 2.817\text{Å}$$

For second order, sin12.01 = 0.2080

$$d = \frac{n\lambda}{2\sin\theta} = \frac{2 \times 0.586 \times 10^{-10}}{2 \times 0.2080} = 2.817\text{Å}$$

For third order, sin18.18 = 0.3120

$$d = \frac{n\lambda}{2\sin\theta} = \frac{3 \times 0.586 \times 10^{-10}}{2 \times 0.3120} = 2.817\text{Å}$$

From the result, it is evident that the interplanar distance doesn't change when the order of the reflection changes.

1.2

ELECTRON THEORY, ENERGY BANDS AND SEMICONDUCTORS (1028–1051)

1028

a. Determine the cubic system in which the monovalent metal, lithium crystallizes given that its lattice constant is 3.51 Å, atomic mass is 6.94 u and mass density is 546 kg/m^3.

b. Find the shape of the Fermi surface and its characteristic dimension, $k_{F.}$ Will there be a distortion of the Fermi surface? (Compare k_F with the distance from the origin to the boundary of the first Brillouin zone in reciprocal space)

c. Also estimate lithium's Fermi energy, its Fermi temperature T_F, and the speed of the fastest moving free electrons.

d. The resistivity of lithium is approximately 10^{-5} Ωcm at 300 K. Find the mean free path and collision time of the free electrons.

(*Wisconsin*)

Sol:

a. Mass of one atom of lithium = Atomic mass/Avogadro number = 1.16×10^{-26} kg ----(1)

Mass contained in a cube of edge 3.51 Å = 546 kg/m^3 × $(3.51 \times 10^{10}$ m$)^3$ = 2.3×10^{-26} kg -----(2)

From (1) and (2), it is evident that there are two atoms per cube. **Hence Li crystallizes in the body-centered cubic system**

b. The Fermi surface is almost spherical with a radius k_F.

To find k_F:

Two electrons (one with up and one with down spin) can exist in a volume of $(2\pi/L)^3$ in reciprocal space. Also, the total number of free electrons per mole (N_0) that form the Fermi sphere of volume $\left(\frac{4\pi}{3}\right)k_F{}^3$ is equal to the Avogadro number (since Li is monovalent). We get the relation

$$\left(\frac{2\pi}{L}\right)^3 :2 = \left(\frac{4\pi}{3}\right)k_F{}^3 : N_0$$

The Fermi radius is hence

$$k_F = \left[3\pi^2\left(\frac{N_0}{V}\right)\right]^{1/3} = [3\pi^2 n]^{1/3}$$

On substituting $V = m/\rho$ and $N_0 = 6.023 \times 10^{23}$ e/mol, $n = 4.74 \times 10^{22}$ e/cm^3

Hence, $k_F = 1.11 \times 10^8$ cm^{-1}

The reciprocal lattice of a body-centered cubic lattice is a face-centered cubic lattice. The distance from the origin of the reciprocal space to the nearest Brillouin zone boundary (110) plane **is $\frac{\pi\sqrt{2}}{a}$ = 1.27×10^8 cm^{-1}**

There will be no distortion of the Fermi sphere as this value is larger than k_F.

c. $E_F = \frac{\hbar^2 k_F{}^2}{2m} = 4.72$ eV; $T_F = \frac{E_F}{K_B} = 64{,}800$ K; $v_F = \frac{\hbar k_F}{m} = 1300$ km/s

d. Knowing $\sigma = \frac{ne^2\tau}{m}$ and $\sigma = \frac{1}{\rho}$

$\tau = 0.76 \times 10^{-14}$ s

1029

Metallic sodium crystalizes in body-centered cubic form, the length of the cube being 4.25×10^{-8} cm. Find the concentration of conduction electrons.

Assume one conduction electron per atom. Adopting the free-electron Fermi gas model for the conduction electrons, derive an expression for the Fermi energy (at 0 K) and show that it depends only on the concentration of conduction electrons, but not on the mass of the crystal.

(*Wisconsin*)

Sol: A bcc lattice contains $\frac{2}{a^3}$ atoms per unit volume, so the concentration of conduction electrons is

$$n = \frac{2}{(4.25 \times 10^{-8})^3} = 2.6 \times 10^{22} \text{ cm}^{-3}.$$

According to Fermi's free-electron gas model, the number of electrons in a volume V with energies in the range E to $E + dE$ is

$$dN = CE^{1/2} f(E) dE$$

with $c = 4\pi V(2m)^{3/2} / \hbar^3$, m being the electron mass. At $T = 0$ K, the distribution function $f(E)$ is

$$f(E) = \begin{cases} 1 & \text{for } E \leq E_F, \\ 0 & \text{for } E > E_F. \end{cases}$$

Hence the total number of electrons in V is

$$N = \int_0^{E_{F0}} CE^{1/2} dE = \frac{2}{3} C E_{F0}^{\frac{3}{2}}.$$

where E_{F0} is the Fermi energy at $T = 0$ K. As $n = \frac{N}{V}$, the Fermi energy at 0 K is

$$E_{F0} = \frac{h^2}{2m}\left(\frac{3n}{8\pi}\right)^{\frac{2}{3}} = \frac{\hbar^2}{2m}(3\pi^2 n)^{\frac{2}{3}}.$$

For metallic sodium we have

$$E_{F0} = \frac{(1.05 \times 10^{-27})^2}{2 \times 9.1 \times 10^{-28}} \times \frac{(3\pi^2 \times 26 \times 10^{21})^{2/3}}{1.6 \times 10^{-12}} = 3.18 \text{ eV}.$$

The expression of E_{F0} shows that it depends only on the concentration of conduction electrons, but not on the mass of the crystal.

1030

Consider a square lattice of lattice parameter "*a*," with identical atoms located in the *xy* plane as shown.

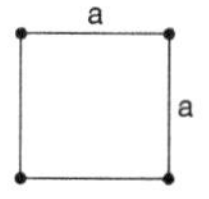

a. Determine the valence electron dispersion relation for this lattice in the tight-binding approximation.

b. Find the energies E_Γ, E_X, and E_M, with the points Γ, X, and M of the Brillouin zone shown in the following. Also find the dispersion relation along the directions $\Gamma \rightarrow$ X, $\Gamma \rightarrow$ M, and X $\rightarrow$ M.

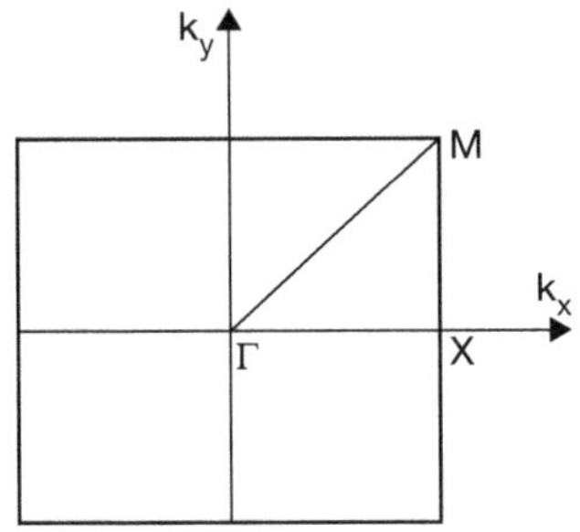

Sol:

a. In the tight-binding approximation,

$$E = -\alpha - \gamma \sum_j e^{-i\bar{k}.\bar{\rho}_j}$$

In which α and γ have positive values and ρ_j represents the vectors connecting an atom at the origin with its nearest neighbors.

An atom at the origin has four nearest neighbors in the real space at a distance a.

$$E = -\alpha - \gamma\left(\cos k_x a + \cos k_x a + \cos k_y a + \cos k_y a\right)$$

$$\boldsymbol{E = -\alpha - 2\gamma\left(\cos k_x a + \cos k_y a\right)}$$

b. The points in the Brillouin zone are $\Gamma = (0,0)$, $X = (\pi/a, 0)$, and $M = (\pi/a, \pi/a)$

$$\boldsymbol{E_\Gamma = -\alpha - 4\gamma} \quad \left(k_x = k_y = 0\right)$$

$$\boldsymbol{E_X = -\alpha} \quad \left(k_x = \pi/a;\ k_y = 0\right)$$

$$\boldsymbol{E_M = -\alpha + 4\gamma} \quad \left(k_x = k_y = \pi/a\right)$$

Along $\Gamma \rightarrow X$, $k_y = 0$. Hence $\boldsymbol{E = -\alpha - 2\gamma - 2\gamma(\cos k_x a)}$

Along $\Gamma \rightarrow M$, $k_x = k_y = \frac{|k|}{\sqrt{2}}$. Hence $\boldsymbol{E = -\alpha - 4\gamma \cos \frac{|k|}{\sqrt{2}} a}$

Along $X \rightarrow M$, $k_x = \pi/a$. Hence $\boldsymbol{E = -\alpha + 2\gamma - 2\gamma \cos k_y a}$

1031

Graphs of the temperature dependence of the electrical resistivity of Cu, Ge, and NaCl are given in Fig. 1.14. They describe "real" high-purity samples: purity > 99.9%, but not 100%. In each case describe briefly the essential mechanism of conduction and explain the temperature dependence illustrated. (Note that the axes of the three graphs are different.)

(Columbia)

Sol:

1. Cu is a metal. The graph of its electric resistivity vs. temperature shows: at $T \approx 0$ K, the resistivity is small and essentially constant; for $0 < T \ll \Theta_D$, $\rho \propto T^5$; for $T > \Theta_D$, $\rho \propto T$. Such behavior is typical of a metal and arises from the relation between the resistivity ρ and the probability P of scattering of the passing electron:

$$\rho = \frac{m}{ne^2} P,$$

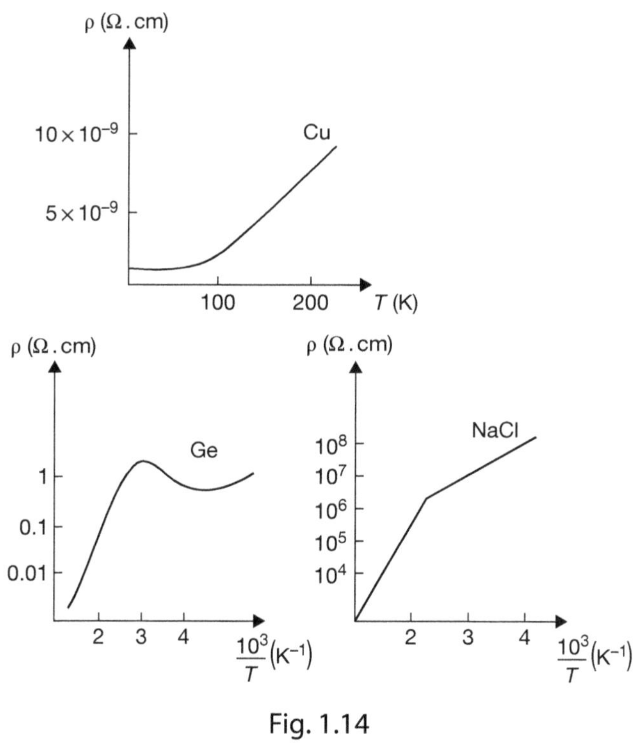

Fig. 1.14

where m is the electron mass and $P = P_{\text{phonon}} + P_{\text{impurity}}$ (Problem **1030** (b)).

a. $T \approx 0$ K

At this temperature, $P_{\text{phonon}} \to 0$ and ρ is contributed almost entirely by P_{impurity}, which is independent of temperature. Thus ρ is at its minimum and essentially constant.

b. $0 < T \ll \Theta_D$, the Debye temperature of Cu.

At such low temperatures, ρ is mainly determined by P_{phonon}, which is proportional to (Problem **1030**)

$$\frac{\text{density of phonons}}{\text{number of collision making up one effective scattering}} \propto \frac{T^3}{T^{-2}} = T^5 .$$

Hence $\rho \propto T^5$.

c. $T > \Theta_D$

ρ is almost entirely contributed by P_{phonon}, which at such high temperatures is proportional to the density of phonons which is $\propto T$. Hence $\rho \propto T$. Note that at such temperatures, the phonons have large momenta and every collision causes some deflection of the passing electrons.

2. Ge is a semiconductor. The graph of its electric resistivity vs. temperature shows: at low temperatures, ρ decreases with increasing temperature; for intermediate temperatures, ρ increases with increasing temperature; at high temperatures, ρ decreases rapidly with increasing temperature. This behavior is explained as follows.

a. At low temperatures, impurity excitation plays the dominant role. The number of excited charge carriers from the impurities increases with temperature, so that ρ decreases with rising temperature.

b. In the middle temperature range, all the impurity atoms are ionized and so the number of charge carriers does not change with temperature. However, scattering of the passing electrons by phonons becomes more frequent as the temperature increases, leading to a lowering of mobility and hence increasing ρ.

c. At high temperatures, intrinsic excitation starts and the concentration of charge carriers rapidly increases with rising temperature, causing ρ to drop quickly.

3. NaCl is an ionic crystal. The semi-log ρ–T curve consists of linear segments with different slopes for low and high temperatures. The reason is as follows.

In an ionic crystal electrical conduction by the holes is the dominant process, the number of holes being proportional to $e^{-u_0/k_B T}$, where u_0 is the activation energy. At low temperatures, conduction by impurity holes is the dominant mechanism; while at high temperatures, the conduction is due mainly to holes resulting from thermal motion of the ions. As their activation energies are different, the semi-log graph shows two lines of different slopes.

1032

A piece of metal with $\mu = 1$ is placed in a static uniform magnetic field $\mathbf{B} = B_0\hat{\mathbf{z}}$. The conduction electrons can be treated as a free electron gas with a scattering time τ and number density n.

a. Derive an expression for the resistivity tensor of this metal.

b. Using the result of (a) derive the low-frequency dispersion relation for waves propagation along $\hat{\mathbf{z}}$. Neglect $\frac{1}{\tau}$ and ω in comparison with the cyclotron frequency ω_c. (Derive a wave equation for $\mathbf{j}$, the current density.)

(*Princeton*)

Sol:

a. In the scattering of a conduction electron of momentum $m\mathbf{v}$, the average change of momentum is $\frac{1}{2}(0 + 2m\mathbf{v}) = m\mathbf{v}$. If on average it suffers $\frac{1}{\tau}$ scatterings per unit time, it is subjected to a damping force $-\frac{m\mathbf{v}}{\tau}$. Hence the equation of motion of a conduction electron is

$$m\frac{d\mathbf{v}}{dt} = -e\mathbf{E} - e\mathbf{v} \times \mathbf{B} - m\frac{\mathbf{v}}{\tau}.$$

As

$$\mathbf{j} = -ne\mathbf{v},$$

the above becomes

$$-\frac{m}{ne}\frac{d\mathbf{j}}{dt} = -e\mathbf{E} + \frac{\mathbf{j} \times \mathbf{B}}{n} + \frac{m\mathbf{j}}{\tau ne}.$$

Consider a solution of the form

$$\mathbf{j} = \mathbf{j}_0 e^{-i\omega t}.$$

Substitution gives

$$\mathbf{E} = \frac{m}{ne^2}\left(\frac{1}{\tau} - i\omega\right)\mathbf{j} + \frac{\mathbf{j} \times \mathbf{B}}{ne},$$

and in particular

$$E_k = \frac{m}{ne^2}\left(\frac{1}{\tau} - i\omega\right) j_k + \frac{\varepsilon_{klq} j_l B_q}{ne}$$
$$= \left[\frac{m}{ne^2}(\frac{1}{\tau} - i\omega)\,\delta_{kl} + \frac{\varepsilon_{klq} B_q}{ne}\right] j_l = \rho_{kl} j_l, \qquad \text{say,}$$

where

$$\delta_{kl} = \begin{cases} 1 & \text{if } k = l, \\ 0 & \text{if } k \neq l, \end{cases}$$
$$\varepsilon_{klq} = \begin{cases} 1 & \text{if } k \neq l \neq q \text{ and } k, l, q \text{ are in an even permutation of 123,} \\ -1 & \text{if } k \neq l \neq q \text{ and } k, l, q \text{ are in an odd permutation of 123,} \\ 0 & \text{if all or any two of } k, l, \text{q are the same,} \end{cases}$$

and Einstein's summation rule is used.

If Ohm's law

$$\mathbf{j} = \sigma \mathbf{E},$$

or

$$\mathbf{E} = \rho \mathbf{j}, \tag{1}$$

holds, a resistivity ρ can be defined. The above shows that ρ has the from of a matrix

$$p = \begin{pmatrix} \frac{m}{ne^2}\left(\frac{1}{\tau} - i\omega\right) & \frac{B_0}{ne} & 0 \\ -\frac{B_0}{ne} & \frac{m}{ne^2}\left(\frac{1}{\tau} - i\omega\right) & 0 \\ 0 & 0 & \frac{m}{ne^2}\left(\frac{1}{\tau} - i\omega\right) \end{pmatrix}.$$

b. In the metal,

$$\mathbf{D} = \epsilon_0 \mathbf{E} + \mathbf{P}, \quad \nabla \cdot \mathbf{D} = 0, \quad \mathbf{B} = \mu_0 \mathbf{H}$$

and Maxwell's equations

$$\nabla \times \mathbf{E} = -\dot{\mathbf{B}}, \quad \nabla \times \mathbf{H} = \dot{\mathbf{D}}$$

give

$$\mu_0 \ddot{\mathbf{D}} = \nabla^2 \mathbf{E} - \nabla(\nabla \cdot \mathbf{E}).$$

An electromagnetic wave propagating along $\hat{\mathbf{z}}$ can be represented by

$$\mathbf{E} = \mathbf{E}_0 e^{i(kz - \omega t)},$$

from which

$$\frac{\partial}{\partial t} = -i\omega, \quad \frac{\partial}{\partial z} = ik, \quad \frac{\partial}{\partial x} = \frac{\partial}{\partial y} = 0.$$

Thus we have

$$-\omega^2\mu_0(\epsilon_0\mathbf{E}+\mathbf{P}) = -k^2\mathbf{E}+k^2E_z\hat{\mathbf{z}}.$$

As

$$\nabla\cdot\mathbf{D} = \frac{\partial D_z}{\partial_z} = ik(\epsilon_0 E_z + P_z) = 0,$$

or

$$E_z = -\frac{P_z}{\epsilon_0},$$

the above can be written as

$$\left(k^2-\frac{\omega^2}{c^2}\right)\mathbf{E} = \mu_0\omega^2\mathbf{P} - \frac{k^2}{\epsilon_0}P_z\hat{\mathbf{z}},$$

or

$$\mathbf{E} = \frac{1}{\epsilon_0\chi}[\mathbf{P}-(\chi+1)P_z\hat{\mathbf{z}}],$$

where $\chi = \left(\frac{ck}{\omega}\right)^2 - 1$. Hence

$$\begin{pmatrix} E_x \\ E_y \\ E_z \end{pmatrix} = \frac{1}{\epsilon_0\chi}\begin{pmatrix} 1 & 0 & 0 \\ 0 & 1 & 0 \\ 0 & 0 & -\chi \end{pmatrix}\begin{pmatrix} P_x \\ P_y \\ P_z \end{pmatrix}. \tag{2}$$

The polarization current density **j** is given by

$$\mathbf{j} = \dot{\mathbf{P}} = -i\omega\mathbf{P}. \tag{3}$$

Writing Eqs. (1) and (2) as matrix equations

$$E = RJ, \quad E = AP = BJ,$$

where E, P, J, are column matrices, R, A, B are square matrices. Equation (3) then gives

$$B = \frac{i}{\omega}A.$$

As

$$(R-B)J = 0,$$

for J not identically zero, we require the determinant

$$|R-B| = 0,$$

i.e.,

$$\left[\frac{m}{ne^2}\left(\frac{1}{\tau}-i\omega\right)+\frac{i}{\epsilon_0\omega}\right]\left\{\left[\frac{m}{ne^2}\left(\frac{1}{\tau}-i\omega\right)-\frac{i}{\epsilon_0\chi\omega}\right]^2+\left(\frac{B_0}{ne}\right)^2\right\} = 0.$$

The first factor of the left-hand side, which involves only the parameters n, τ of the electron gas and the angular frequency ω of the propagating wave, is generally not zero. Hence

$$\left[\frac{m}{ne^2}\left(\frac{1}{\tau} - i\omega\right) - \frac{i}{\epsilon_0 \chi \omega}\right]^2 = -\left(\frac{B_0}{ne}\right)^2 .$$

Defining the plasma frequency

$$\omega_{\mathrm{P}} = \sqrt{\frac{ne^2}{\epsilon_0 m}}$$

and the cyclotron frequency

$$\omega_{\mathrm{c}} = \frac{eB_0}{m},$$

we can write the above as

$$\left[\left(\frac{1}{\tau} - i\omega\right) - \frac{i\omega_{\mathrm{p}}^2}{\chi\omega}\right]^2 = -\omega_{\mathrm{c}}^2 .$$

As $\frac{1}{\tau}$ and ω can be neglected in comparison with ω_{c} this gives

$$-\frac{i\omega_{\mathrm{p}}^2}{\chi\omega} = \pm i\omega_{\mathrm{c}} .$$

Hence the low-frequency dispersion relations are

$$-\omega_{\mathrm{p}}^2 = \pm\omega\,\omega_{\mathrm{c}}\chi = \pm\omega\,\omega_{\mathrm{c}}\left(\frac{c^2k^2}{\omega^2} - 1\right),$$

or

$$\frac{c^2k^2}{\omega^2} = 1 \mp \frac{\omega_{\mathrm{p}}^2}{\omega\,\omega_{\mathrm{c}}} .$$

1033

The Drude-Lorentz formula for the dielectric constant of a solid is

$$\varepsilon(\omega) = 1 + \frac{\omega_{\mathrm{p}}^2}{(\omega_0^2 - \omega^2) - i\omega\tau^{-1}} .$$

Here ω_{p} is the plasma frequency, ω_0 is the energy gap for interband transitions and τ is the scattering time of the electron.

a. At room temperature a reasonable value for Cu is $\tau = 10^{-14}$sec. Give order of magnitude estimates of ω_{p} and ω_0 for this metal. You may want to make use of the characteristic "color" of the metal in determining ω_0. Plot the real and imaginary parts of $\varepsilon(\omega)$ as a function of ω (in eV).

b. At room temperature, calculate $\sigma(\omega)$, the complex frequency-dependent conduction of Cu.

c. What is $\sigma(\omega)$ for perfectly pure, defect-free Cu at zero temperature?

(*Chicago*)

Sol:

a. Suppose each copper atom contributes one free electron. The concentration of free electrons is then

$$n = \frac{6.02 \times 10^{23} \times 8.9}{63.5} = 8.4 \times 10^{22}\mathrm{cm}^{-3} = 8.4 \times 10^{28}\mathrm{m}^{-3}.$$

For the electron, $\frac{e^2}{\epsilon_0 m} = 3 \times 10^3 \mathrm{m}^2\mathrm{s}^{-2}$, so that the plasma frequency is (Problem **1032**)

$$\omega_\mathrm{p} = \sqrt{\frac{ne^2}{\epsilon_0 m}} = 1.6 \times 10^{16}\mathrm{s}^{-1}.$$

Hence

$$\hbar\,\omega_\mathrm{p} = 6.6 \times 10^{-14} \times 1.6 \times 10^{16} = 10.5\,\mathrm{eV} \approx 10\,\mathrm{eV}.$$

The color of copper is redish, corresponding to a wavelength $\lambda_0 \approx 6 \times 10^{-7}$ m, or $\omega_0 \approx 3 \times 10^{15}$ s^{-1}. Hence

$$\hbar\,\omega_0 \approx 2\,\mathrm{eV}.$$

At room temperature, $\tau = 10^{-14}$ s, or, $\hbar\tau^{-1} = 6.6 \times 10^{-2}$ eV $\approx 10^{-1}$ eV, and the real and imaginary parts of $\varepsilon(\omega)$ are

$$\mathrm{Re}\,\varepsilon(\omega) = 1 + \frac{\omega_\mathrm{p}^2(\omega_0^2 - \omega^2)}{(\omega_0^2 - \omega^2)^2 + \omega^2\tau^{-2}} \approx 1 + \frac{10^2 \cdot (4 - \omega^2)}{(4 - \omega^2)^2 + \omega^2 \cdot 10^{-2}},$$

$$\mathrm{Im}\,\varepsilon(\omega) = \frac{\omega_\mathrm{p}^2\,\omega\tau^{-1}}{(\omega_0^2 - \omega^2)^2 + \omega^2\tau^{-2}} \approx \frac{20}{(4 - \omega^2)^2 + \omega^2 \cdot 10^{-2}}.$$

These are plotted in Fig. 1.15.

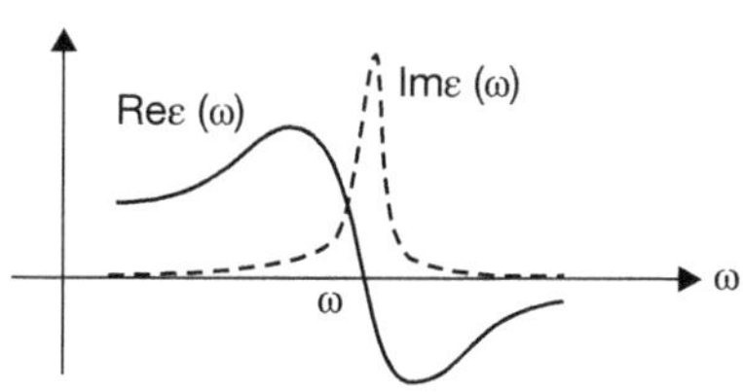

Fig. 1.15

b. Applying Ohm's law $\mathbf{J} = \sigma\mathbf{E}$ and the expression for the polarization current density $\mathbf{J} = \dot{\mathbf{P}} = -i\omega\mathbf{P}$, we find the electric susceptibility

$$\chi \equiv \frac{P}{\epsilon_0 E} = \frac{i\sigma}{\epsilon_0 \omega}.$$

Then

$$\mathbf{D} = \epsilon \mathbf{E} = \epsilon_0 \mathbf{E} + \mathbf{P}$$

gives the complex dielectric constant of Cu

$$\varepsilon(\omega) = 1 + \chi = 1 + \frac{i\sigma(\omega)}{\epsilon_0 \omega}.$$

The Drude-Lorentz formula then gives

$$\frac{i\sigma}{\epsilon_0 \omega} = \frac{\omega_p^2}{(\omega_0^2 - \omega^2)^2 - i\omega\tau^{-1}},$$

or

$$\sigma(\omega) = \frac{\epsilon_0 \omega \omega_p^2 \left[\omega\tau^{-1} - i(\omega_0^2 - \omega^2)\right]}{\left[\omega_0^2 - \omega^2\right]^2 + \omega^2 \tau^{-2}}.$$

c. At zero temperature, $\tau \to \infty$ and the complex conductivity is

$$\sigma(\omega) = -\frac{i\epsilon_0 \omega \omega_p^2}{\omega_0^2 - \omega^2}.$$

1034

At time $t = 0$ a charge distribution $\rho(\mathbf{r})$ exists within an idealized homogeneous conductor whose permittivity ϵ and conductivity σ are constants. Obtain $\rho(\mathbf{r}, t)$ for subsequent times.

(*Wisconsin*)

Sol: Use of Maxwell's equation

$$\nabla \cdot \mathbf{D} = \epsilon \nabla \cdot \mathbf{E} = \rho$$

and Ohm's law

$$\mathbf{j} = \sigma \mathbf{E}$$

in the continuity equation which expresses conservation of charge,

$$\frac{\partial \rho}{\partial t} + \nabla \cdot \mathbf{j} = 0,$$

gives

$$\frac{\partial \rho}{\partial t} + \frac{\sigma}{\epsilon}\rho = 0.$$

With the initial condition $\rho(\mathbf{r}, 0) = \rho(\mathbf{r})$, integration gives

$$\rho(\mathbf{r}, t) = \rho(\mathbf{r}) e^{-\sigma t/\epsilon}.$$

1035

Consider a two-dimensional square lattice.

a. The kinetic energy of a free electron at a corner of the first Brillouin zone is larger than that of an electron at the midpoint of a side face of the zone by a factor b. Calculate b.

b. The crystal potential of the corresponding material is

$$V(x, y) = -2V_0\left(\cos\frac{2\pi x}{a} + \cos\frac{2\pi y}{a}\right),$$

where V_0 is a constant and a is the lattice spacing. Find approximately the energy gap at the midpoint of the zone face.

c. Assuming that your result in (b) is exact and that the material is divalent, write down a condition for the system to be metallic.

(*Princeton*)

Sol:

a. The kinetic energy of a free electron is

$$E = \frac{-p^2}{2m} = \frac{\hbar^2 k^2}{2m} \propto k^2.$$

The first Brillouin zone for a two-dimensional square lattice of lattice constant a is a square in the **k**-space of edge $\frac{2\pi}{a}$, as shown in Fig. 1.16. As

$$\overline{\Gamma M} = \sqrt{2}\,\overline{\Gamma X},$$

$$b = \frac{E_M}{E_X} = \frac{\overline{\Gamma M}^2}{\overline{\Gamma X}^2} = 2.$$

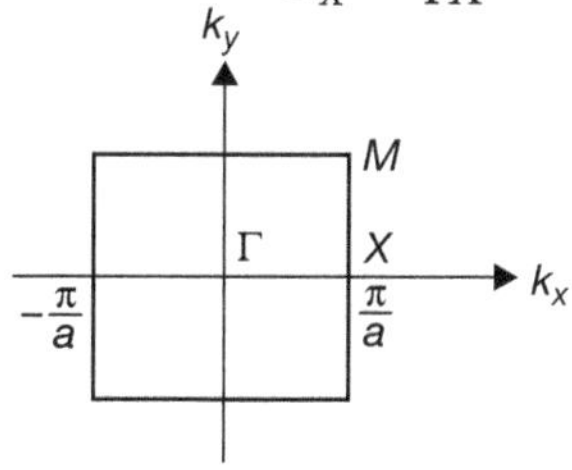

Fig. 1.16

b. We can write

$$V(x, y) = -V_0\left(e^{i2\pi x/a} + e^{-i2\pi x/a} + e^{i2\pi y/a} + e^{-i2\pi y/a}\right).$$

Let the basis vectors of the lattice be $\mathbf{b}_1$, $\mathbf{b}_2$. The midpoint of a side face of the zone in the reciprocal lattice has vector $\mathbf{k} = \frac{2\pi}{a}\mathbf{b}_1$. The corresponding Fourier coefficient in $V(x,y)$ is $-V_0$, so the energy gap is

$$\Delta_x = 2|V_x| = 2|V_0|.$$

c. For a divalent material, the condition for it to be metallic is that the energy bands overlap. Using the above result, this means that

$$E_X + \Delta_x \le E_M,$$

i.e.,

$$\frac{1}{2m}\hbar^2\left(\frac{\pi}{a}\right)^2 + 2|V_0| \le 2 \cdot \frac{\hbar^2}{2m}\left(\frac{\pi}{a}\right)^2,$$

giving

$$|V_0| \le \frac{\pi^2\hbar^2}{4ma^2}.$$

1036

Consider an electron of charge e in a one-dimensional lattice with energy levels

$$\varepsilon(k) = -2T\cos(ka),$$

where a is the lattice constant. A small uniform electric field **E** is imposed parallel to the lattice. Describe qualitatively the motion of the electron in **k**-space and in real space in both the absence and presence of scattering. What is meant by small **E** and what else can happen in a real (multiband) crystal when **E** is no longer small?

(Chicago)

Sol: In the absence of scattering, the equation of motion of an electron represented by a Bloch wave in an applied electric field **E** is

$$\frac{d\mathbf{p}}{dt} = \hbar\frac{d\mathbf{k}}{dt} = \mathbf{F} = -e\mathbf{E}.$$

Integration gives

$$\mathbf{k} = \mathbf{k}(0) - \frac{e\mathbf{E}t}{\hbar},$$

showing that **k** varies linearly with t. In **k**-space all the electrons move with the same velocity opposite to the direction of the electric field, as shown in Fig. 1.17 (a). When an electron reaches the boundary of the first Brillouin zone, say at point A, it is reflected and reappears at A′ on the other side of the origin. The states at A and A′ are perfectly identical. In this way the electron moves periodically in **k**-space.

Under the action of the applied electric field **E**, the electron state changes continually and so does its velocity (identical with the group velocity) $v = \frac{\partial \omega}{\partial k} = \frac{1}{\hbar}\frac{\partial \varepsilon(k)}{\partial k}$, ε being the electron energy, as shown in

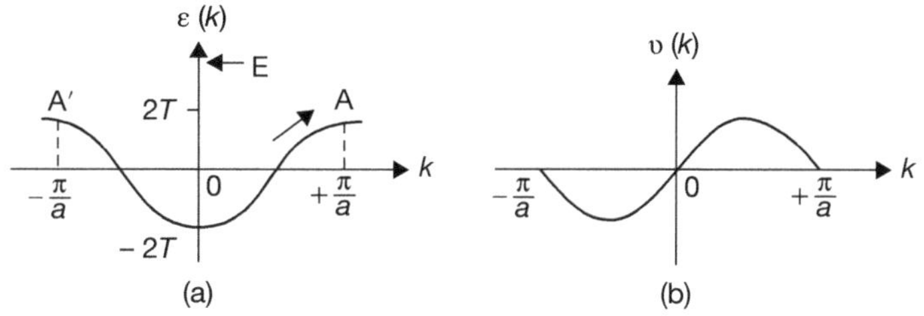

Fig. 1.17

Fig. 1.17 (b). The electron velocity thus alternates between positive and negative values. The motion of the electron in real space is also periodic.

In the presence of scattering, such oscillation as mentioned above is not observed. This is because in a crystal there are various mechanisms of scattering and the equation of motion given above only applies between scatterings. As the time between two scatterings is very short, the electron wave vector only moves along a small portion of the reciprocal lattice vector before the electron is scattered away, making it impossible to execute the oscillation described.

Small **E** means that it does not allow an electron to gain sufficient energy to transit to a higher energy band. A real crystal has a multitude of bands and if the electric field is large enough transitions between bands will take place. The condition for such transitions not to take place is

$$eEa \ll \left[\varepsilon_{\text{gap}}(k)\right]^2 / \varepsilon_{\text{F}}.$$

1037

Consider a one-dimensional metal with atoms regularly spaced a distance b apart in the tight-binding approximation, with one electron per atom. The atomic wave functions are of the form $\psi(r - R_i)$ where R_i is the position of the nucleus of the ith atom. ($R_i = ib$, the nuclei will be considered to be fixed in space.) Assume further

$$\int \psi^*(r - R_i) H \psi(r - R_i) d^3 r = -E_0,$$
$$\int \psi^*(r - R_i) H \psi(r - R_{i+1}) d^3 r = -V,$$
$$\int \psi^*(r - R_i) H \psi(r - R_{i+j}) d^3 r = 0 \quad \text{for } j \geq 2,$$

where H is the Hamiltonian. Calculate the following:

a. The electronic band structure (ε_k vs. k).

b. The electronic density of states.

c. The electronic contribution to the cohesive energy (relative to $-E_0$).

d. The electronic contribution to the specific heat at temperature T. (Assume $k_B T \ll V$).

(Princeton)

Sol:

a. According to the Bloch theorem the motion of an electron in a lattice is described by a traveling wave $e^{i\mathbf{k}\cdot\mathbf{r}}$ with an amplitude oscillating periodically from cell to cell. In the tight-binding method the modulation function is the atomic orbital $\psi(\mathbf{r} - \mathbf{R}_l)$ which is large in the neighborhood of an atom at $\mathbf{R}_l$ but decays rapidly away from it. For a one-dimensional lattice the electron wave function is thus

$$\phi_k(\mathbf{r}) = \frac{1}{\sqrt{N}} \sum_l e^{ikR_l} \psi(r - R_l),$$

where N is the number of atoms in the lattice. Then the energy of the electron is

$$\begin{aligned} E(k) &= \langle \phi_k | H | \phi_k \rangle \\ &= \frac{1}{N} \sum_{l,l'} e^{ik(R_l - R_{l'})} \langle \psi(r - R_l) | H | \psi(r - R_{l'}) \rangle, \end{aligned}$$

with the summation taken over all the atoms in the lattice. Note that for each particular choice of l the sum over l' yields the same result. As only nearest-neighbor interactions are important, we have

$$\begin{aligned} E(k) &= \langle \psi(r - R_l) | H | \psi(r - R_l) \rangle \\ &\quad + e^{-ikb} \langle \psi(r - R_l) | H | \psi(r - R_{l+1}) \rangle \\ &\quad + e^{ikb} \langle \psi(r - R_l) | H | \psi(r - R_{l-1}) \rangle \\ &= -E_0 - V(e^{-ikb} + e^{ikb}) \\ &= -E_0 - 2V \cos(kb). \end{aligned}$$

b. The number of states with wave vectors having magnitude smaller than or equal to k for a one-dimensional lattice of length L is (Problem **1014**)

$$N = \frac{Lk}{\pi},$$

whence the density of state per unit length of the lattice is

$$\rho(k) = \frac{1}{L}\frac{dN}{dk} = \frac{1}{\pi}.$$

For electron states, as there are two spin directions the density of state per unit length is

$$\rho(k) = \frac{2}{\pi}.$$

As

$$\rho(E)dE = \rho(k)dk,$$

we have

$$\begin{aligned}\rho(E) = \rho(k)\left(\frac{dE}{dk}\right)^{-1} &= [\pi bV\sin(kb)]^{-1}\\ &= \left[\pi bV\sqrt{1-\left(\frac{E_0+E}{2V}\right)^2}\right]^{-1}.\end{aligned}$$

c. As each atom contributes one electron, there are N electrons in the lattice and $L = Nb$, L being the length of the lattice. Then

$$N = L\int_0^{E_m} f(E)\rho(E)dE,$$

where $f(E)$ is the Fermi-Dirac distribution function

$$f(E) = \frac{1}{e^{\beta(E-E_F)}+1},$$

with $\beta = (k_B T)^{-1}$ and E_F being the Fermi energy, which applies to particles of spin $\frac{1}{2}$. At $T = 0$ K, $f(E) = 1$ for $E \ll E_F$ and $f(E) = 0$ for $E > E_F$. For $k_B T \ll E_F$, $E_F(T) \approx E_F(0)$ and only electrons with energies within about $k_B T$ of E_F are effected by the distribution function, some electrons transiting from below E_F to above E_F. Thus, unless the T-dependence is specifically required, we may approximate by taking $f(E) \approx 1$ for low temperatures. Hence

$$\begin{aligned}N &= L\int_0^{E_m} \rho(E)dE = L\int_0^{k_m} \rho(k)dk\\ &= Nb\int_0^{k_m} \frac{2}{\pi}dk = \frac{2Nb}{\pi}k_m,\end{aligned}$$

giving $k_m = \frac{\pi}{2b}$, and

$$E_F \approx E_m = -E_0 - 2V\cos(k_m b) = -E_0 - 2V.$$

The total energy of the electrons is

$$\begin{aligned} E_t &= L\int_0^{k_m} E(k)\rho(k)dk \\ &= L\int_0^{k_m} [-E_0 - 2V\cos(kb)]\frac{2}{\pi}dk \\ &= \frac{2Nb}{\pi}\left[-E_0 - \frac{2V}{b}\sin(kb)\right]_0^{\pi/2b} \\ &= N\left(-E_0 - \frac{4V}{\pi}\right). \end{aligned}$$

Thus each electron on average contributes $-E_0 - \frac{4V}{\pi}$ to the total energy, and the cohesive energy per electron relative to $-E_0$ is therefore $\frac{4V}{\pi}$.

d. As only electrons lying below and above but not near the Fermi level $E_F = -E_0$ contribute to the specific heat, we just have to calculate

$$\varepsilon_t(T) = \frac{L}{\pi bV}\int_{-2V}^{2V}\frac{\varepsilon d\varepsilon}{\left(e^{\beta\varepsilon}+1\right)\sqrt{1-\left(\frac{\varepsilon}{2V}\right)^2}},$$

where $\varepsilon = E - E_F = E + E_0$.

Let

$$f(\varepsilon) = \frac{1}{e^{\beta\varepsilon}+1}, \quad F(\varepsilon) = \sqrt{1-\left(\frac{\varepsilon}{2V}\right)^2}$$

and consider

$$\begin{aligned} I = \int_{-2V}^{2V} f(\varepsilon)F'(\varepsilon)d\varepsilon &= [f(\varepsilon)F(\varepsilon)]_{-2V}^{2V} - \int_{-2V}^{2V} f'(\varepsilon)F(\varepsilon)d\varepsilon \\ &= -\int_{-2V}^{2V} f'(\varepsilon)F(\varepsilon)d\varepsilon. \end{aligned}$$

Expand $F(\varepsilon)$ around $\varepsilon = 0$ according to Taylor's theorem:

$$F(\varepsilon) = F(0) + \varepsilon F'(0) + \frac{1}{2}\varepsilon^2 F''(0) + \cdots.$$

As $F(0) = 1$, $F'(0) = 0$, $F''(0) = -\frac{1}{4V^2}$, we have, to second order in ε,

$$I \approx -\int_{-2V}^{2V} f'(\varepsilon)d\varepsilon + \frac{1}{8V^2}\int_{-2V}^{2V}\varepsilon^2 f'(\varepsilon)d\varepsilon\ .$$

The first integral gives

$$[f(\varepsilon)]_{-2V}^{2V} = \frac{1}{e^{2\beta V}+1} - \frac{1}{e^{-2\beta V}+1} \approx -1$$

as it is assumed that $k_B T \ll V$, or $\beta V \gg 1$. Setting $\beta\varepsilon = x$, the second integral becomes

$$-\frac{1}{8V^2\beta^2}\int_{-\infty}^{\infty}\frac{x^2 e^x\, dx}{(e^x+1)^2} = -\frac{\pi^2}{24V^2\beta^2}.$$

Thus

$$I \approx 1 - \frac{\pi^2}{24V^2}(k_B T)^2,$$

and

$$\begin{aligned} \varepsilon_t(T) &= \frac{L}{\pi b V}(-4V^2)I \\ &\approx N\left[-\frac{4V}{\pi} + \frac{\pi}{6}\frac{(k_B T)^2}{V}\right]. \end{aligned}$$

Hence the specific heat per electron is

$$C_e = \frac{d\varepsilon_t}{dT} \approx \frac{\pi}{3}\frac{k_B^2 T}{V}.$$

1038

The method of orthogonalized plane waves (OPW, closely associated with the method of pseudo-potential) is very often used to make band structure calculations for electron states in metals. It explains successfully why the nearly free electron approximation can be used although the actual lattice potential acting on the electrons is not weak at all.

Discuss the essential physical ideas and mathematical steps in this method by including, among others, descriptions of a

a. construction of Bloch function out of core states and its properties,

b. construction of the orthogonalized plane wave states and their properties,

c. construction of the pseudopotential and its non-uniqueness,

d. physical explanation why the pseudopotential can be treated as small.

(SUNY, Buffalo)

Sol: In a metal the valence electrons are nearly free, but inside a core the wave function oscillates very rapidly. In regions outside the cores, the electron wave is essentially a plane wave

$$|\mathbf{k}\rangle = \frac{1}{\sqrt{N\Omega}} e^{ik\cdot\mathbf{r}},$$
$$|\mathbf{k}+\mathbf{k}_i\rangle = \frac{1}{\sqrt{N\Omega}} e^{i(\mathbf{k}+\mathbf{k}_i)\cdot\mathbf{r}},$$

where N is the number of unit cells in the crystal, Ω is the volume of a unit cell, $\mathbf{k}_i$ is a reciprocal lattice vector.

a. In the core zone the electrons are tightly bound. The tight-binding approximation gives the wave function as

$$|\mathbf{\Phi}_{j\mathbf{k}}(r)\rangle = \frac{1}{\sqrt{N}} \sum_l e^{i\mathbf{k}\cdot\mathbf{R}_l} \varphi_j(r - \mathbf{R}_l),$$

where $\varphi_j(\mathbf{r} - \mathbf{R}_l)$ is the jth state orbital of the atom located at $\mathbf{R}_l$ and the summation extends over all the atoms in the lattice.

b. Taking a plane wave with wave vector $\mathbf{k}$ as basis, construct a function $\chi_i(\mathbf{k}, \mathbf{r})$ orthogonal to all the electron wave functions with wave vector $\mathbf{k}$ in the core zone:

$$\chi_i(\mathbf{k}, \mathbf{r}) = |\mathbf{k}+\mathbf{k}_i\rangle - \sum_{j=1}^{N} \mu_{ij} |\mathbf{\Phi}_{j\mathbf{k}}\rangle,$$

where the coefficient μ_{ij} is determined by the orthogonal condition

$$\int \mathbf{\Phi}_{j\mathbf{k}} \chi_i(\mathbf{k}, \mathbf{r}) d^3\mathbf{r} = 0.$$

The wave function χ_i $(\mathbf{k}, \mathbf{r})$, called an orthogonalized plane wave function, is orthogonal to the electron wave function in the core zone. It behaves as a plane wave at large distances from an atom, but oscillates rapidly near a core. This makes such functions suitable for construction of the wave function of a valence electron in the crystal, in the form a linear combination of several orthogonalized plane waves:

$$\psi_{\mathbf{k}} = \sum_i C_i \chi_i.$$

c. As

$$\chi_i = |\mathbf{k}+\mathbf{k}_i\rangle - \sum_j |\mathbf{\Phi}_{j\mathbf{k}}\rangle \langle \mathbf{\Phi}_{j\mathbf{k}} || \mathbf{k}+\mathbf{k}_i\rangle,$$

denoting

$$\hat{p} = \sum_j |\mathbf{\Phi}_{j\mathbf{k}}\rangle \langle \mathbf{\Phi}_{j\mathbf{k}}|$$

we have

$$\psi_{\mathbf{k}} = (1-\hat{p})|\mathbf{k}+\mathbf{k}_i\rangle.$$

Introducing the pseudo wave function

$$\varphi = \sum_i C_i|\mathbf{k} + \mathbf{k}_i\rangle$$

and substituting it in the Schrödinger equation, we have

$$-\frac{\hbar^2}{2m}\nabla^2\varphi + W\varphi = E\varphi,$$

where

$$\begin{aligned} W &= V(\mathbf{r}) - \left[-\frac{\hbar^2}{2m}\nabla^2 + V(\mathbf{r})\right]\hat{p} + E\hat{p} \\ &= V(\mathbf{r}) + \sum_j (E - E_j)|\mathbf{\Phi}_{j\mathbf{k}}\rangle\langle\mathbf{\Phi}_{j\mathbf{k}}| \end{aligned}$$

is known as a pseudopotential.

d. The pseudopotential is not unique. For if we substitute

$$W' = V(\mathbf{r}) + \sum_j f(E, E_j)|\mathbf{\Phi}_{j\mathbf{k}}\rangle\langle\mathbf{\Phi}_{j\mathbf{k}}|,$$

where $f(E, E_j)$ is an arbitrary function, in the Schrödinger equation

$$-\frac{\hbar^2}{2m}\nabla^2\varphi + W'\varphi = E'\varphi,$$

and take the inner product of its both sides with $\psi^*_{\mathbf{k}}$, we can show that $E' = E$. Thus W' is also a pseudopotential.

e. Now write

$$W = V(\mathbf{r}) + \sum_j (E - E_j)|\mathbf{\Phi}_{j\mathbf{k}}\rangle\langle\mathbf{\Phi}_{j\mathbf{k}}| = V(\mathbf{r}) + V_B,$$

where V(r), the real crystal potential, is an attractive potential, and

$$V_B = \sum_j (E - E_j)|\mathbf{\Phi}_{j\mathbf{k}}\rangle\langle\mathbf{\Phi}_{j\mathbf{k}}|.$$

In a crystal the energy E of a valence electron is higher than that of an inner electron, E_j, so that V_B has the property of repulsing electrons and is a repulsive potential. Thus V_B tends to cancel $V(\mathbf{r})$. This cancellation of the crystal potential by atomic functions is usually appreciable, leading to a very weak and quite smooth pseudopotential W.

1039

Consider a one-dimensional solid of length $L = Na$ made up of N diatomic molecules, the interatomic spacing within a molecule is $b\left(b < \frac{a}{2}\right)$. The centers of adjacent molecules are a distance a apart. We represent the potential energy as a sum of delta functions centered on each atom:

$$V = -A \sum_{n=0}^{N-1} \left[\delta\left(x - na + \frac{b}{2}\right) + \delta\left(x - na - \frac{b}{2}\right) \right],$$

with A a positive quantity and $n = 0, 1, 2, \ldots, N-1$. The potential is shown in Fig. 1.18.

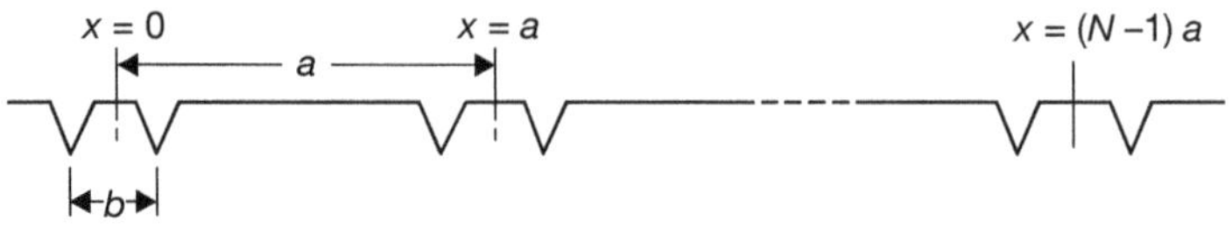

Fig.1.18

a. Consider free electrons in this solid (i.e., neglect V for the moment) and periodic boundary conditions. Derive the allowed values of the electron wave vectors k, and normalize the wave function.

b. Expressing the potential as a Fourier series

$$V = \sum_q V_q e^{iqx},$$

find the allowed values of q and the coefficients V_q.

c. Assuming A to be small, show that for certain values of k there are energy gaps. Derive a general formula for the gaps and show in particular that the gap in energy at the top of the first zone is proportional to $\cos\left(\frac{\pi b}{a}\right)$.

d. Derive an expression for the number of states there are in the first zone. If each atom has one electron, will the substance be a conductor or an insulator?

e. Suppose $b = a/2$. Show what happens to the results of the previous sections and give a brief explanation.

(*Princeton*)

Sol:

a. For a one-dimensional solid the Hamiltonian of a free electron, mass m, is

$$H = \frac{p^2}{2m} = -\frac{\hbar^2}{2m}\nabla^2 = -\frac{\hbar^2}{2m}\frac{d^2}{dx^2}.$$

The Schrödinger equation

$$H\psi = E\psi$$

then becomes

$$\frac{d^2\psi}{dx^2}+\frac{2mE}{\hbar^2}\psi=0.$$

Consider a plane wave solution

$$\psi=A_0e^{ikx}.$$

Substitution gives

$$k=\sqrt{\frac{2mE}{\hbar^2}}.$$

Normalization

$$\int_0^L\psi^*\psi dx=1,$$

L being the length of the crystal, then gives

$$A_0=\frac{1}{\sqrt{L}}.$$

The periodic boundary condition $\psi(0)=\psi(L)$ requires

$$e^{ikL}=1,$$

yielding

$$k=\frac{2n\pi}{L},\quad n=0,\pm1,\pm2,\ldots.$$

This gives the allowed electron wave vectors k, for which the normalized wave function is

$$\psi=\frac{1}{\sqrt{L}}e^{ikx}.$$

b. As $V(x)$ has period a, in the Fourier series

$$V(x)=\sum_q V_q e^{iqx},$$

we require

$$e^{iqx}=e^{iq(x+a)},$$

or

$$q=\frac{2n\pi}{a},\quad n=0,\pm1,\pm2,\ldots,$$

which are the allowed values of q. Consider

$$\begin{aligned}\int_0^a V(x)e^{-iq'x}dx &= \sum_q V_q\int_0^a e^{i\frac{2\pi}{a}(n-n')x}dx\\ &=\sum_q V_q a\,\delta_{qq'}=V_{q'}a,\end{aligned}$$

where $q' = \frac{2n'\pi}{a}$ is a particular q. Thus the coefficients are given by

$$V_q = \frac{1}{a}\int_0^a V(x)e^{-iqx}\,dx.$$

In the neighborhood of $x = na$,

$$V(x) = -A\left[\delta\left(x - na + \frac{b}{2}\right) + \delta\left(x - na - \frac{b}{2}\right)\right],$$

so that

$$\begin{aligned} V_q &= -\frac{A}{a}\left[\int_0^a \delta\left(x - na + \frac{b}{2}\right)e^{-iqx}\,dx + \right. \\ &\qquad \left. \int_0^a \delta\left(x - na - \frac{b}{2}\right)e^{-iqx}\,dx\right] \\ &= -\frac{A}{a}\left[e^{-iq\left(na-\frac{b}{2}\right)} + e^{-iq\left(na+\frac{b}{2}\right)}\right] \\ &= -\frac{A}{a}e^{iqna}\cdot 2\cos\left(\frac{bq}{2}\right) \\ &= -\frac{2A}{a}\cos\left(\frac{bq}{2}\right) \end{aligned}$$

as $e^{iqna} = e^{i2n^2\pi} = 1$.

c. For the one-dimensional periodic potential

$$V(x) = \sum_q V_q e^{iqx}$$

with $q = \frac{2n\pi}{a}$, $n = 0, \pm 1, \pm 2, \ldots$, the Schrodinger equation for an electron moving in a one-dimensional crystal

$$[H_0 + V(x)]\psi = E\psi,$$

where $H_0 = -\frac{\hbar^2}{2m}\frac{d^2}{dx^2}$, can be solved by expressing the solution as a series

$$\psi = \sum_k C_k \varphi_k,$$

where φ_k are the solutions of the free-electron Schrödinger equation

$$H_0\varphi_k = E_k^0\varphi_k,$$

i.e.,

$$\varphi_k = \frac{1}{\sqrt{L}} e^{ikx},$$

with energy

$$E_k^0 = \frac{\hbar^2 k^2}{2m},$$

where $k = \frac{2\pi n}{L}$, $n = 0, \pm 1, \pm 2, \ldots$.

Substitution in the Schrödinger equation gives

$$\sum_k C_k E_k^0 \varphi_k + \sum_k \sum_q C_k V_q e^{iqx} \varphi_k = \sum_k C_k E \varphi_k.$$

As

$$e^{iqx} \varphi_k = \frac{1}{\sqrt{L}} e^{i(q+k)x} = \varphi_{q+k},$$

the wave vectors that make up $\psi(x)$ differ from one another by a reciprocal lattice vector q and the above can be written as

$$\sum_k C_k (E - E_k^0) \varphi_k = \sum_k \sum_q C_k V_q \varphi_{q+k}.$$

Then as the summation over q and k go from $-\infty$ to ∞, the above can be rewritten as

$$\sum_k C_k (E - E_k^0) \varphi_k = \sum_k \sum_q C_{k-q} V_q \varphi_k.$$

Equating the coefficients of $\varphi_{k'}$ gives

$$C_{k'} (E - E_{k'}^0) = \sum_q C_{k'-q} V_q.$$

Letting $k' = k - K$, $q = K' - K$, where K is an arbitrary reciprocal lattice vector, in the above, we have

$$C_{k-K} (E - E_{k-K}^0) = \sum_{K'} C_{k-K'} V_{K'-K} \tag{1}$$

For a free electron, $V(x) = 0$ and the right-hand side vanishes. This shows that

$$C_{k-K} = 0 \qquad \text{for } E \neq E_{k-K}^0,$$
$$C_{k-K} \neq 0 \qquad \text{only if } E = E_{k-K}^0,$$

for the free-electron states. For A small the periodic potential can be taken as small perturbation. These coefficients are then used to express the wave function of wave vector k for an electron in the potential field:

$$\psi_k = \sum_{k'} C_{k'} \varphi_{k'}.$$

Suppose for certain k degeneracy occurs: $E_{k-K_1}^0 = E_{k-K_2}^0 = \ldots$. For a one-dimensional lattice, there is only two-fold degeneracy. Suppose $E_k^0 = E_{k-K_1}^0$. Then only C_k, $C_k - K_1$ are nonvanishing and Eq. (1) gives

$$C_k(E_k - E_k^0) = C_k V_0 + C_{k-K_1} V_{K_1}, \tag{2}$$

$$C_{k-K_1}(E_k - E_k^0) = C_k V_{K_1} + C_{k-K_1} V_0 \cdot \tag{3}$$

As $V(x)$ is real

$$V(x) = V^*(x) = \sum_q V_q e^{iqx} = \sum_q V_q^* e^{-iqx} = \sum_q V_q^* e^{iqx}$$

since q goes from $-\infty$ to ∞. Hence $V_q = V_{-q}^* = V_{-q}$ as V_q is real. Equations (2) and (3) now become

$$(E_k - E_k^0 - V_0) C_k - V_{K_1} C_{k-K_1} = 0,$$
$$V_{K_1} C_k - (E_k - E_k^0 - V_0) C_{k-K_1} = 0.$$

For C_k, C_{k-k1} be nonvanishing we require

$$(E_k - E_k^0 - V_0)^2 - V_{K_1}^2 = 0$$

or

$$E_k = E_k^0 + V_0 \pm |V_{K_1}| \cdot$$

This equation means that an energy gap

$$E_g = \Delta E_k = 2|V_{K_1}|$$

exists at

$$k = \frac{1}{2}(k + K_1 - k) = \frac{K_1}{2}.$$

For a linear crystal the first Brillouin zone is given by $-\frac{\pi}{a} \leq k \leq \frac{\pi}{a}$. At the top of the zone $k = \frac{\pi}{a} = \frac{K_1}{2}$,yielding $K_1 = \frac{2\pi}{a}$ and so $q = -K_1 = -\frac{2\pi}{a}$. Then we have from (a)

$$|V_{K_1}| = |V_g| = \frac{2A}{a} \cos\left(\frac{\pi b}{2}\right).$$

Thus

$$E_g(k) \propto \cos\left(\frac{\pi b}{2}\right).$$

d. The first zone consists of states with wave vectors in

$$-\frac{\pi}{a} \leq k \leq \frac{\pi}{a}$$

with $k = \dfrac{2n\pi}{L}$, $n = 0, \pm 1, \pm 2\ldots$. The number of states in the first zone is thus

$$\frac{2\pi}{a}\left(\frac{2\pi}{L}\right)^{-1} = \frac{L}{a} = N.$$

Since each state corresponds to two spin orientations, there are $2N$ states.

The solid has N diatomic molecules, or $2N$ atoms, and each atom contributes one electron. Thus there are $2N$ electrons which completely fill up the $2N$ states in the first zone. The solid is therefore an insulator.

e. It can be seen from Fig. 1.18 that if $b = \frac{a}{2}$, the period of the potential becomes $\frac{a}{2}$. The Brillouin zones will expand and the above results will change as follows:

a. Remains the same.

b. $q = \frac{4\pi}{a}$, $V_0 = \frac{2A}{a}$, $V_n = (-1)^{n+1}\frac{2A}{a}$.

c. $E_g \propto |V_n|$.

d. The first zone of the k-space will have $2N$ states, or $4N$ states taking spin into account. The $2N$ electrons available are not sufficient to fill up all the states. The material is therefore a conductor.

1040

Consider a one-dimensional periodic potential $U(x)$ viewed as an array of identical potential barriers $V(x)$ of width a, centered at the points $x = \pm na$ where n is an integer. The barrier $V(x)$, schematically drawn in Fig. 1.19, may be characterized by a transmission coefficient $t(k)$ and a reflection coefficient $r(k)$ for an electron incident on the barrier with energy $E = \frac{\hbar^2 k^2}{2m}$.

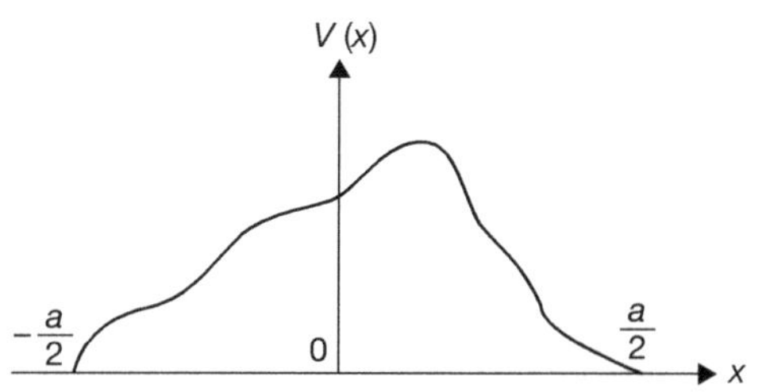

Fig. 1.19

a. For the single barrier case, write down the most general solution for the wave function of an electron with energy $E = \frac{\hbar^2 k^2}{2m}$.

b. How is the energy of the Bloch electron related to its wave vector k for the crystal Hamiltonian case, i.e., $U(x) = \sum_{n=-\infty}^{n=+\infty} V(x + na)$. Verify that this gives the right answer in the free-electron case $V = 0$.

c. By recalling some simple properties of t and r, demonstrate that Bloch waves occur only in bands of energies.

d. Suppose the barriers are weak, i.e., $|t| = 1$ and $|r| = 0$, find a simple expression for the width of the energy gaps.

(Princeton)

Sol:

a. An electron of energy E coming from the left of the single barrier can be represented by a wave

$$\psi_1(x) = \begin{cases} e^{ikx} + r_1 e^{-ikx}, & x \le -\frac{a}{2}, \\ t_1 e^{ikx} & x \ge \frac{a}{2}, \end{cases}$$

k being given by the relation $E = \frac{\hbar^2 k^2}{2m}$. An electron coming from the right can be represented by a wave

$$\psi_r(x) = \begin{cases} e^{-ikx} + r_r e^{ikx}, & x \ge \frac{a}{2}, \\ t_r e^{-ikx} & x \le -\frac{a}{2}. \end{cases}$$

If $V(x)$ is not symmetrical, the reflection and transmission coefficients r, t for the two cases are generally different. The general wave function for these regions is a linear combination of $\psi_1(x)$ and $\psi_r(x)$:

$$\psi(x) = A\,\psi_1(x) + B\,\psi_r(x),$$

A, B being real constants.

Within the barrier $-\frac{a}{2} \le x \le \frac{a}{2}$, the wave function satisfies the one-dimensional Schrödinger equation

$$\left[-\frac{\hbar}{2m}\frac{d^2}{dx^2} + V(x)\right]\psi = E\psi.$$

b. For a periodic crystal potential

$$U(x) = \sum_{n=-\infty}^{\infty} V(x + na),$$

the Hamiltonian within the region $-\frac{a}{2} \le x \le \frac{a}{2}$ is identical with that for a single barrier and we can make use of the Bloch theorem to find the general solution of the Schrödinger equation:

$$\psi(x) = A\,\psi_1(x) + B\,\psi_r(x),$$

with

$$\psi(x + a) = e^{iKa}\,\psi(x),$$

where K is a positive real constant, and hence, on differentiation with respect to x,

$$\psi'(x+a) = e^{iKa}\psi'(x).$$

Thus at $x = -\frac{a}{2}$ we have

$$A\psi_l\left(\tfrac{a}{2}\right) + B\psi_r\left(\tfrac{a}{2}\right) = e^{iKa}\left[A\psi_l\left(-\tfrac{a}{2}\right) + B\psi_r\left(-\tfrac{a}{2}\right)\right],$$
$$A\psi_l'\left(\tfrac{a}{2}\right) + B\psi_r'\left(\tfrac{a}{2}\right) = e^{iKa}\left[A\psi_l'\left(-\tfrac{a}{2}\right) + B\psi_r'\left(-\tfrac{a}{2}\right)\right],$$

or

$$A\left[t_l\beta - \alpha(1 + r_l\beta)\right] + B(1 + r_r\beta - t_r\alpha\beta) = 0,$$
$$A\left[t_l\beta - \alpha(1 - r_l\beta)\right] + B(-1 + r_r\beta + t_r\alpha\beta) = 0,$$

with

$$\alpha = e^{iKa}, \quad \beta = e^{ika}.$$

Taking the sum and difference of the last two equations we have

$$(t_l\beta - \alpha)A + r_r\beta B = 0,$$
$$-r_l\alpha\beta A + (1 - t_r\alpha\beta)B = 0.$$

For nontrivial solutions of A, B, the determinant of their coefficients must vanish, giving

$$(t_l t_r - r_l r_r)e^{ika} + e^{-ika} = t_l e^{-iKa} + t_r e^{iKa}.$$

This gives the relation between the kinetic energy of the electron $E - V = \frac{\hbar^2 K^2}{2m}$, and the wave vector k.

A crystal is usually symmetric so that $V(x) = V(-x)$ and hence $r_l = r_r = r$, $t_l = t_r = t$. The above relation then becomes

$$(t^2 - r^2)e^{ika} + e^{-ika} = 2t\cos(Ka). \tag{1}$$

If the electron is completely free, $r = 0$, $t = 1$, and the above gives

$$\cos(ka) = \cos(Ka),$$

or

$$k = K.$$

This is the right answer since for $V = 0$ we have $E = \frac{\hbar^2 K^2}{2m}$.

c. Consider an electron traveling from left to right through a single barrier, assumed symmetric. The continuity of the probability current density $-\frac{i\hbar}{2m}(\psi^*\psi' - \psi\psi'^*)$applied to the regions $x \le -\frac{a}{2}$ and $x \ge \frac{a}{2}$ gives

$$\begin{aligned}\psi_1^*\psi_1' - \psi_1\psi_1'^* &= (\alpha^{-1} + r^*\alpha)\frac{\partial}{\partial x}(\alpha + r\alpha^{-1}) \\ &\quad -(\alpha + r\alpha^{-1})\frac{\partial}{\partial x}(\alpha^{-1} + r^*\alpha) \\ &= t^*\alpha^{-1}\frac{\partial}{\partial x}(t\alpha) - t\alpha\frac{\partial}{\partial x}(t^*\alpha^{-1})\,,\end{aligned}$$

where $\alpha = e^{ikx}$, and hence

$$1 - rr^* = tt^*\,,$$

or

$$|r|^2 + |t|^2 = 1$$

as expected from energy conservation.

Consider next two simultaneously traveling electrons of the same energy, one from left to right and the other from right to left, crossing the barrier. The probability current density is then zero everywhere, in particular at $x \leq -\frac{a}{2}$. Thus

$$\begin{aligned}&(\psi_1 + \psi_\mathrm{r})^*\frac{\partial}{\partial x}(\psi_1 + \psi_\mathrm{r}) - (\psi_1 + \psi_\mathrm{r})\frac{\partial}{\partial x}(\psi_1 + \psi_\mathrm{r})^* \\ &= (\alpha^{-1} + r^*\alpha + t^*\alpha)\frac{\partial}{\partial x}(\alpha + r\alpha^{-1} + t\alpha^{-1}) \\ &\quad -(\alpha + r\alpha^{-1} + t\alpha^{-1})\frac{\partial}{\partial x}(\alpha^{-1} + r^*\alpha + t^*\alpha) \\ &= 0\,,\end{aligned}$$

or

$$1 - rr^* - tt^* - r^*t - rt^* = 0.$$

Hence

$$r^*t + rt^* = 0\,,$$

which means that m^*t is purely imaginary. Let

$$t = |t|e^{i\delta}, \quad r = i|r|e^{i\delta}$$

and substitute these in Eq. (1), which then gives

$$\frac{\cos(ka + \delta)}{|t|} = \cos(Ka)\,. \tag{2}$$

With increasing electron energy, the transmission coefficient increases, i.e., $|t|$ increases toward unity as k increases. Then the variation of the left-hand side of Eq. (2) with k is as shown in Fig. 1.20. However, as $\cos(Ka)$ is confined to the region between —1 and 1, certain values of k are forbidden by Eq. (2). These forbidden gaps are shown as shadowed areas in the figure. Hence the possible Bloch waves occurs only in bands of energies.

d. Consider an energy gap in the k-space with boundaries k_1, k_2 and midpoint at k_0 and width 2ϵ. It can be seen from Fig. 1.20 and Eq. (2) that

$$\left|\cos(k_0 a + \delta)\right| = 1\,,$$
$$\left|\cos(k_0 a + \delta + \epsilon a)\right| = |t|,$$

or

$$k_0 a + \delta = n\pi,$$

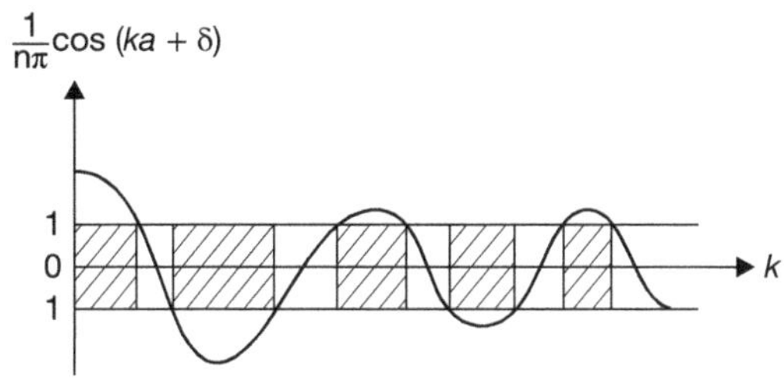

Fig.1.20

where n is zero or an integer, and

$$\left|\sin(n\pi + \epsilon a)\right| = \sqrt{1 - |t|^2} = |r|.$$

For a weak barrier, $|r| \to 0$, $|t| \to 1$, $|\delta| \to 0$, and as can be seen from Fig. 1.20, $\epsilon a \to 0$, so that

$$k_0 = \frac{1}{2}(k_2 + k_1) \approx \frac{n\pi}{2a},$$

and

$$\left|\sin(n\pi)\cos(\epsilon a) + \sin(\epsilon a)\cos(n\pi)\right| \approx \epsilon a = |r|,$$

i.e.,

$$k_2 - k_1 = \frac{|r|}{a}.$$

Hence the energy gap has width

$$\Delta E = \frac{\hbar^2}{2m}(k_2^2 - k_1^2) = \frac{\hbar^2}{m}k_0(k_2 - k_1) = \frac{n\pi}{2}\frac{\hbar^2}{ma^2}|r|.$$

1041

a. For a hexagonal close-packed structure with periodic potential $V(\mathbf{r}) = \Sigma_{\mathbf{G}} V_{\mathbf{G}} e^{i\mathbf{G}\cdot\mathbf{r}}$ (**G** being a reciprocal lattice vector), show that $V_{\mathbf{G}} = 0$ for the **G** corresponding to the first Brillouin zone face in the c direction (that is, the normal to the face is in the (0 0 1) direction). What does this imply for the first-order energy gap across this zone face?

b. Explain why it is not possible to obtain an insulator made up of monovalent atoms in an hcp structure.

c. Na metal has bcc structure and has one valence electron per atom. Find a formula for and estimate the threshold wavelength (in Å) for interband transitions (from the lowest to the next higher energy band) in Na metal. Make use of the fact that k_F is appreciably less than k_{ZB} (the k-vector to the zone boundary nearest the origin). The size of the cube edge a in Na is 4.23 Å.

(Princeton)

Sol:

a. The coefficients in the expansion

$$V(\mathbf{r}) = \sum_{\mathbf{G}} V_{\mathbf{G}} e^{i\mathbf{G}\cdot\mathbf{r}}$$

are

$$V_{\mathbf{G}} = \frac{1}{v}\int_{\text{cell}} V(\mathbf{r}) e^{-i\mathbf{G}\cdot\mathbf{r}}\, d\mathbf{r},$$

where v is the volume of a unit cell.

If there are n identical atoms in the unit cell with positions $\mathbf{d}_1, \mathbf{d}_2, \ldots, \mathbf{d}_j, \ldots, \mathbf{d}_n$, the periodic potential can be expressed as

$$V(\mathbf{r}) = \sum_{\mathbf{R}}\sum_{j=1}^{n} \Phi(\mathbf{r} - \mathbf{R} - \mathbf{d}_j),$$

where $\mathbf{R}$ is the position of a lattice point. Hence

$$\begin{aligned} V_{\mathbf{G}} &= \frac{1}{v}\int_{\text{cell}} e^{-i\mathbf{G}\cdot\mathbf{r}} \sum_{\mathbf{R}}\sum_{j=1}^{n} \Phi(\mathbf{r} - \mathbf{R} - \mathbf{d}_j) d\mathbf{r} \\ &= \frac{1}{v}\sum_{j=1}^{n}\sum_{\mathbf{R}} \int_{\text{cell}} e^{-i\mathbf{G}\cdot\mathbf{r}}\, \Phi(\mathbf{r} - \mathbf{R} - \mathbf{d}_j) d\mathbf{r} \\ &= \frac{1}{v}\sum_{j=1}^{n} \int_{\text{wholes pace}} e^{-i\mathbf{G}\cdot\mathbf{r}}\, \Phi(\mathbf{r} - \mathbf{d}_j) d\mathbf{r} \\ &= \frac{1}{v}\int_{\text{wholes pace}} e^{-i\mathbf{G}\cdot\mathbf{r}}\, \Phi(\mathbf{r}) \left(\sum_{j=1}^{n} e^{i\mathbf{G}\cdot\mathbf{d}_j}\right) d\mathbf{r}. \end{aligned}$$

In the expansion the sum $\sum_{j=1}^{n} e^{i\mathbf{G}\cdot\mathbf{d}_j}$ is the geometric structure factor. A unit cell of the hexagonal close-packed (hcp) structure contains two atoms at (0 0 0) and $\left(\frac{2}{3}\frac{1}{3}\frac{1}{2}\right)$. Thus for $\mathbf{G} = \frac{2\pi}{c}\mathbf{b}_3$,

$$\sum_{j=1}^{n} e^{i\mathbf{G}\cdot\mathbf{d}_j} = 1 + e^{\frac{i2\pi}{c}\cdot\frac{c}{2}} = 0,$$

and hence

$$V_{\mathbf{G}} = 0.$$

The first-order energy gap across this zone face is therefore

$$2|V_{\mathbf{G}}| = 0.$$

b. In the hcp structure, each energy band contains $2 \times 2 \times N = 4N$ states, where N is the number of unit cells. However, for a solid made up of monovalent atoms, the total number of valence electrons in $2N$, not enough to fill up the energy band completely. Hence it cannot be an insulator.

c. Conservation of lattice momentum requires

$$\mathbf{k}' = \mathbf{k} + \mathbf{k}_{ph} + \mathbf{G}.$$

The wavelength of visible light is about 5000 Å, giving $k_{ph} = \frac{2\pi}{\lambda} = 10^5\text{cm}^{-1}$. The linear size of the Brillouin zone π/a is of the order 10^8 cm^{-1}. Hence $\mathbf{k}_{ph}$ can affect the wave vector in the above formula only to a very small extent. Because of translation symmetry, a change of the wave vector by $\mathbf{G}$ does not generate any effect. Thus interband transitions can be considered as transitions between two energy bands with identical wave vectors, as shown by Fig. 1.21.

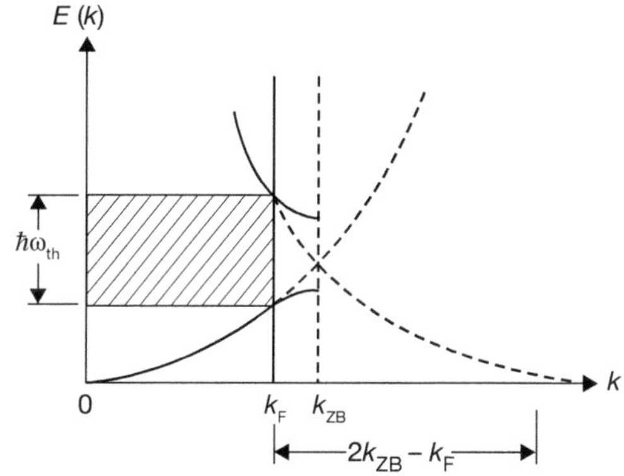

Fig. 1.21

Since k_F is much less than k_{ZB}, one can make an estimation of the threshold wavelength for interband transitions using the free-electron approximation. From Fig. 1.21 it can be seen that

$$\hbar\,\omega_{\rm th} = \hbar\frac{2\pi c}{\lambda_{\rm th}} = \frac{\hbar^2}{2m}(2\,k_{\rm ZB} - k_{\rm F})^2 - \frac{\hbar^2}{2m}k_{\rm F}^2 = \frac{2\hbar^2}{m}k_{\rm B}(k_{\rm B} - k_{\rm F}),$$

which gives the threshold wavelength

$$\lambda_{\rm th} = \frac{\pi mc}{\hbar\,k_{\rm ZB}(k_{\rm ZB} - k_{\rm F})},$$

where

$$k_{\rm F} = (3\pi^2 n)^{1/3} = \left(3\pi^2\frac{2}{a^3}\right)^{1/3} = \frac{1}{a}(6\pi^2)^{1/3}.$$

The reciprocal lattice of a bcc lattice with edge a is an fcc lattice with edge $\frac{4\pi}{a}$. Thus $k_{\rm ZB}$ is a quarter of the diagonal of the fee face,

$$k_{\rm ZB} = \frac{1}{4}\sqrt{2}\,\frac{4\pi}{a} = \frac{\sqrt{2}\,\pi}{a}.$$

Hence

$$\begin{aligned}\lambda_{\rm th} &= \frac{mc}{\hbar}\frac{a^2}{2\pi - \sqrt{2}(6\pi^2)^{1/3}}\\ &= \frac{mc^2}{\hbar c}\frac{a^2}{2\pi - \sqrt{2}(6\pi^2)^{1/3}}.\end{aligned}$$

With $\hbar c = 1.97 \times 10^3$ eV·Å, $mc^2 = 0.511$ MeV, $a = 4.23$ Å, we obtain

$$\lambda_{\rm th} = 6034\ \text{Å}.$$

1042

Figure 1.22 is an energy versus wave vector diagram for electrons in a one-dimensional solid.

a. If n is the number density for electrons and p is that for holes, what can be inferred about p/n?

b. Does this material have an even or odd number of conduction electrons per unit cell? Justify your answer.

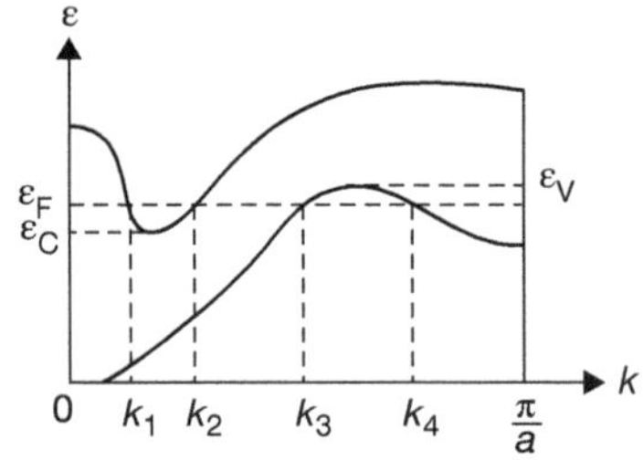

Fig. 1.22

c. Which is greater, the effective mass of the electron or that of the hole? In terms of quantities in the diagram, derive approximate expressions for the effective masses.

(*Princeton*)

Sol:

a. $p/n = 1$.

b. Figure 1.22 shows that there is no overlapping of the energy bands so that the conduction band is completely filled. Since each unit cell can take two electrons, this material contains an even number of conduction electrons per unit cell.

c. The region (k_1, k_2) corresponds to conduction electrons and (k_3, k_4) to holes. The figure indicates that the energy curve varies more smoothly in (k_3, k_4) than in (k_1, k_2). Hence the effective mass, which is given by

$$m^* = \left|\hbar^2 \left(\frac{d^2 \epsilon}{dk^2}\right)^{-1}\right|,$$

is greater for the holes than for the electrons.

Approximating the energy curve in the regions (k_1, k_2), (k_3, k_4) by parabolas

$$\begin{aligned} \varepsilon - \varepsilon_C &= a\left(k - \frac{k_1 + k_2}{2}\right)^2 \\ \varepsilon_V - \varepsilon &= b\left(k - \frac{k_3 + k_4}{2}\right)^2, \end{aligned}$$

where

$$\begin{aligned} a &= 4(\varepsilon_F - \varepsilon_C)(k_2 - k_1)^{-2} \\ b &= 4(\varepsilon_v - \varepsilon_F)(k_4 - k_3)^{-2}, \end{aligned}$$

respectively, we find

$$m_e^* = \frac{\hbar^2}{2a} = \frac{\hbar^2 (k_2 - k_1)^2}{8(\varepsilon_F - \varepsilon_C)}$$

as the effective mass of an electron, and

$$m_h^* = \frac{\hbar^2}{2b} = \frac{\hbar^2 (k_4 - k_3)^2}{8(\varepsilon_v - \varepsilon_F)}$$

as the effective mass of a electron hole. As $|k_2 - k_1| < |k_4 - k_3|$, $0 < \varepsilon_V - \varepsilon_F < \varepsilon_F - \varepsilon_C$,

$$m_e^* < m_h^*.$$

1043

Consider an intrinsic semiconductor whose electronic density of states function $N(E)$ is depicted in Fig. 1.23.

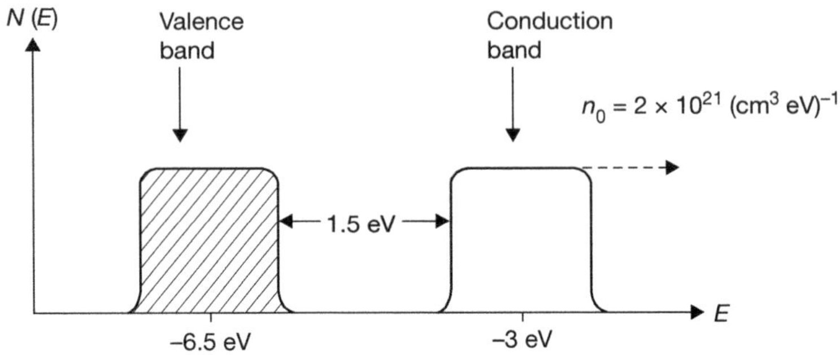

Fig. 1.23

1. Where is the Fermi level with respect to the valence and conduction bands?
2. Write the Fermi distribution function assuming that $N(E)$ already contains the spin degeneracy factor of 2.
3. Estimate the density of conduction band electrons at room temperature.

(*Wisconsin*)

Sol:

1. The number of electrons in the conduction band, whose bottom is ε_C, and the number of holes in the valence band, whose top is ε_V, are respectively

$$n = \int_{\varepsilon_C}^{\infty} N(E) f(E) dE,$$

$$p = \int_{-\infty}^{\varepsilon_V} N(E)(1 - f(E))\, dE,$$

where

$$f(E) = \frac{1}{e^{(E-E_F)/k_B T} + 1}$$

is the Fermi-Dirac distribution function giving the probability that an energy level E is occupied. For an intrinsic semiconductor, $n = p$, from which the Fermi level E_F can be derived. For example consider the case where $E - E_F \gg k_B T$, then $f(E) \approx e^{-(E\sim E_F)k_B T}$, whence

$$n \approx \int_{\varepsilon_C}^{\infty} n_0 e^{-(E-E_F)/k_B T} dE = n_0 k_B T e^{-(\varepsilon_C - E_F)/k_B T},$$

$$p \approx n_0 k_B T e^{-(E_F - \varepsilon_V)/k_B T},$$

where n_0 is the density of electron states. From $n = p$ we have

$$E_F = \frac{\varepsilon_C + \varepsilon_V}{2}.$$

2. The function $f(E)$ above gives the distribution of the electrons among the energy states E of the conduction band and is known as the Fermi-Dirac distribution function.

3. As

$$\varepsilon_C - E_F \approx \frac{E_g}{2} = 0.75\,\text{eV}$$

and at room temperature $k_B T = \frac{1}{40}$ eV, we have the density of conduction electrons

$$n \approx n_0 k_B T \exp\left(-\frac{\varepsilon_C - E_F}{k_B T}\right) = 4.68 \times 10^6\ \text{cm}^{-3}.$$

1044

Consider an intrinsic semiconductor. Let ε be the energy of an electron. Let $g_C(\varepsilon)$ be the density of states in the conduction band, and $g_V(\varepsilon)$ be the density of states in the valence band (see Fig. 1.24). Assume $\varepsilon_C - \varepsilon_F \gg k_B T$, $\varepsilon_F - \varepsilon_V \gg k_B T$, and

$$g_C(\varepsilon) = C_1(\varepsilon - \varepsilon_C)^{1/2},$$
$$g_V(\varepsilon) = C_2(\varepsilon_V - \varepsilon)^{1/2},$$

where ε_C represents the energy of the bottom of the conduction band and ε_V the top of the valence band. The Fermi energy is ε_F.

a. Find an expression for n, the number of electrons in the conduction band, in terms of k_B, T, C_1, ε_C, ε_F and a dimensionless definite integral.

b. Find an expression for p, the number of holes in the valence band, in terms of k_B, T, C_2, ε_V, ε_F and a dimensionless definite integral.

c. Find an explicit expression for $\varepsilon_F(T)$.

d. Which, if any, of the results of (a), (b) or (c) remain true if the material is doped with donor atoms? Explain

(*Princeton*)

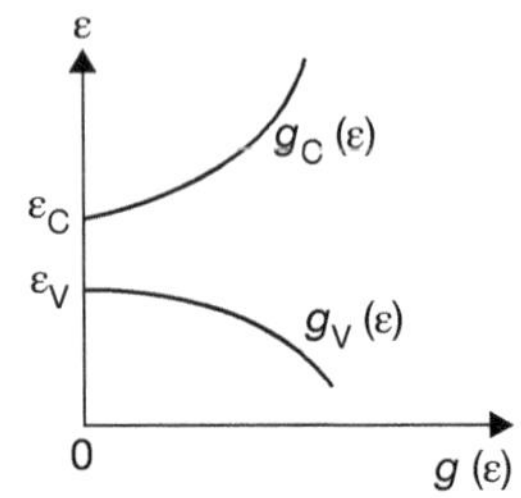

Fig.1.24

Sol:

a. The concentration of electrons in the conduction band is given by

$$n = \int_{\varepsilon_C}^{\infty} f(\varepsilon) g_C(\varepsilon) d\varepsilon$$

with

$$f(\varepsilon) = \frac{1}{e^{\beta(\varepsilon - \varepsilon_F)} + 1},$$

where $\beta = (k_B T)^{-1}$. Since $\varepsilon - \varepsilon_F \geq \varepsilon_C - \varepsilon_F \gg k_B T$,

$$f(\varepsilon) \approx e^{-\beta(\varepsilon - \varepsilon_F)}.$$

Setting $x = \beta(\varepsilon — \varepsilon_C)$, we have

$$n \approx C_1 \int_{\varepsilon_C} e^{-\beta(\varepsilon - \varepsilon_F)} (\varepsilon - \varepsilon_C)^{\frac{1}{2}} d\varepsilon$$

$$= C_1 \beta^{-3/2} e^{-\beta(\varepsilon_C - \varepsilon_F)} \int_0^{\infty} e^{-x} x^{1/2} dx$$

$$= C_1 (k_B T)^{3/2} e^{-(\varepsilon_C - \varepsilon_F)/k_B T} \int_0^{\infty} e^{-x} x^{1/2} dx.$$

b. The probability of hole occupation of a state is $1 - f(\varepsilon)$. As $\varepsilon_F - \varepsilon \geq \varepsilon_F - \varepsilon_V \gg k_B T$,

$$1 - f(\varepsilon) \approx \frac{e^{\beta(\varepsilon - \varepsilon_F)}}{e^{-\beta(\varepsilon_F - \varepsilon)} + 1} \approx e^{\beta(\varepsilon - \varepsilon_F)}.$$

Setting $x = -\beta(\varepsilon - \varepsilon_V)$, we have the concentration of holes in the valence band

$$p = C_2\beta^{-3/2}e^{\beta(\varepsilon_V-\varepsilon_F)}\int_0^\infty e^{-x}x^{1/2}\,dx$$

$$= C_2(k_B T)^{3/2}e^{-(\varepsilon_F-\varepsilon_V)/k_BT}\int_0^\infty e^{-x}x^{1/2}\,dx.$$

c. For an intrinsic semiconductor, $n = p$. Hence

$$C_1 e^{-(\varepsilon_C-\varepsilon_F)/k_BT} = C_2 e^{-(\varepsilon_F-\varepsilon_V)/k_BT},$$

giving

$$\varepsilon_F = \frac{1}{2}\left(\varepsilon_C + \varepsilon_V - k_B T \ln\frac{C_1}{C_2}\right).$$

d. If the material is doped, the results of (a) and (b) still hold true so long as it remains in the equilibrium state. However, because of doping the condition $n = p$ is no longer valid and the Fermi level ε_F as given in (c) will shift.

1045

At room temperature, $k_BT/e = 26$ mV. A sample of cadmium sulfide displays a mobile carrier density of 10^{16} cm^{-3} and a mobility coefficient $\mu = 10^2$ cm^2/volt sec.

a. Calculate the electrical conductivity of this sample.

b. The carriers are continuously trapped into immobile sites and then being thermally reionized into mobile states. If the average free life-time in a mobile state is 10^{-5} second, what is the rms distance a carrier diffuses between successive trappings?

c. If the charge carriers have an effective mass equal to 0.1 times the mass of a free electron, what is the average time between successive scatterings?

(*Wisconsin*)

Sol:

a. The electrical conductivity is given by $\sigma = ne\mu$. With $n = 10^{22}$ m^{-3}, $e = 1.6 \times 10^{-19}$ C, $\mu = 10^{-2}$ m^{2}V^{-1}s^{-1}, we have for the material $\sigma = 16\ \Omega^{-1}m^{-1}$.

b. The law of equipartition of energy

$$\frac{1}{2}m\bar{v}_x^2 = \frac{1}{2}m\bar{v}_y^2 = \frac{1}{2}m\bar{v}_z^2 = \frac{1}{2}k_B T$$

gives

$$\frac{1}{2}m\bar{v}^2 = \frac{3}{2}k_B T,$$

or

$$\bar{v}^2 = \sqrt{\frac{3k_B T}{m}}.$$

The rms distance l between successive trappings is given by

$$l^2 = \bar{v}^2 t^2.$$

Hence

$$l = \sqrt{\frac{3k_B T}{m}}\,t = \sqrt{3\left(\frac{k_B T}{e}\right)\frac{e}{m}}\,t.$$

With $\frac{k_B T}{e} = 26 \times 10^{-3}\,\text{V}$, $\frac{e}{m} = 1.76 \times 10^{11}\,\text{Ckg}^{-1}$, $t = 10^{-5}$ s, we have $l = 1.17$ m.

c. The free electron model of metals gives

$$\sigma = \frac{ne^2\langle\tau\rangle}{m^*},$$

where m^* is the effective mass of an electron. Then the average time between successive scatterings is

$$\langle\tau\rangle = \frac{0.1\sigma}{ne}\left(\frac{m}{e}\right) = 5.7 \times 10^{-15}\,\text{s}.$$

1046

Determine the form of the temperature dependence of the electrical conductivity $\frac{ne^2\tau}{m}$ (where τ = effective collision time) of a semiconductor in a temperature domain where the density n of free carriers is constant and the dominant scattering mechanism is Rutherford scattering from a constant and small number of charged impurities.

(*Wisconsin*)

Sol: The differential cross section for Rutherford scattering of a particle of charge e, mass m, and velocity v by an impurity nucleus of charge Ze per unit solid angle is

$$\frac{d\phi}{d\Omega}=\left(\frac{Ze^2}{8\pi\epsilon m v^2}\right)^2\mathrm{cosec}^4\left(\frac{\theta}{2}\right)=R^2\mathrm{cosec}^4\left(\frac{\theta}{2}\right),$$

where ϵ is the permittivity of the semiconductor material and θ is the scattering angle as shown in Fig. 1.25. The probability per unit time of scattering through a scattering angle between θ and $\theta + d\theta$ is

$$dP = N_I v d d\phi = 2\pi N_I v R^2 \frac{\sin\theta d\theta}{\sin^4\left(\frac{\theta}{2}\right)}$$

$$= 8\pi N_I v R^2 \frac{d(1-\cos\theta)}{(1-\cos\theta)^2},$$

where N_I is the number of impurity atoms per unit volume.

In the expression for electrical conductivity,

$$\sigma = \frac{ne^2\tau}{m},$$

the effective collision time, or the relaxation time, τ, is identical with the mean free time between collisions τ_c only if the scattering is isotropic, which is not the case for Rutherford scattering. Instead,

$$\tau = \tau_c(1-\langle\cos\theta\rangle)^{-1},$$

where $\langle\cos\theta\rangle$ is the average of the cosine of the scattering angle θ. Thus

$$\frac{1}{\tau} = \frac{\langle 1-\cos\theta\rangle}{\tau_c} \approx \int(1-\cos\theta)dP$$

$$= 8\pi N_1 v R^2 \int_{\theta_{min}}^{\theta_{max}} \frac{d(1-\cos\theta)}{(1-\cos\theta)}$$

$$= \frac{Z^2e^4N_I}{8\pi\epsilon^2m^2v^3}\ln\left(\frac{1-\cos\theta_{max}}{1-\cos\theta_{min}}\right).$$

Hence, treating the logarithmic function as approximately constant,

$$\frac{1}{\tau}\propto N_I v^{-3}.$$

Since each degree of freedom of the carrier has an average energy $\frac{1}{2}k_B T$,

$$v=\sqrt{\frac{3k_BT}{m}}\propto T^{\frac{1}{2}}$$

and

$$\frac{1}{\tau}\propto T^{-\frac{3}{2}}.$$

It follows that the electric conductivity

$$\sigma=\frac{ne^2\tau}{m}\propto T^{\frac{3}{2}}.$$

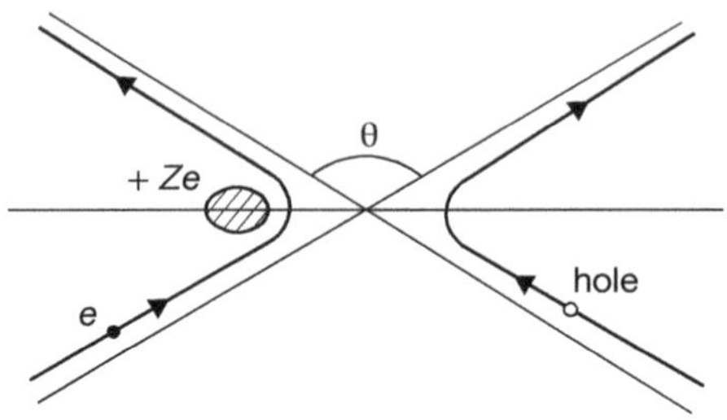

Fig. 1.25

1047

Derive an expression for the Hall coefficient for an intrinsic semiconductor in terms of the electron and hole mobilities and the density of carriers.

(*Wisconsin*)

Sol: Consider a rectangular sample of the intrinsic semiconductor parallel to the xy plane, to which an electric field E_x and a magnetic field B_z are applied as shown in Fig. 1.26. The electric field causes the electrons to drift in the $-x$ direction, and the holes to drift in the x direction, assuming the end surfaces perpendicular to x-axis are joined in a closed circuit. Because of the drift, the magnetic field exerts Lorentz forces $q\mathbf{v} \times \mathbf{B}$ on the carriers, causing both the electrons and holes to deflect in the $-y$ direction, toward the D surface. While the electrons and holes tend to cancel each other, the cancellation is not complete with the result that a net charge accumulates on D and an equal and opposite charge on C. This produces an electric field, the Hall field ε_y, in the y direction.

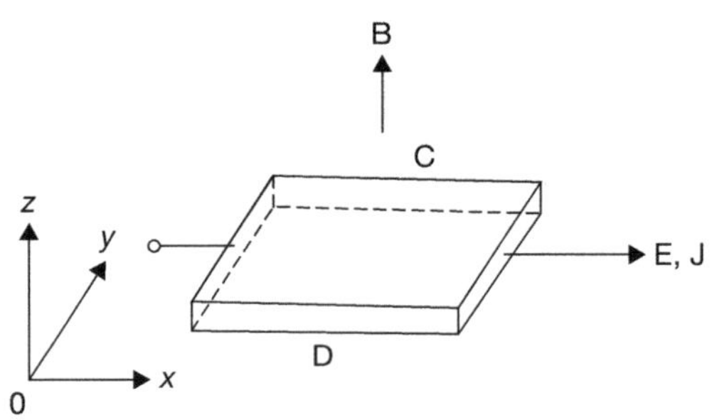

Fig.1.26

The electric conductivity of a semiconductor is

$$\sigma = nq\mu_n + pq\mu_p,$$

where n, p are the electron and hole concentrations respectively and μ_n, μ_p are their mobilities defined as velocity per unit field strength in the direction of the field. In the steady state there is no net current in the y direction. However the hole and electron currents are not zero individually. Let their densities be $(J_p)_y$, $(J_n)_y$ respectively. Then

$$\begin{aligned}(J_p)_y &= pq\mu_p\varepsilon_y - pq\mu_p(v_p)_x B_z \\ &= pq\mu_p\varepsilon_y - pq\mu_p^2 E_x B_z, \\ (J_n)_y &= nq\mu_n\varepsilon_y - nq\mu_n(v_n)_x B_z \\ &= nq\mu_n\varepsilon_y + nq\mu_n^2 E_x B_z.\end{aligned}$$

In the steady state,

$$J_y = (J_P)_y + (J_n)_y = 0,$$

giving

$$\varepsilon_y = \left(\frac{p\mu_p^2 - n\mu_n^2}{p\mu_p + n\mu_n}\right) E_x B_z.$$

The Hall coefficient, defined as $\frac{\varepsilon_y}{J_x B_z}$, is then

$$R_H = \frac{1}{q}\frac{(p\mu_p^2 - n\mu_n^2)}{(p\mu_p + n\mu_n)^2},$$

since $J_x = \sigma E_x$. For an intrinsic semiconductor, $n = p$ and

$$R_H = \frac{1}{qn}\frac{(\mu_p - \mu_n)}{(\mu_p + \mu_n)},$$

1048

Assume that the E vs. **k** relationship for electrons in the conduction band of a hypothetical tetravalent n-type semiconductor can be approximated by

$$E = ak^2 + \text{constant}.$$

The cyclotron resonance for electrons in a field $B = 0.1$ Weber/m^2 occurs at an angular rotation frequency $\omega_c = 1.8 \times 10^{11}$ rad s^{-1}.

a. Find the value of a.

b. Assume that the semiconductor is doped with pentavalent donors. Estimate the number of donors per m^3, given that the Hall coefficient at room temperature is $R_H = 6.25 \times 10^{-6}$ m^3 coul^{-1} and the relative dielectric constant is $\varepsilon_r = 15$.

(Wisconsin)

Sol:

a. An electron in the conduction band behaves as if it had an effective mass m^* given by

$$\frac{1}{m^*} = \frac{1}{\hbar}\frac{\partial^2 E}{\partial k^2} = \frac{2a}{\hbar^2}.$$

The cyclotron resonance frequency for the electron is

$$\omega_c = \frac{eB}{m^*}.$$

Hence

$$a = \frac{\hbar^2}{2}\frac{\omega_c}{eB} = \frac{(1.05 \times 10^{-34})^2 \times 1.8 \times 10^{11}}{2 \times 1.6 \times 10^{-19} \times 0 \cdot 1} = 6.2 \times 10^{-38} \mathrm{Jm}^2.$$

b. For an n-type semiconductor, $n \gg p$ and (Problem **1047**)

$$R_H \approx -\frac{1}{ne}.$$

At room temperature, the impurity atoms are almost entirely ionized. Hence the impurity number density is

$$N_1 \approx n = -\frac{1}{R_H e} = \frac{1}{6.25 \times 10^{-6} \times 1.6 \times 10^{-19}} = 10^{24} \mathrm{m}^{-3}.$$

1049

Optical excitation of intrinsic germanium creates an average density of 10^{12} conduction electrons per cm^3 in the material at liquid nitrogen temperature. At this temperature, the electron and hole mobilities are equal, $\mu = 0.5 \times 10^4$ cm^2/volt sec. The germanium dielectric constant is 20.

a. If 100 volts is applied across a 1 cm cube of crystal under these conditions, about how much current is observed?

b. What is the approximate value of the diffusion coefficient for the electrons and holes? (At room temperature, $\frac{k_B T}{e} = 26\,\mathrm{mV}$)

(Wisconsin)

Sol:

a. For an intrinsic semiconductor, $n = p$ and the current density is

$$j = \sigma E = (n\mu_n + p\mu_p)eE = 2n\mu_n eE\,,$$

with $n = 10^{18}\ \mathrm{m^{-3}}$, $\mu_n = 0.5\ \mathrm{m^2V^{-1}s^{-1}}$, $E = \dfrac{100}{10^{-2}} = 10^4\mathrm{Vm^{-1}}$, $V = 100$ V, $e = 1.6 \times 10^{-19}$ C, the current density is

$$j = 1.6 \times 10^3\,\mathrm{Am^{-2}}$$

and the current is

$$I = jS = 1.6 \times 10^3 \times 10^4 = 0.16\,\mathrm{A} = 160\,\mathrm{mA}.$$

b. The diffusion coefficient D is given by the Einstein relation

$$\frac{D}{\mu} = \frac{k_B T}{e}.$$

As at liquid nitrogen temperature $T_D = 77$ K, $\mu_n = \mu_p = \mu$, we have

$$D_n = D_p = D.$$

The diffusion coefficient for the electrons and holes is then

$$D = \frac{k_B T_D}{e}\mu = \frac{1.38 \times 10^{-23} \times 77 \times 0.5}{1.6 \times 10^{-19}} = 3.3 \times 10^{-3}\mathrm{m^2s^{-1}}.$$

1050

The low temperature ($T \sim 4$ K) optical absorption spectrum of very pure InP, a direct gap semiconductor, is shown in Fig. 1.27, where intensity of absorption is plotted as a function of phonon energy (full lines). There are a few lower intensity peaks between C and A; these have not been shown for simplicity.

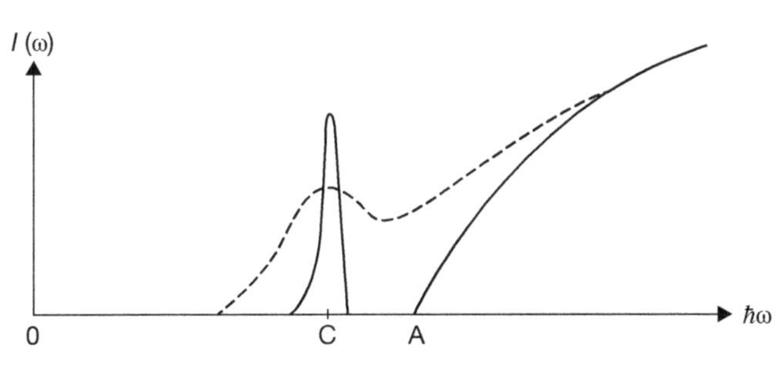

Fig.1.27

a. What is the physical origin of the absorption continuum with threshold at A, and of the peak at C?

The energies E_A and E_C are 1.400 eV and 1.386 eV respectively. The static dielectric constant of InP is 9.6, and the hole in InP is much lighter than the electron $(m_h^* \ll m_e^*)$. Prom the above information calculate

b. the direct band gap in InP,

c. the hole mass m_h^*.

At room temperature the sharp line C is broadened out and is not clearly seen (dotted lines).

d. Explain this effect.

(*Princeton*)

Sol:

a. The absorption continuum with threshold at A is caused by the absorption of a photon from the incident beam by a valence electron, which then jumps into the conduction band. The process can take place if $h\nu \geq E_g$, the energy gap. The lower photon energy limit corresponds to the threshold. The excited electron in the conduction band and the hole left in the valence band both behave as free particles.

If the incident photon energy $h\nu < E_g$, it is not sufficient to excite a valence electron into the conduction band. However, it may still be sufficient to excite an electrons from some atom of the crystal leaving behind a hole, the electron-hole pair forming a bound state, called an exciton, in which the two particles revolve around their center of mass. The photon involved in exciton absorption has energy $h\nu = E_g - E_{ex}$, where E_{ex} is the exciton binding energy, usually about 0.01 eV. The exciton spectrum consists of a sharp line, slightly below the threshold of the absorption continuum, shown as the sharp peak at C in Fig. 1.27. The line may often be broadened by interaction of the exciton with impurities and other effects.

b. $E_g = E_A = 1.400$ eV.

c. Consider the exciton as analogous to the hydrogen atom. Then the energy of the nth state is

$$E_n = E_g - \frac{\mu e^4}{2(4\pi\epsilon_0)^2\epsilon^2\hbar^2 n^2},$$

where

$$\mu = \frac{m_e^* m_h^*}{m_e^* + m_h^*} \approx m_h^*,$$

as $m_h^* \ll m_e^*$, is the reduced mass of the exciton system. For the ground state, $n = 1$, $E_1 = E_C = 1.386$ eV, $E_g = 1.400$ eV, we have

$$m_h^* c^2 = 2(E_g - E_c)\epsilon^2 \left(\frac{e^2}{4\pi\epsilon_0 \hbar c}\right)^{-2}.$$

With

$$E_g - E_c = 0.014 \text{ eV}, \quad \epsilon = 9.6, \quad \frac{e^2}{4\pi\epsilon_0 \hbar c} = \frac{1}{137},$$

we find

$$m_h^* c^2 = 4.85 \times 10^4 \text{eV} = 9.5 \times 10^{-2} \text{m}_e c^2 ,$$

or

$$m_h^* = 9.5 \times 10^{-2} \text{m}_e.$$

d. At room temperature, lattice vibration is appreciable and the electron in an exciton system interacts with the optical branch of the lattice vibration, causing broadening of the exciton line, which may merge with the characteristic absorption band although the peak of the exciton line may still be discernable.

1051

In crystalline silicon, the energy gap is 1.14 eV, the hole effective mass is roughly $m_h = 0.3$ m, and the electron effective mass is $m_e = 0.2$ m.

a. Making (and justifying) appropriate approximations, find an expression for the function $f(T)$ in the law of mass action:

$$np = f(T),$$

where n and p are the electron and hole concentrations and T is the temperature. (You may need to know $\int_0^\infty x^{1/2} e^{-x} dx = \left(\frac{\pi}{4}\right)^{1/2}$).

b. What concentration N_d of pentavalent As donors must be added to make the extrinsic conductivity 10^4 times greater than the intrinsic conductivity at room temperature? (Neglect acceptor impurities and take the static dielectric constant to be $\varepsilon = 11.8$)

c. The work function ϕ_s of a semiconductor is the difference in energy between an electron at rest in vacuum and the Fermi energy in the semiconductor. If a metal with work function ϕ_m is used to make contact with the semiconductor, use band diagrams of the materials at the junction to show that the junction can be rectifying or ohmic, depending on the relative magnitudes of ϕ_m and ϕ_s.

(*Princeton*)

Sol:

a. Near the bottom of the conduction band of a semiconductor an electron of wave vector k has energy

$$E(k) = E_g + \frac{\hbar^2 k^2}{2m_e},$$

where E_g is the energy gap and the top of the valence band is taken to be the zero-energy level, as shown in Fig. 1.28.

The number of states per unit volume of the semiconductor in the range k, $k + dk$ is

$$\frac{4\pi k^2\, dk}{(2\pi)^3} = \frac{1}{4\pi^2}\left(\frac{2m_e}{\hbar^2}\right)^{\frac{3}{2}} (E - E_g)^{\frac{1}{2}}\, dE\,.$$

As each k corresponds to two spin states, the density of states per unit interval is

$$g(E) = \frac{1}{2\pi^2}\left(\frac{2m_e}{\hbar^2}\right)^{\frac{3}{2}} (E - E_g)^{\frac{1}{2}}.$$

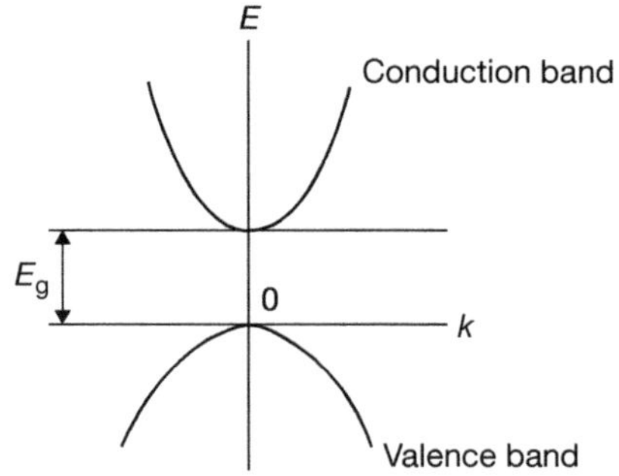

Fig.1.28

When the system is in thermal equilibrium at temperature T, the electrons follow the Fermi distribution function

$$f(E) = \frac{1}{1 + e^{(E-E_{\mathrm{F}})/k_{\mathrm{B}}T}}.$$

In a semiconductor in Fermi level E_{F} usually lies within the forbidden band between the valence and conduction bands, and its distance from either is far greater than $k_{\mathrm{B}}T$. Thus $E - E_{\mathrm{F}} \gg k_{\mathrm{B}}T$ and we can take

$$f(E) \approx e^{(E_{\mathrm{F}}-E)/k_{\mathrm{B}}T}.$$

The electron concentration of the conduction band is therefore

$$\begin{aligned} n &= \int_{E_{\mathrm{g}}}^{\infty} f(E)g(E)dE \\ &\approx \frac{1}{2\pi^2}\left(\frac{2m_{\mathrm{e}}}{\hbar^2}\right)^{\frac{3}{2}} \int_{E_{\mathrm{g}}}^{\infty} (E - E_{\mathrm{g}})^{\frac{1}{2}} e^{(E_{\mathrm{F}}-E)/k_{\mathrm{B}}T} dE \\ &= \frac{1}{2\pi^2}\left(\frac{2m_{\mathrm{e}}k_{\mathrm{B}}T}{\hbar^2}\right)^{\frac{3}{2}} e^{-(E_{\mathrm{g}}-E_{\mathrm{F}})/k_{\mathrm{B}}T} \int_{0}^{\infty} x^{\frac{1}{2}} e^{-x} dx \end{aligned}$$

with $x = \dfrac{E - E_{\mathrm{g}}}{k_{\mathrm{B}}T}$. Hence

$$n \approx N_{\mathrm{e}} e^{-(E_{\mathrm{g}}-E_{\mathrm{F}})/k_{\mathrm{B}}T}$$

with

$$N_{\mathrm{e}} = 2\left(\frac{m_{\mathrm{e}}k_{\mathrm{B}}T}{2\pi\hbar^2}\right)^{\frac{3}{2}}.$$

Similarly, the hole concentration in the valence band is

$$p = \int_{-\infty}^{0} f_{\mathrm{h}}(E)\, g_{\mathrm{h}}(E)dE,$$

where

$$\begin{aligned} f_{\mathrm{h}}(E) &= 1 - f(E) \approx e^{-(E_{\mathrm{F}}-E)/k_{\mathrm{B}}T} \\ g_{\mathrm{h}}(E) &= \frac{1}{2\pi^2}\left(\frac{2m_{\mathrm{h}}}{\hbar^2}\right)^{\frac{3}{2}} (-E)^{\frac{1}{2}}. \end{aligned}$$

Hence

$$p \approx N_{\mathrm{h}} e^{-E_{\mathrm{F}}/k_{\mathrm{B}}T}$$

with

$$N_h = 2\left(\frac{m_h k_B T}{2\pi\hbar^2}\right)^{\frac{3}{2}}.$$

Therefore

$$\begin{aligned} f(T) &= np \approx N_e N_h e^{-E_g/k_B T} \\ &= 4(m_e m_h)^{\frac{3}{2}}\left(\frac{k_B T}{2\pi\hbar^2}\right)^3 e^{-E_g/k_B T}. \end{aligned}$$

b. The expression of $np = f(T)$ shows that, at a definite temperature and for a nondegenerate semiconductor, the product of the carrier concentrations at thermal equilibrium is independent of the impurity present. Hence it is also true for an intrinsic semiconductor for which $n = p = n_i$. Thus for an intrinsic semiconductor, the conductivity is

$$\sigma_i = e n_i(\mu_e + \mu_h) \approx 2e n_i \mu_e.$$

The donor As is pentavalent while the acceptor Si is tetravalent so each donor atom contributes one free electron at room temperature. Thus $n \approx N_d$. Neglecting the effect of the acceptor impurities, the conductivity of the material is

$$\sigma = e\mu_e N_d.$$

For

$$\frac{\sigma}{\sigma_i} = \frac{N_d}{2n_i} = 10^4,$$

we require

$$N_d = 2 \times 10^4 n_i.$$

From (a)

$$\begin{aligned} n_i &= \sqrt{f(T)} = 2(m_e m_h)^{\frac{3}{4}}\left(\frac{k_B T}{2\pi\hbar^2}\right)^{\frac{3}{2}} e^{-E_g/k_B T} \\ &= 2(0.2 \times 0.3)^{0.75}\left(\frac{k_B T mc^2}{2\pi\hbar^2 c^2}\right)^{1.5} e^{-E_g/k_B T}. \end{aligned}$$

With

$$\begin{aligned} k_B T &\approx \frac{1}{40}\,\text{eV at room temperature,} \\ E_g &= 1.14\,\text{eV for Si,} \\ mc^2 &= 0.511\,\text{MeV}, \quad c = 3 \times 10^8\,\text{ms}^{-1}, \\ \hbar &= 6.582 \times 10^{-16}\,\text{eVs}, \end{aligned}$$

we have

$$n_{\text{i}} = 2(0.2 \times 0.3)^{0.75}\left(\frac{0.511 \times 10^6}{80\pi}\right)^{1.5}$$
$$\left(\frac{1}{6.58 \times 10^{-16} \times 3 \times 10^8}\right)^3 e^{-22.8}$$
$$= 3.62 \times 10^{14}\text{m}^{-3},$$

and hence

$$N_{\text{d}} = 7.2 \times 10^{18}\text{m}^{-3} .$$

c. If $\mathbf{\Phi}_{\text{m}} < \mathbf{\Phi}_{\text{s}}$, the energy band diagrams before and after the metal makes contact with the *n*-type semiconductor are as shown in Fig. 1.29 (a) and (b) respectively.

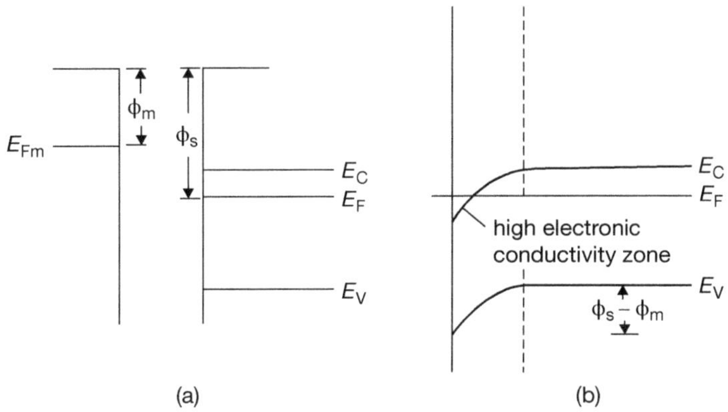

Fig1.29

After contact, because $\Phi_{\text{m}} < \Phi_{\text{s}}$, the electrons in the metal flow into the semiconductor. The energy band near the surface of the semiconductor bends down to form a high-conductivity antiblocking layer whose effect on the contact resistance between the semiconductor and the metal is very small. Hence such a junction is ohmic.

If $\mathbf{\Phi}_{\text{m}} > \mathbf{\Phi}_{\text{s}}$, the energy band diagrams before and after the metal makes contact with the *n*-type semiconductor are as shown in Fig. 1.30 (a) and (b) respectively.

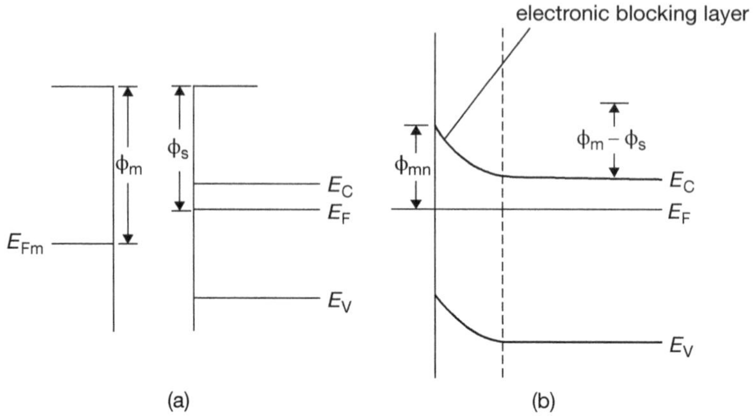

Fig. 1.30

After contact, as $\boldsymbol{\Phi}_m > \boldsymbol{\Phi}_s$, the electrons in the semiconductor flow into the metal, and the energy band near the surface of the semiconductor bends up to form a high-resistance zone, i.e., a blocking layer or surface potential barrier. When a forward bias is applied, the metal becomes positive and the semiconductor negative, the surface potential barrier decreases, and the electron current flowing from the semiconductor to the metal increases significantly, forming a large forward current. Conversely when a reverse bias is applied, the surface potential barrier will increases and the electron current flowing from the semiconductor to the metal will decrease appreciably, approaching zero. Thus, only a few electrons flow from the metal to the semiconductor, forming only a small reverse current. Hence such a junction acts as a rectifier.

1.3

ELECTROMAGNETIC PROPERTIES, OPTICAL PROPERTIES AND SUPERCONDUCTIVITY (1052–1076)

1052

The orienting tendency of an electric field E on a permanent dipole is opposed by thermal agitation.

a. Using classical statistical mechanics, calculate the total polarization of a gas of N independent permanent dipoles of moment P.

b. Show that for small fields, the orientational polarization per dipole is inversely proportional to the temperature.

c. Discuss the effect of this phenomenon on the dielectric constant of water, a polarizable molecule, at high frequencies and at low frequencies.

(*Princeton*)

Sol:

a. In a gas of independent dipoles, the latter are able to rotate freely. The potential energy of a dipole of moment $\mathbf{p}$ in an applied electric field $\mathbf{E}$ is

$$V = -\mathbf{p} \cdot \mathbf{E} = -pE\cos\theta,$$

where θ is the angle it makes with the direction of the field. Classical statistical mechanics gives that the probability of finding the dipole along the θ direction is given by the Boltzmann distribution function

$$f(\theta) = e^{-V/k_B T} = \exp\left(\frac{pE\cos\theta}{k_B T}\right).$$

The average value of p along the direction of the field is then

$$\langle p\cos\theta\rangle = \frac{\displaystyle\int p\cos\theta f(\theta)d\Omega}{\displaystyle\int f(\theta)d\Omega}$$

with $d\Omega = 2\pi\sin\theta d\theta$. Thus

$$\begin{aligned}\langle p\cos\theta\rangle &= \frac{\displaystyle\int_0^\pi p\cos\theta\exp\left(\frac{pE\cos\theta}{k_B T}\right)2\pi\sin\theta d\theta}{\displaystyle\int_0^\pi \exp\left(\frac{pE\cos\theta}{k_B T}\right)2\pi\sin\theta d\theta}\\ &= \frac{\displaystyle p\int_{-1}^{1} e^{\beta x}xdx}{\displaystyle\int_{-1}^{1} e^{\beta x}dx}\\ &= \frac{p}{\beta}(\beta\coth\beta - 1),\end{aligned}$$

where $x = \cos\theta$, $\beta = \dfrac{pE}{k_B T}$.

As there are N independent dipoles in the gas the total polarization is

$$P = N\langle p\cos\theta\rangle = \frac{Nk_B T}{E}\left(\frac{pE}{k_B T}\coth\frac{pE}{k_B T} - 1\right).$$

b. For small fields, $\beta = \dfrac{pE}{k_B T} \ll 1$ and

$$\coth\beta = \frac{1}{\beta} + \frac{\beta}{3} - \frac{\beta^3}{45} + \ldots \approx \frac{k_B T}{pE} + \frac{pE}{3k_B T}.$$

By definition, the orientational polarizability per dipole is

$$\alpha = \frac{P}{EN} \approx \frac{p^2}{3k_B T} \propto T^{-1}.$$

c. If the electric field oscillates, the dipoles, following the field, will flip back and forth as the field reverses its direction during each cycle. However, the dipole may experience friction due to its collision with other molecules in the gas, causing some loss of energy known as dielectric loss. This means that a relaxation time τ is involved, as well as a phase lag between the field and the polarization, and that the dielectric constant $\epsilon = \epsilon_0 + an$, where ϵ_0 is the vacuum permittivity and n the number of dipoles per unit volume, is complex:

$$\epsilon = \epsilon_1 + i\epsilon_2 .$$

The variations of ϵ_1, ϵ_2 with $\omega\tau$ are shown in Fig. 1.31. The imaginary part, ϵ_2, is proportional to the dielectric loss.

At low frequencies ($\omega\tau \ll 1$), dielectric loss is small and $\epsilon \approx \epsilon_1 \approx \epsilon_r$, the static permittivity. As the frequency increases to the vicinity of $\omega\tau \approx 1$, dielectric loss becomes appreciable and the real part of the dielectric constant decreases with increasing frequency in that region. At high frequencies ($\omega\tau \gg 1$), the flippings of the dipoles are unable to follow the rapid oscillations of the field and dielectric loss again becomes small. In this region the real part ϵ_1 tends to a constant ϵ_∞, the dielectric constant for high frequencies.

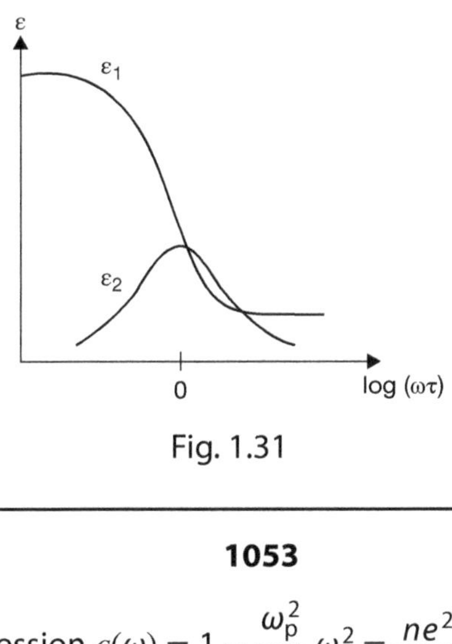

Fig. 1.31

1053

a. Derive the expression $\epsilon(\omega) = 1 - \frac{\omega_p^2}{\omega^2}, \omega_p^2 = \frac{ne^2}{\epsilon_0 m}$ for the dielectric constant as a function of ω for a free electron gas of number density n.

b. Show clearly that metals are opaque to light for which ω is less than ω_p.

c. Calculate the wavelength cutoff for Na metal if the volume of a primitive unit cell in Na is 35×10^{-30} m^3.

(*Wisconsin*)

Sol:

a. Take the x-axis parallel to the applied electric field $\mathbf{E} = \mathbf{E}_0 e^{i\omega t}$. The displacement of a free electron from its neutral position, taken as the origin, is given by

$$m\frac{d^2x}{dt^2} = -eE_0 e^{i\omega t}.$$

Try a solution of the form $x = x_0 e^{i\omega t}$. Substitution gives

$$x_0 = \frac{eE_0}{m\omega^2}.$$

The dipole moment per unit volume, or the polarization, of the electron gas is by definition

$$P = -nex = -\frac{ne^2}{m\omega^2}E_0 e^{i\omega t} = -\frac{ne^2}{m\omega^2}E.$$

The (relative) dielectric constant ϵ is then

$$\epsilon = \frac{D}{\epsilon_0 E} = \frac{\epsilon_0 E + P}{\epsilon_0 E} = 1 - \frac{ne^2}{\epsilon_0 m}\frac{1}{\omega^2} = 1 - \frac{\omega_p^2}{\omega^2}$$

where $\omega_p = \sqrt{\frac{ne^2}{\epsilon_0 m}}$ is the plasma (angular) frequency of the gas.

b. The refractive index of a substance is $n = \sqrt{\epsilon}$. For light of frequency $\omega < \omega_p$ incident on a metal of plasma frequency ω_p, n is imaginary. This means that such light cannot be transmitted through the metal but is totally reflected from its surface. Thus metals are opaque to light for which $\omega < \omega_p$.

c. The sodium atom has valency one and in the metallic state each atom contributes one conduction electron. As each primitive cell contains one atom, the number density of conduction electrons is

$$n = \frac{1}{35 \times 10^{-30}} = 2.86 \times 10^{28}\,\text{m}^{-3}.$$

Then

$$\begin{aligned}\omega_p^2 &= 4\pi c^2 n\left(\frac{e^2}{4\pi\epsilon_0 mc^2}\right)\\ &= 4\pi \times (3 \times 10^8)^2 \times 2.86 \times 10^{28} \times 2.82 \times 10^{-15}\\ &= 9.12 \times 10^{31}\,\text{s}^{-2},\end{aligned}$$

and the cutoff angular frequency is

$$\omega = \omega_p = 9.55 \times 10^{15}\,\text{s}^{-1}.$$

Hence the cutoff wavelength is

$$\lambda = \frac{2\pi c}{\omega} = 2.0 \times 10^{-7}\,\text{m} = 2000\,\text{Å}.$$

1054

A solid is composed of a collection of identical atoms, each of which can be modeled as an electron of charge q and mass m coupled by a spring of constant k to a fixed atomic site. Energy dissipation is modeled by a damping force in each atom equal to $-b\mathbf{v}$, where $\mathbf{v}$ is the electron velocity. The number density of these atoms in the solid N. A plane polarized electromagnetic wave of frequency ω propagates in the solid. Treat the oscillators classically and nonrelativistically. For uniformity of notation, use $\omega_0^2 = \kappa/m$, $\gamma = b/m$ and $\omega_p^2 = 4\pi Nq^2/m$.

a. Compute the index of refraction $n(\omega)$.

b. Sketch the real and imaginary parts of $n(\omega)$ in the low density limit ($\omega_p/\omega_0 \ll 1$ and $\omega_p/\gamma \ll 1$).

c. At resonance (i.e., $\omega = \omega_0$), compute the attenuation length in the solid in the low density limit.

In the vicinity of resonance, the group velocity v_g, defined as $d\omega/dk$, can exceed the speed of light. Explain why this is not a violation of causality.

(*MIT*)

Sol:

a. Let $\mathbf{r}$ be the displacement of an electron of charge q from the fixed atomic site and $\mathbf{E} = \mathbf{E}_0 e^{-i\omega t}$ be the electric field intensity of the traversing wave at the location of the atom. The equation of motion of the electron is

$$m\ddot{\mathbf{r}} + b\dot{\mathbf{r}} + \kappa\mathbf{r} = q\mathbf{E}.$$

Try a solution $\mathbf{r} = \mathbf{r}_0 e^{-i\omega t}$. Substitution gives

$$\mathbf{r} = \frac{q\mathbf{E}}{-m\omega^2 - i\omega b + \kappa} = \frac{q}{m}\frac{\mathbf{E}}{\omega_0^2 - \omega^2 - i\omega\gamma}.$$

The electric dipole moment per unit volume, or the polarization, of the solid is then

$$\mathbf{P} = nq\mathbf{r} = \frac{Nq^2}{m}\frac{E}{\omega_0^2 - \omega^2 - i\omega\gamma}.$$

As $\mathbf{P}$ and $\mathbf{E}$ are parallel, the relative dielectric constant is

$$\epsilon = \frac{\epsilon_0 E + P}{\epsilon_0 E} = 1 + \frac{\omega_p^2}{\omega_0^2 - \omega^2 - i\omega\gamma},$$

Where $\omega_p^2 = \dfrac{Nq^2}{\epsilon_0 m}$ (in SI units). The index of refraction n is now complex:

$$n \equiv \sqrt{\epsilon} = \eta + i\kappa$$

With η, k given by

$$\eta^2 - \kappa^2 = 1 + \frac{(\omega_0^2 - \omega^2)\,\omega_p^2}{(\omega_0^2 - \omega^2)^2 + \omega^2\gamma^2},$$

$$2\eta\kappa = \frac{\omega\,\omega_p^2\,\gamma}{(\omega_0^2 - \omega^2)^2 + \omega^2\gamma^2}.$$

b. As $\omega_p \propto \sqrt{N}$, in the low density limit, $\eta^2 - k^2 \sim 1$, $2\eta k \sim 0$. This means that the imaginary part of n, k, is small and the real part, η, is close to unity at low densities. The variations $\eta(\omega)$, $k(\omega)$ for low densities are sketched in Fig. 1.32. Note that k is maximum at $\omega = \omega_0$. The corresponding maximum absorption is known as resonance absorption. Also $\eta > 1$ for $\omega < \omega_0$ and $\eta < 1$ for $\omega > \omega_0$.

c. At resonance, $\omega \approx \omega_0$ and

$$\eta^2 - \kappa^2 = 1, \quad 2\eta\kappa = \frac{\omega_p^2}{\omega_0\,\gamma}.$$

In the low density limit, $\omega_p \ll \gamma$ and the above set of equations has solution

$$\eta \approx 1, \quad \kappa \approx \frac{\omega_p^2}{2\,\omega_0\,\gamma}.$$

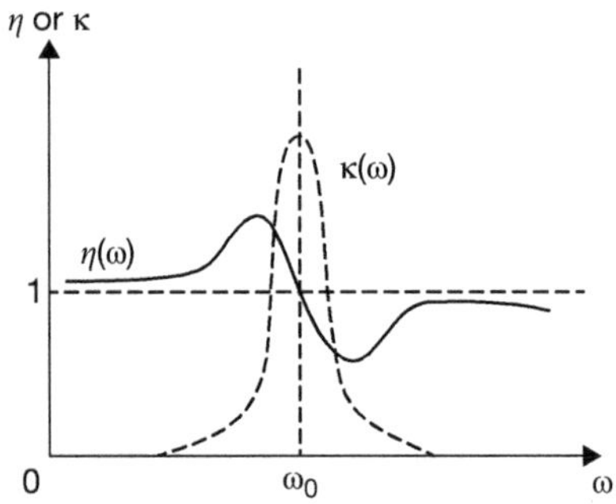

Fig. 1.32

The wave function has an exponential factor

$$e^{i(knr-\omega t)} = e^{-\kappa kr}e^{i(k\eta r-\omega t)},$$

where $k = \omega_0/c$, showing that the intensity has an attenuation factor $e^{-2k\omega_0 r/c}$. The attenuation length l is then given by

$$\frac{2\kappa\,\omega_0}{c}l = 1,$$

or

$$l = \frac{c}{2\kappa\omega_0} = \frac{c\gamma}{\omega_p^2}.$$

d. The index of refraction is by definition

$$n = \frac{c}{v} = \frac{ck}{\omega},$$

giving

$$k = \frac{n(\omega)\omega}{c}.$$

The group velocity v_g is then

$$v_g = \left(\frac{dk}{d\omega}\right)^{-1} = \left(\frac{\omega}{c}\frac{dn}{dw} + \frac{n}{c}\right)^{-1}.$$

In the low density limit, near resonance $\omega \approx \omega_0$, $\kappa \ll 1$ so that

$$n = \eta + i\kappa \approx \eta,$$

$$v_g \simeq \frac{c}{\omega_0\left(\frac{d\eta}{d\omega}\right)_{\omega_0} + \eta}.$$

As $\left(\frac{d\eta}{d\omega}\right)_{\omega_0} = \frac{1}{2\eta}\left(\frac{d\eta^2}{d\omega}\right)_{\omega_0} = -\frac{\omega_p^2}{\omega_0\gamma^2}$, $\eta \approx 1$ for low densities,

$$v_g \approx \frac{c}{1 - \frac{\omega_p^2}{\gamma^2}} > c,$$

which shows that in the vicinity of resonance the group velocity exceeds the space of light. However, this is not a violation of causality because when anomalous dispersion occurs the group velocity can no longer be identified with the velocity of energy propagation, which is always less than c.

1055

Let a dielectric be described as N valence electrons per unit volume, bound to ions harmonically with angular frequency ω_0. It is in a uniform magnetic field **B** directed along the z-axis. Monochromatic light of frequency ω is sent through the dielectric along the z-direction.

a. If the light is circularly polarized, find the indices of refraction.

b. If the light is linearly polarized, calculate the angle through which the polarization rotates as the wave propagates a distance z in the dielectric (you may assume that $\Delta n \ll n$, where $\Delta n = n_R - n_L$ and $n = (n_R + n_L)/2$).

(*Wisconsin*)

Sol:

a. The equation of motion for a valence electron is

$$m\frac{d^2\mathbf{r}}{dt^2} = -e\mathbf{E} - e\mathbf{V}\times\mathbf{B} - m\omega_0^2\mathbf{r},$$

where $\mathbf{r}$ is the displacement of the electron from its equilibrium position. The collective motion of the N valence electrons generates a polarization

$$\mathbf{P} = -Ne\mathbf{r}.$$

The equation of the polarization is accordingly

$$-\frac{m}{Ne}\frac{d^2\mathbf{P}}{dt^2} = -e\mathbf{E} - e\left(-\frac{1}{Ne}\frac{d\mathbf{P}}{dt}\right)\times\mathbf{B} + \frac{m\omega_0^2}{Ne}\mathbf{P}.$$

As $\mathbf{E} \sim e^{i\omega t}$, we have $\mathbf{P} \sim e^{i\omega t}$ and $\frac{d}{dt} = i\omega$. Substitution in the above gives

$$\frac{m\omega^2}{Ne}\mathbf{P} = -e\mathbf{E} - \frac{i\omega\mathbf{B}\times\mathbf{P}}{N} + \frac{m\omega_0^2}{Ne}\mathbf{P},$$

or

$$\mathbf{E} = \frac{\omega_0^2 - \omega^2}{\epsilon_0\,\omega_\mathrm{p}^2}\mathbf{P} + \frac{i\omega}{Ne}\mathbf{B}\times\mathbf{P},$$

where $\omega_\mathrm{p}^2 = \frac{Ne^2}{m\epsilon_0}$. As $\mathbf{B} = (0, 0, B)$, the above can be written as

$$\begin{pmatrix} E_x \\ E_y \\ E_z \end{pmatrix} = \begin{pmatrix} \frac{\omega_0^2 - \omega^2}{\epsilon_0\,\omega_\mathrm{p}^2} & \frac{i\omega}{Ne}B & 0 \\ -\frac{i\omega}{Ne}B & \frac{\omega_0^2 - \omega^2}{\epsilon_0\,\omega_\mathrm{p}^2} & 0 \\ 0 & 0 & \frac{\omega_0^2 - \omega^2}{\epsilon_0\,\omega_\mathrm{p}^2} \end{pmatrix} \begin{pmatrix} P_x \\ P_y \\ P_z \end{pmatrix}.$$

Maxwell's equations for free space

$$\nabla\times\mathbf{E} = -\dot{\mathbf{B}}, \qquad \nabla\times\mathbf{H} = \dot{\mathbf{D}}, \qquad \nabla\cdot\mathbf{D} = 0$$

with $\mathbf{B} = \mu_0\mathbf{H}$, $\mathbf{D} = \epsilon_0\mathbf{E} + \mathbf{P}$ give

$$\mu_0\epsilon_0\ddot{\mathbf{E}} - \nabla^2\mathbf{E} = \frac{1}{\epsilon_0}\nabla(\nabla\cdot\mathbf{P}) - \mu_0\ddot{\mathbf{P}}.$$

With $\mathbf{E} = E_0 e^{i(\omega t - k\cdot\mathbf{r})}$, $\mathbf{P} = P_0 e^{i(\omega t - k\cdot\mathbf{r})}$, $\mathbf{k} = k\mathbf{e}_z$, we have

$$\frac{\partial^2}{\partial t^2} = -\omega^2, \qquad \nabla^2\mathbf{E} = -k^2\mathbf{E},$$

$$\nabla\cdot\mathbf{P} = -i\mathbf{k}\cdot\mathbf{P} = -ikP_z, \qquad \nabla(\nabla\cdot\mathbf{P}) = -k^2P_z\mathbf{e}_z,$$

and thus

$$\epsilon_0(k^2c^2 - \omega^2)\mathbf{E} = -k^2c^2P_z\mathbf{e}_z + \omega^2\mathbf{P}$$

since $\mu_0\epsilon_0 = c^{-2}$. Letting $\chi = \frac{k^2c^2}{\omega^2} - 1$, we can write the above as

$$\epsilon_0\chi\mathbf{E} = P_x\mathbf{e}_x + P_y\mathbf{e}_y - \chi P_z\mathbf{e}_z\ ,$$

or

$$\begin{pmatrix} E_x \\ E_y \\ E_z \end{pmatrix} = \frac{1}{\epsilon_0\chi}\begin{pmatrix} 1 & 0 & 0 \\ 0 & 1 & 0 \\ 0 & 0 & -\chi \end{pmatrix}\begin{pmatrix} P_x \\ P_y \\ P_z \end{pmatrix}.$$

The above two matrix equations combine to give

$$\begin{pmatrix} \frac{\omega_0^2-\omega^2}{\epsilon_0\omega_{\mathrm{p}}^2} - \frac{1}{\epsilon_0\chi} & \frac{i\omega}{Ne}B & 0 \\ -\frac{i\omega}{Ne}B & \frac{\omega_0^2-\omega^2}{\epsilon_0\omega_{\mathrm{p}}^2} - \frac{1}{\epsilon_0\chi} & 0 \\ 0 & 0 & \frac{\omega_0^2-\omega^2}{\epsilon_0\omega_{\mathrm{p}}^2} + \frac{1}{\epsilon_0} \end{pmatrix}\begin{pmatrix} P_x \\ P_y \\ P_z \end{pmatrix} = 0\ .$$

In general $\frac{\omega_0^2-\omega^2}{\epsilon_0\,\omega_{\mathrm{p}}^2} + \frac{1}{\epsilon_0} \neq 0$, hence $P_z = 0$ and so $E_z = 0$. This means that an incident plane wave will remain a plane wave. The condition for nonvanishing solutions is that determinant of the square matrix in the equation for **P** must vanish, giving

$$\frac{\omega_0^2-\omega^2}{\epsilon_0\,\omega_{\mathrm{p}}^2} - \frac{1}{\epsilon_0\chi\pm} = \pm\frac{\omega B}{Ne}.$$

Thus there are two refractive indices

$$n_\pm = \frac{ck_\pm}{\omega} = \sqrt{1+\chi\pm} = \sqrt{1 + \frac{Ne\omega_p^2}{Ne(\omega_0^2-\omega^2) \mp \epsilon_0\,\omega_{\mathrm{p}}^2\,\omega B}}.$$

Corresponding to $n_\pm$ we have

$$\frac{\omega B}{Ne}P_x = \mp\frac{i\omega B}{Ne}P_y, \quad \text{or} \quad P_y = \pm iP_x$$

and hence the amplitudes

$$P_\pm = C\pm\begin{pmatrix} 1 \\ \pm i \\ 0 \end{pmatrix},$$

where C_+, C_- are constants. Similarly,

$$E_\pm = \frac{C_\pm}{\epsilon_0\chi\pm}\begin{pmatrix} 1 \\ \pm i \\ 0 \end{pmatrix}.$$

Since the factors $\pm i$ correspond to additional phase angles $\pm\frac{\pi}{2}$, the above shows that E_+, E_- are right-hand and left-hand circularly polarized waves with indices of refraction n_+, n_- respectively. (Right-hand rotation means that looking against the direction of light, **E** rotates clockwise. For E_+, $E_y = iE_x$, or $E_x = \frac{E_0}{2}\cos\omega t = \frac{E_0}{2}\cos(-\omega t)$, $E_y = \frac{E_0}{2}\cos\left(\omega t + \frac{\pi}{2}\right) = \frac{E_0}{2}\sin(-\omega t)$.)

b. The linearly polarized light can be decomposed into right-and left-hand circulating polarized components

$$\mathbf{E} = \mathbf{E}_+ + \mathbf{E}_- = \frac{\mathbf{E}_0}{2}\begin{pmatrix}1\\ i\end{pmatrix} e^{i(\omega t - k+z)} + \frac{\mathbf{E}_0}{2}\begin{pmatrix}1\\ -i\end{pmatrix} e^{i(\omega t - k-z)}$$

$$= \mathbf{E}_0\begin{pmatrix}\cos\left(\frac{k_+ - k_-}{2}z\right)\\ -\sin\left(\frac{k_+ - k_-}{2}z\right)\end{pmatrix} e^{i\left(\omega t - \frac{k_+ + k_-}{2}z\right)}.$$

This shows that when looking against the direction of light propagation, the plane of polarization turns anticlockwise with an angle

$$\theta = \frac{k_+ - k_-}{2}z = \frac{\omega z}{2c}(n_+ - n_-) = \frac{\omega z}{2c}\Delta n,$$

after traveling the medium for a distance z. Assuming

$$\frac{\epsilon_0\,\omega_\mathrm{p}^2\,\omega B}{Ne\left(\omega_0^2 - \omega^2\right)} \ll 1,$$

we have

$$n_\pm = \sqrt{1 + \frac{\omega_\mathrm{p}^2}{\omega_0^2 - \omega^2}\left(1 \pm \frac{\epsilon_0\,\omega_\mathrm{p}^2\,\omega B}{Ne(\omega_0^2 - \omega^2)}\right)}$$

$$\approx \sqrt{1 + \frac{\omega_\mathrm{p}^2}{\omega_0^2 - \omega^2}} \pm \frac{\epsilon_0\,\omega_\mathrm{p}^4\,\omega B}{2Ne(\omega_0^2 - \omega^2)^2}\left(1 + \frac{\omega_\mathrm{p}^2}{\omega_0^2 - \omega^2}\right)^{-\frac{1}{2}}.$$

Note that this approximation agrees with the condition $\Delta n \ll \frac{1}{2}(n_+ + n_-)$.

As

$$\Delta n \approx \frac{\epsilon_0\,\omega_\mathrm{p}^4\,\omega B}{Ne(\omega_0^2 - \omega^2)\sqrt{(\omega_0^2 - \omega^2)^2 + \omega_\mathrm{p}^2(\omega_0^2 - \omega^2)}},$$

we obtain

$$\theta = \frac{\omega_\mathrm{p}^2\omega^2\,\omega_\mathrm{c} z}{2c(\omega_0^2 - \omega^2)\sqrt{(\omega_0^2 - \omega^2)^2 + \omega_\mathrm{p}^2(\omega_0^2 - \omega^2)}},$$

where $\omega_\mathrm{c} = \frac{eB}{m}$.

1056

Circularly polarized light propagates along the *z*-axis through a fully ionized medium in which a static magnetic field *B* is also directed along the *z*-axis. The index of refraction for right-hand circularly polarized radiation of angular frequency ω is

$$n_R^2 = 1 - \frac{4\pi Ne^2/m}{\omega^2 + \frac{eB\omega}{mc}},$$

where *N* is the electron density and *e*, *m* and *c* have their usual meaning.

a. What is the corresponding expression for left-hand circularly polarized light?

b. If $B = 0$, what is the phase velocity?

c. if $B = 0$, what is the group velocity? You may assume that $\omega \gg \sqrt{\frac{4\pi Ne^2}{m}}$ but not $\omega \to \infty$.

(*Wisconsin*)

Sol:

a. The equation of motion of an electron in the medium is

$$m\frac{d^2\mathbf{r}}{dt^2} = -e(\mathbf{E} + \mathbf{v} \times \mathbf{B}) ,$$

where $\mathbf{r}$ is the displacement of the electron from the original position. The force on the electron due to the magnetic field associated with the traversing light has been omitted as it is smaller than the force due to the associated electric field by a factor v/c. With $\mathbf{E} = E_x\mathbf{e}_x + E_y\mathbf{e}_y$, $\mathbf{B} = B\mathbf{e}_z$, $\mathbf{r} = x\mathbf{e}_x + y\mathbf{e}_y + z\mathbf{e}_z$, the above becomes

$$\begin{aligned} m\ddot{x} &= -eE_x - eB\dot{y}, \\ m\ddot{y} &= -eE_y + eB\dot{x}, \\ m\ddot{z} &= 0. \end{aligned}$$

The last equation shows that $\dot{z} =$ constant so that any velocity in the z direction is not associated with the oscillation caused by the traversing light and can hence be neglected. We can also take $z = 0$.

For right-hand circularly polarized light, we have

$$E_x = E_0 e^{i(\omega t - kz)}, \quad E_y = iE_x,$$

and so

$$x = x_0 e^{i(\omega t - kz)}, \quad y = y_0 e^{i(\omega t - kz)}.$$

The equation of motion can thus be written as

$$-m\omega^2 x = -eE_x - i\omega eBy,$$
$$-m\omega^2 y = -eE_y + i\omega eBx.$$

In terms of the polarization vector $\mathbf{P} = -Ne\mathbf{r}$, the above become

$$-\frac{m\omega^2}{Ne}P_x = eE_x - \frac{i\omega B}{N}P_y,$$
$$-\frac{m\omega^2}{Ne}P_y = eE_y + \frac{i\omega B}{N}P.$$

Maxwell's equation for a source-free isotropic medium,

$$\nabla + \mathbf{E} = -\dot{\mathbf{B}}, \quad \nabla \times \mathbf{H} = \dot{\mathbf{D}},$$

with

$$\mathbf{B} = \mu_0 \mathbf{H}, \quad \mathbf{D} = \epsilon_0 \mathbf{E} + \mathbf{P}$$

give

$$\mu_0 \ddot{\mathbf{P}} - \frac{1}{\epsilon_0}\nabla(\nabla \cdot \mathbf{P}) = \nabla^2 \mathbf{E} - \frac{1}{c^2}\ddot{\mathbf{E}}.$$

We also have $\nabla \cdot \mathbf{D} = 0$, which gives

$$\nabla \cdot \mathbf{P} = \frac{1}{\epsilon_0}\nabla \cdot \mathbf{E} = \frac{1}{\epsilon_0}\frac{\partial E_z}{\partial z} = 0$$

since $\mathbf{E}$ depends on z only and $E_z = 0$. This also implies that $P_z = 0$. Noting that

$$\nabla^2 \mathbf{E} = -k^2 \mathbf{E}, \quad \frac{\partial^2}{\partial t^2} = -\omega^2,$$

we finally have

$$\mathbf{P} = \chi \mathbf{E},$$

where

$$\chi = \epsilon_0 \left(\frac{c^2 k^2}{\omega^2} - 1 \right).$$

The differential equations for P_x, P_y can now be written as

$$\left(\frac{\omega^2}{Ne} + \frac{e}{m\chi} \right) E_x - \frac{i\omega B}{Nm} E_y = 0,$$
$$\frac{i\omega B}{Nm} E_x + \left(\frac{\omega^2}{Ne} + \frac{e}{m\chi} \right) E_y = 0.$$

For nonvanishing solutions, we require that the determinant of the coefficients of E_x, E_y be zero. This gives

$$\left(\frac{\omega^2}{Ne} + \frac{e}{m\chi} \right)^2 = \left(\frac{\omega B}{Nm} \right)^2,$$

which has two solutions

$$\frac{\omega^2}{Ne} + \frac{e}{m\chi\pm} = \pm\frac{\omega B}{Nm}.$$

The top sign gives $Ey = -iE_x$, which is the left-hand circularly polarized light. The refraction index n_R for the right-hand circularly polarized light is then given by

$$\begin{aligned} n_R^2 &= \frac{\epsilon_0 E + P}{\epsilon_0 E} = 1 + \frac{\chi-}{\epsilon_0} \\ &= 1 - \frac{\omega_p^2}{\omega^2 + \omega\,\omega_g}, \end{aligned}$$

where $\omega_p^2 = \frac{Ne^2}{m\epsilon_0}$, $\omega_g = \frac{eB}{m}$. Note that in Gaussian units,

$$\omega_p^2 = \frac{4\pi Ne^2}{m}, \quad \omega_g = \frac{eB}{mc},$$

and the expression for n_R^2 is identical with that given in the problem. For the left-hand circularly polarized light,

$$n_L^2 = 1 + \frac{\chi+}{\epsilon_0} = 1 - \frac{\omega_p^2}{\omega^2 - \omega\,\omega_g}.$$

b. If $B = 0$,

$$n^2 = 1 - \frac{\omega_p^2}{\omega^2} = 1 + \frac{\chi}{\epsilon_0} = \frac{c^2k^2}{\omega^2},$$

giving

$$c^2k^2 = \omega^2 - \omega_p^2 .$$

The phase velocity is then

$$v_p = \frac{\omega}{k} = \frac{c}{\sqrt{1 - \left(\frac{\omega_p}{\omega}\right)^2}}.$$

c. As $2c^2kdk = 2\omega d\omega$, the group velocity is

$$v_g = \frac{d\omega}{dk} = \frac{c^2k}{\omega} = c\sqrt{1 - \left(\frac{\omega_p}{\omega}\right)^2}.$$

1057

Below are listed some measured properties of NaCl at room temperature: static dielectric constant $\varepsilon = 5.90$, side of unit cell $a = 5.6$ Å, atomic mass of Na = 23 amu (1 amu = 1.66×10^{-24} g), atomic mass of Cl = 35.5 amu.

a. Polaritons results from the coupling between phonons and photons in ionic crystals. To see this, derive the dispersion relation $\omega(k)$ and draw the dispersion diagram for transverse electromagnetic modes propagating through an ionic crystal (e.g. NaCl). Neglect atomic polarization.

b. For what band of photon wavelengths would you expect NaCl to reflect incident radiation? (That is, obtain and evaluate expressions for the shortest and longest wavelengths of this band.)

(*Princeton*)

Sol:

a. Figure 1.33 (a) shows optical and acoustic branches of the dispersion relation for a one-dimensional diatomic lattice as shown in Fig. 1.33 (b) without any coupling between them. Near $k = 0$, the acoustic branch is linear, $\omega = vk$. As $v \approx 10^{-5}c$, the corresponding wave for light, $\omega = ck$ will practically coincide with the ω-axis. Thus only the coupling between the optical mode of $k = 0$ and light is possible.

Neglecting electronic polarizability, we assume polarization to be caused entirely by ionic motion. Consider the motion of a pair of ions in an applied electric field **E** with $k \approx 0$. The relative displacement **W** of a positive from a negative ion has equation of motion

$$\mu \ddot{\mathbf{W}} = -2K\mathbf{W} + q\mathbf{E},$$

where $\mu = \dfrac{Mm}{M+m}$ is the reduced mass of the pair of ions and K is the force constant between them, q is the magnitude of the charge of each ion. When the ions are stationary, $\ddot{\mathbf{W}} = 0$ and $\mathbf{W} = \dfrac{q}{2k}\mathbf{E}$. If V is the volume per unit pair, and $\rho_0 = q/V$ is the charge density, the polarization is by definition

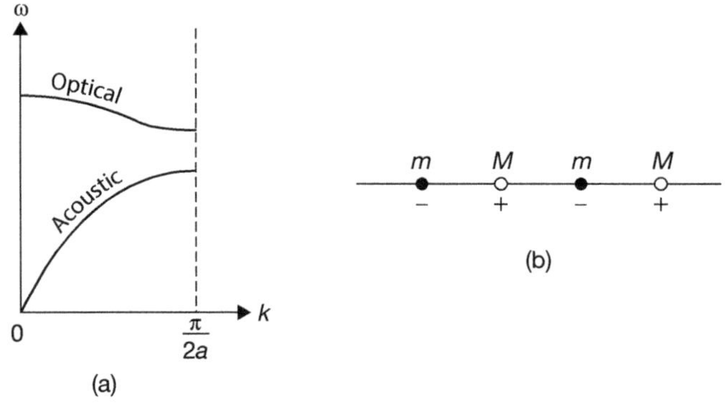

Fig. 1.33

$$\mathbf{P} = \rho_0 \mathbf{W} = \frac{q^2}{2KV}\mathbf{E} = \epsilon_0 \chi \mathbf{E},$$

where ϵ_0 is the free-space permittivity and χ the electric susceptibility related to the relative dielectric constant ε by $\chi = \varepsilon - 1$. The force constant is then

$$K = \frac{q^2}{2\chi\epsilon_0 V} = \frac{q^2}{2(\varepsilon - 1)\epsilon_0 V}.$$

Setting $\mathbf{Q} = \frac{\rho_0}{\epsilon_0}\mathbf{W}$, we have from the equation of motion

$$\ddot{\mathbf{Q}} + \omega_0^2 \mathbf{Q} - \omega_0^2 \chi \mathbf{E} = 0, \quad (1)$$

where

$$\omega_0 = \sqrt{\frac{2K}{\mu}} = \sqrt{\frac{q^2}{\mu(\varepsilon - 1)\epsilon_0 V}}$$

is the angular frequency of the optical mode when $k = 0$, which is characteristic of the optical mode. To find the interaction of the optical mode with the traversing electromagnetic waves and derive the coupled mode, we have to combine this characteristic equation with Maxwell's equations

$$\nabla \times \mathbf{E} = -\dot{\mathbf{B}},$$
$$\nabla \times \mathbf{H} = -\dot{\mathbf{D}} + \mathbf{J},$$

with $\mathbf{B} = \mu_0 \mathbf{H}$, $\mathbf{D} = \epsilon_0 \mathbf{E}$, $\mathbf{J}_0 = \rho_0 \dot{\mathbf{W}} = \epsilon_0 \dot{\mathbf{Q}}$ being the current density. Furthermore, as the medium is charge-free (apart from the ions), $\nabla \cdot \mathbf{E} = \frac{1}{\epsilon_0} \nabla \cdot \mathbf{D} = 0$. Hence

$$\nabla \times (\nabla \times \mathbf{E}) = -\nabla^2 \mathbf{E} = -\mu_0 \nabla \times \dot{\mathbf{H}} = -\mu_0 \epsilon_0 (\ddot{\mathbf{E}} + \ddot{\mathbf{Q}}),$$

or

$$\nabla^2 \mathbf{E} = \frac{1}{c^2}\frac{\partial^2}{\partial t^2}(\mathbf{E} + \dot{\mathbf{Q}}). \quad (2)$$

Taking the direction of propagation of the electromagnetic wave as the x direction, we have

$$\mathbf{E} = \mathbf{E}_0 e^{i(kx - \omega t)}, \quad \mathbf{Q} = \mathbf{Q}_0 e^{i(kx - \omega t)}.$$

Substituting these in Eqs. (1) and (2) results in

$$\begin{cases} (\omega_0^2 - \omega^2)\mathbf{Q}_0 = \chi \omega_0^2 \mathbf{E}_0, \\ k^2 c^2 \mathbf{E}_0 = \omega^2 (\mathbf{E}_0 + \mathbf{Q}_0) \end{cases},$$

which give the dispersion relation of the coupled mode

$$\frac{c^2 k^2}{\omega^2} = \frac{\epsilon \omega_0^2 - \omega^2}{\omega_0^2 - \omega^2}.$$

It is seen that for $k \to 0$, $\omega \to 0$ or $\omega \to \sqrt{\epsilon}\,\omega_0$; for $k \to \infty$, $\omega \to \omega_0$ or $\omega \to \infty$. The relation consists of two branches as shown in Fig. 1.34.

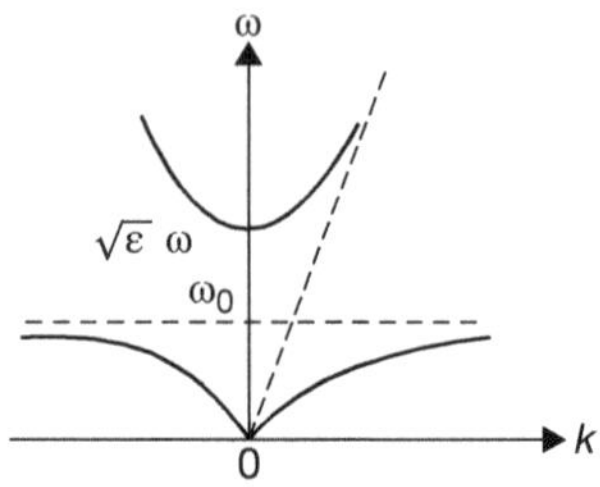

Fig. 1.34

b. As shown in Fig. 1.34, in the band between ω_0 and $\sqrt{\varepsilon}\,\omega_0$, coupling between the phonons and photons is forbidden and the incident light will be completely reflected. The shortest and longest wavelengths of this band are

$$\lambda_{\min} = \frac{2\pi c}{\sqrt{\varepsilon}\,\omega_0} = \frac{2\pi c}{q}\sqrt{\epsilon_0 \mu V\left(1 - \frac{1}{\varepsilon}\right)},$$

$$\lambda_{\max} = \frac{2\pi c}{\omega_0} = \frac{2\pi c}{q}\sqrt{\epsilon_0 \mu V(\epsilon - 1)}\,.$$

For NaCl, $V = \frac{1}{4}a^3$, $\mu = \frac{M_{\text{Na}} m_{\text{Cl}}}{M_{\text{Na}} + m_{\text{Cl}}}$, $q = e$, with the data given we obtain

$$\lambda_{\min} = 32.2\,\mu\text{m},$$

$$\lambda_{\max} = 78.3\,\mu\text{m}.$$

1058

Suppose you are given an alkali halide with $\varepsilon(0) = 5.9$, $\varepsilon(\infty) = 2.25$, reflectivity $= 0$ at $\lambda = 3.06\ \mu$m, and $\varepsilon(\omega)$ real.

a. Find ω_L, ω_T. (Recall that ω_T is the TO phonon frequency in the absence of any coupling to external fields and defines a pole of $\varepsilon(\omega)$, whereas ω_L defines a zero of $\varepsilon(\omega)$.)

b. Sketch the reflectivity vs. ω.

(Princeton)

Sol:

a. For an ionic crystal, if viscous damping can be neglected, the dielectric constant is real and the relative dielectric constant can be expressed as

$$\varepsilon(\omega) = \varepsilon(\infty) + \frac{[\varepsilon(0) - \varepsilon(\infty)]\,\omega_\text{T}^2}{\omega_\text{T}^2 - \omega^2}.$$

The refractive index of a substance is $n = \sqrt{\varepsilon(\omega)}$ and for normal incidence the reflectivity is

$$R = \left(\frac{n-1}{n+1}\right)^2 = \left(\frac{\sqrt{\varepsilon(\omega)}-1}{\sqrt{\varepsilon(\omega)}+1}\right)^2 .$$

If $R = 0$ at $\omega = \omega_0$, then $\varepsilon(\omega_0) = 1$, or

$$1 = \varepsilon(\infty) + \frac{[\varepsilon(0) - \varepsilon(\infty)]\,\omega_T^2}{\omega_T^2 - \omega_0^2},$$

giving

$$\omega_T = \left[\frac{\varepsilon(\infty) - 1}{\varepsilon(0) - 1}\right]^{1/2} \omega_0 .$$

As $\omega_0 = \frac{2\pi c}{\lambda_0} = \frac{2\pi \times 3 \times 10^8}{30.6 \times 10^{-6}} = 6.16 \times 10^{13}\text{s}^{-1}$, $\varepsilon(0) = 5.9$, $\varepsilon(\infty) = 2.25$,

$$\omega_T = \sqrt{\frac{1.25}{4.9}} \times 6.16 \times 10^{13} = 3.11 \times 10^{13}\text{s}^{-1} .$$

At $\omega = \omega_L$, $\varepsilon(\omega) = 0$, i.e.,

$$\varepsilon(\infty) + \frac{[\varepsilon(0) - \varepsilon(\infty)]\,\omega_T^2}{\omega_T^2 - \omega_L^2} = 0,$$

giving the Lyddane–Sachs–Teller relation

$$\frac{\omega_L^2}{\omega_T^2} = \frac{\varepsilon(0)}{\varepsilon(\infty)} .$$

Hence

$$\omega_L = \left[\frac{\varepsilon(0)}{\varepsilon(\infty)}\right]^{1/2} \omega_T = 5.04 \times 10^{13}\text{s}^{-1} .$$

b. We have

for $\omega = 0$, $\varepsilon(\omega) = \varepsilon(0) = 5.9$, $R = \frac{\left(\sqrt{\varepsilon(0)} - 1\right)^2}{\left(\sqrt{\varepsilon(0)} + 1\right)^2} = 0.17$;

for $\omega \to \infty$, $\varepsilon(\omega) \to \varepsilon(\infty) = 2.25$, $R = \frac{(\sqrt{\varepsilon(\infty)} - 1)^2}{(\sqrt{\varepsilon(\infty)} + 1)^2} = 0.04$;

for $\omega = \omega_0 = 6.16 \times 10^{13}\text{s}^{-1}$,

$$\varepsilon(\omega) = 1, \quad R = 0 ;$$

for $R = 1$, $\varepsilon(\omega) = 0$ and $\omega = \sqrt{\frac{\varepsilon(0)}{\varepsilon(\infty)}}\,\omega_T = \omega_L = 5.04 \times 10^{13}\text{s}^{-1}$.

Furthermore, for $\omega_T < \omega < \omega_L$, $\varepsilon(\omega) < 0$, which means that n is imaginary, total reflection occurs and $R = 1$. Hence we can sketch the R vs. ω curve, which is shown in the semi-log graph Fig. 1.35.

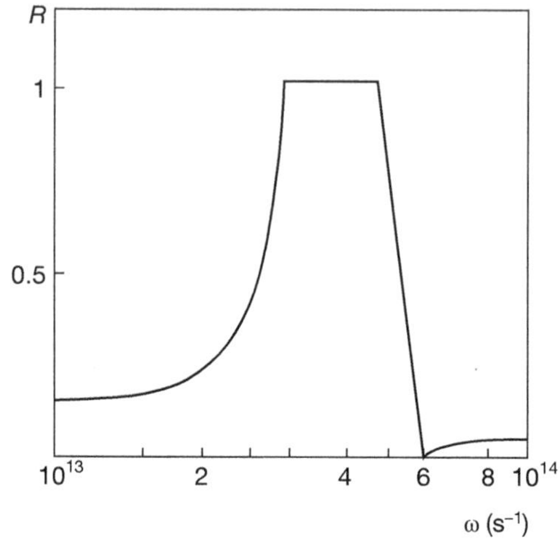

Fig. 1.35

1059

A rough sketch of the absorption coefficient for electromagnetic radiation is given for a nonmagnetic pure solid substance. The substance absorbs light in a narrow range in the infrared (0.15 eV) and then does not absorb light until the photon energy is 2.3 eV. Briefly explain this pattern of absorption and answer the following.

Fig. 1.36

a. Is this material a metal or an insulator? Why?

b. Is there more than one atom per unit cell in the crystal structure of this material? Why?

(Wisconsin)

Sol: The absorption of light in the infrared is caused by the interaction of photons with lattice vibration, or more specifically, by the interaction of long-wavelength optical phonons with photons. Locally the long wavelength optical vibration of the lattice causes the atoms or molecules to behave as oscillating electric dipoles. In their interaction with the electromagnetic field of the traversing light, absorption of the latter occurs. The strong absorption above 2.3 eV is due to excitation of electrons from valence band into conduction band.

a. This material is an insulator (probably a semiconductor as the energy gap ΔE is 2.3 eV).

b. There are more than two atoms per unit cell, otherwise the optical branch would have been absent and there would be no absorption peak at 0.15 eV.

1060

A paramagnetic salt containing ions with magnetic moment $\mu_{mj} = \pm\mu_B$ is placed in an external magnetic B_o. Each ion in the salt has one unpaired electron in the s-state.

a. Determine the number of energy states and the energy difference between the highest and lowest energy state.

b. For 20% of the ions to occupy the lowest energy state, what must be the applied magnetic field?

Sol:

A. Each ion in this paramagnetic system has one unpaired electron in the s-state, which implies that the orbital angular momentum (L) is equal to 0. Consequently, the total angular momentum (J) is the same as the spin quantum number (S), which is ½.

The formula to calculate the number of energy levels in this case is 2J + 1. Since J is ½, there are two energy states: one aligned parallel to the field and one aligned antiparallel to the field. Each of these states has an energy of $\pm\mu_B B_o$. Hence, the energy difference between the two states is $\Delta E = +\mu_B B_o - (-\mu_B B_o) = 2\mu_B B_o$.

B. For a two-level system, such as a paramagnetic ion with a magnetic moment in an external magnetic field, the ratio of the populations of the higher and lower energy states is given by:

$$\frac{N_2}{N_1} = e^{-\left(\frac{\Delta E}{k_B T}\right)}$$

where N_2 and N_1 are the populations of the higher and lower energy states, respectively, ΔE is the energy difference between the two states, k is the Boltzmann constant, and T is the temperature.

We want x percent of the ions to be in the lowest energy state, which means that N_1 should be x percent of the total, and N_2 should be (1 – x) percent of the total.

$$\frac{1-x}{x} = e^{-\left(\frac{\Delta E}{k_B T}\right)}$$

$$-\left(\frac{\Delta E}{k_B T}\right) = \left|\ln\left(\frac{1-x}{x}\right)\right| = \ln(1-x) - \ln x$$

$$\frac{2\mu_B B_o}{k_B T} = -\ln(1-x) + \ln x = \left|\ln\left(\frac{x}{1-x}\right)\right|$$

$$B_o = \frac{k_B T}{2\mu_B}\left|\ln\left(\frac{x}{1-x}\right)\right| = \frac{k_B T}{2\mu_B}\left|\ln\left(\frac{0.2}{0.8}\right)\right| = 0.693\frac{k_B T}{\mu_B}$$

1061

Consider a crystal containing a paramagnetic impurity whose energy levels are $\pm\varepsilon$ in a given magnetic field.

a. Write the contribution of this impurity to the partition function of the crystal in the presence of the field.

b. Write the corresponding expression of the impurity's contribution to the specific heat and plot it as a function of temperature.

c. Compare the temperature dependence of this specific heat with that of the phonon specific heat (Debye theory) in the limit of very low temperatures and give a physical interpretation of their difference.

d. Perform the comparison (c) for the limit of very high temperatures.

(Chicago)

Sol:

a. The contribution of the impurity to the partition function of the crystal is

$$z = e^{-\varepsilon/k_B T} + e^{\varepsilon/k_B T} = 2\cosh\left(\frac{\varepsilon}{k_B T}\right).$$

b. The mean energy of the crystal due to the impurity is

$$\bar{E} = -\frac{d\ln z}{d\left(\frac{1}{k_B T}\right)} = -\varepsilon\tanh\left(\frac{\varepsilon}{k_B T}\right),$$

and hence the specific heat due to the impurity is

$$c_V = \frac{d\bar{E}}{dT} = k_B\left(\frac{\varepsilon}{k_B T}\right)^2 \mathrm{sech}^2\left(\frac{\varepsilon}{k_B T}\right).$$

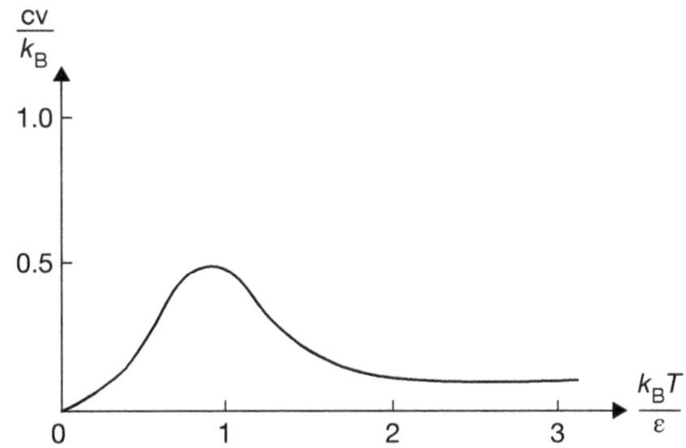

Fig. 1.37

c. In the limit of very low temperatures, $\text{sech}^2\left(\frac{\varepsilon}{k_B T}\right) \sim 4e^{-\frac{2\varepsilon}{k_B T}}$. Hence as $T \to 0$, $c_v \sim 4k_B\left(\frac{\varepsilon}{k_B T}\right)\exp\left(\frac{-2\varepsilon}{k_B T}\right) \to 0$. The more accurate Debye's model gives $c_v \sim T^3$ for $T \ll \theta_D$, the Deybe temperature. The reason is that at low temperatures, only those lattice vibration modes for which $\hbar\omega < k_B T$ can be excited and the contribution of such modes to the specific heat is proportional to T^3.

d. In the limit of very high temperatures $k_B T \gg \varepsilon$, $\text{sech}^2\left(\frac{\varepsilon}{k_B T}\right) \to 1$ and hence $c_v \sim T^{-2}$. However, in Debye's model c_v approaches a constant at high temperatures. This is because at a high temperatures $T \gg \theta_D$ every mode is excited and has an energy $k_B T$, giving rise to a constant c_v. Using the above information, we sketch the curve of c_v/k_B vs. $\frac{k_B T}{\varepsilon}$ in Fig. 1.37.

1062

a. Show that the zero-temperature spin susceptibility of an independent electron gas is

$$\chi = \mu_B^2 g(\varepsilon_F),$$

where μ_B is the Bohr magneton and $g(\varepsilon_F)$ is the density of states (per unit energy) at the Fermi energy. "Independent" is intended to imply that electron-electron interactions are also ignored.

b. Evaluate $g(\varepsilon_F)$ for a free electron gas in which electron-ion interactions are also ignored. (A derivation is required here; do not simply state the standard result.)

c. The spin susceptibility has a weak temperature dependence of the form

$$\chi(T) = \chi(0)(1 + \alpha T^2).$$

Would you expect α to be positive or negative? (Qualitative arguments only are expected.)

(Princeton)

Sol:

a. The energy of an electron in a magnetic field of flux density **B** is

$$\varepsilon(k) - \boldsymbol{\mu} \cdot \mathbf{B} = \varepsilon(k) \mp \mu_B B,$$

where $\boldsymbol{\mu}$, is the magnetic dipole moment due to the spin of the electron, $\varepsilon(k)$ denotes electron energy in the absence of the magnetic field, and the $\pm$ signs correspond to dipole moments parallel and antiparallel to the field respectively. Let $g(\varepsilon)$ be density of states of independent electrons per unit volume taking spin into account. Setting $\varepsilon = \varepsilon(k)$, we have for the total number of electrons with parallel and antiparallel spins respectively

$$N_+ = V \int_0^{\varepsilon_+(k)_{\max}} \frac{1}{2} g(\varepsilon) d\varepsilon,$$

$$N_- = V \int_0^{\varepsilon_-(k)_{\max}} \frac{1}{2} g(\varepsilon) d\varepsilon.$$

At $T = 0$, $\varepsilon_F = \varepsilon_+(k)_{\max} - \mu_B B = \varepsilon_-(k)_{\max} + \mu_B B$, where ε_F is the Fermi energy, and thus

$$N_+ = V \int_0^{\varepsilon_F + \mu_B B} \frac{1}{2} g(\varepsilon) d\varepsilon,$$

$$N_- = V \int_0^{\varepsilon_F - \mu_B B} \frac{1}{2} g(\varepsilon) d\varepsilon.$$

If $\mu_B B \gg \varepsilon_F$, we can take $\epsilon \approx \epsilon_F$ under the integral sign and find

$$N_+ - N_- \approx \frac{1}{2} V \cdot 2\mu_B B \cdot g(\varepsilon_F) = V\mu_B B g(\varepsilon_F).$$

The magnetization M is by definition

$$M = \frac{1}{V}(N_+ - N_-)\mu_B = \mu_B^2 g(\varepsilon_F) B.$$

The spin susceptibility is therefore

$$\chi = \frac{M}{H} = \frac{\mu_0 M}{B} = \mu_0 \mu_B^2 g(\varepsilon_F),$$

or, in Gaussian units where $\mu_0 = 1$,

$$\chi = \mu_B^2 g(\varepsilon_F),$$

provided that B is small,

b. For a free electron

$$\varepsilon = \frac{\hbar^2 k^2}{2m}, \quad \frac{dk}{d\varepsilon} = \frac{m}{\hbar^2 k}.$$

The number of electron states per unit volume in the energy range ε, $\varepsilon + d\varepsilon$ corresponds to the density of states in k-space in a spherical shell of radius k and thickness dk:

$$\rho(\varepsilon)d\varepsilon = 2 \times \frac{4\pi k^2 \, dk}{(2\pi)^3} = \frac{k^2}{\pi^2} \frac{dk}{d\varepsilon} d\varepsilon,$$

where the factor 2 is due to the fact that each energy state comprises two spin states. Hence

$$\rho(\varepsilon) = \frac{mk}{\pi^2 \hbar^2} = \frac{1}{2\pi^2} \left(\frac{2m}{\hbar^2} \right)^{\frac{3}{2}} \varepsilon^{\frac{1}{2}}.$$

c. When $k_B T \ll \varepsilon_F$, the excited electrons will lie within a range $k_B T$ above the Fermi surface. For those electrons whose magnetic moments are exactly parallel to the applied magnetic field, $\varepsilon_+(k)_{max}$ is greater and so the fraction of such excited electrons is relatively small and the reduction of their chemical potential is less. In order to maintain equilibrium, some of the electrons whose magnetic moments are almost parallel to the magnetic field will orientate so that their moments are more parallel to the magnetic field. Hence we expect $\alpha < 0$.

1063

Consider a conduction electron gas at absolute zero temperature in a weak magnetic field B. The concentration of spin-up, N_+, and spin-down, N_-, electrons may be parameterized by the quantity x

$$N_+ = \frac{1}{2} N(1 + x), \quad N_- = \frac{1}{2} N(1 - x),$$

where N is the total number of electrons.

a. Calculate the total energy of the gas and evaluate the magnetization M.

b. We may approximate the effect of exchange interactions among the conduction electrons if we assume that only electrons with parallel spins interact with each other, with energy $-V$ where $V > 0$. How is the magnetization modified?

(*Princeton*)

Sol:

a. The energy of a conduction electron is

$$\varepsilon = \frac{p^2}{2m} \pm \mu B,$$

where $E = \frac{p^2}{2m}$ is the kinetic energy of the electron and μ is its magnetic moment. The signs +, − are for spins parallel and antiparallel to B respectively. The density of states is (problem **1062**)

$$\rho(E) = C\sqrt{E}dE,$$

where C is a constant, and the densities of electrons with magnetic moments parallel and antiparallel to B are respectively

$$N_+ = C\int_0^{E_+} \rho(E)dE, \quad N_- = C\int_0^{E_-} \rho(E)dE.$$

At absolute-zero temperature, all the states below a certain energy ε_0 are filled by electrons of both spin directions. As

$$\varepsilon_0 = E_+ - \mu B = E_- + \mu B,$$

we have

$$E_+ = \varepsilon_0 + \mu B, \qquad E_- = \varepsilon_0 - \mu B,$$

$$N_+ \equiv \frac{1}{2}N(1+x) = C\int_0^{\varepsilon_0+\mu B} E^{\frac{1}{2}}dE = \frac{2}{3}C(\varepsilon_0 + \mu B)^{3/2},$$

$$N_- \equiv \frac{1}{2}N(1-x) = C\int_0^{\varepsilon_0-\mu B} E^{\frac{1}{2}}dE = \frac{2}{3}C(\varepsilon_0 - \mu B)^{3/2}.$$

Suppose E_F is the Fermi energy in the absence of any applied magnetic field, then

$$N = 2C\int_0^{E_F} E^{\frac{1}{2}}dE = \frac{4}{3}CE_F^{3/2},$$

giving $C = \frac{3}{4}NE_F^{-\frac{3}{2}}$. Hence

$$\varepsilon_0 + \mu B = E_F(1+x)^{2/3},$$
$$\varepsilon_0 - \mu B = E_F(1-x)^{2/3},$$

whose difference gives for $x \ll 1$

$$x \approx \frac{3\mu B}{2E_F}.$$

The total energies of the electrons with spin + and spin − are respectively

$$\begin{aligned}
E_{+\text{tot}} &= C\int_0^{\varepsilon_0+\mu B}(E-\mu B)E^{\frac{1}{2}}dE \\
&= \frac{2}{5}C(\varepsilon_0+\mu B)^{5/2} - \mu BN_+ , \\
&= \frac{2}{5}CE_F^{\frac{2}{5}}(1+x)^{\frac{5}{3}} - \frac{1}{2}\mu BN(1+x) \\
&= \frac{3}{10}E_F N(1+x)^{\frac{5}{3}} - \frac{1}{2}\mu BN(1+x) ,
\end{aligned}$$

$$\begin{aligned}
E_{-\text{tot}} &= C\int_0^{\varepsilon_0-\mu B}(E+\mu B)E^{\frac{1}{2}}dE \\
&= \frac{2}{5}C(\varepsilon_0-\mu B)^{5/2} + \mu BN_- , \\
&= \frac{3}{10}E_F N(1-x)^{\frac{5}{3}} + \frac{1}{2}\mu BN(1-x).
\end{aligned}$$

Hence, the total energy of the gas is

$$\begin{aligned}
E_{\text{tot}} &= \frac{3}{10}NE_F\left[(1+x)^{5/3}+(1-x)^{5/3}\right] - \mu BNx \\
&\approx \frac{3}{5}NE_F - \mu BNx \\
&\approx \frac{3}{5}NE_F - \frac{2\mu^2B^2}{2E_F}N.
\end{aligned}$$

Note that for a weak field it is justified to take $\mu B \ll E_F$, or $x \ll 1$.

As the magnetic moment arises from electron spin, $\mu = \mu B$, the Bohr magneton. The total magnetic moment vM, where M is the magnetization and v the volume of the electron gas, is then

$$vM = (N_+ - N_-)\mu_B = Nx\mu_B$$

or

$$M = nx\mu_B = \frac{3n\mu_B^2 B}{2E_F},$$

where $n = N/v$ is the number of electrons per unit volume of the gas.

b. There are N_+ electrons of spin +, making up $\frac{1}{2}N_+(N_+ - 1) \approx \frac{1}{2}N_+^2$ pairs, each of which contributes an interaction energy $-V$. Thus, taking into account the interaction energies we have

$$E_{+\text{tot}} \approx \frac{3}{10}E_F N(1+x)^{5/3} - \frac{1}{2}\mu_B BN(1+x) + \frac{1}{2}N_+^2(-V)$$
$$= \frac{3}{10}E_F N(1+x)^{5/3} - \frac{1}{2}\mu_B BN(1+x) - \frac{1}{8}VN^2(1+x)^2,$$

and similarly

$$E_{-\text{tot}} \approx \frac{3}{10}E_F N(1-x)^{5/3} + \frac{1}{2}\mu_B BN(1-x) - \frac{1}{8}VN^2(1-x)^2.$$

The total energy of the electron gas is then

$$E_{\text{tot}} = \frac{3}{10}E_F N\left[(1+x)^{\frac{5}{3}} + (1-x)^{\frac{5}{3}}\right] - \mu_B BNx$$
$$-\frac{V}{8}N^2[(1+x)^2 + (1-x)^2].$$

At equilibrium, $\frac{\partial E_{\text{tot}}}{\partial x} = 0$, giving

$$\frac{1}{2}E_F\left[(1+x)^{\frac{2}{3}} - (1-x)^{\frac{2}{3}}\right] - \mu_B B - \frac{V}{4}N[(1+x) - (1-x)] = 0,$$

or

$$x \approx \frac{6\mu_B B}{4E_F - 3VN}$$

for $x \ll 1$.

Hence the magnetization is

$$M = nx\mu_B = \frac{6n\mu_B^2 B}{4E_F - 3VN}.$$

Note that for $V = 0$ this reduces to the expression for non-interacting electrons.

1064

Consider an ideal gas of stable vector mesons (massive particles with spin 1) with magnetic moment μ.

a. Compute the paramagnetic susceptibility per unit volume in a weak magnetic field.

b. Perform the corresponding calculation with classical dipoles.

(Chicago)

Sol:

a. For an ideal gas, in which the particles are non-interacting, the partition function of a single particle is

$$z_1 = e^{-\beta\varepsilon} + 1 + e^{\beta\varepsilon},$$

where $\varepsilon = \mu B$, $\beta = \frac{1}{k_B T}$. The probabilities for its magnetic moment to be parallel, perpendicular and antiparallel to the magnetic field **B** are respectively

$$P_+ = \frac{e^{\beta\varepsilon}}{z_1}, \quad P_0 = \frac{1}{z_1}, \quad P_- = \frac{e^{-\beta\varepsilon}}{z_1}.$$

Hence the mean magnetic moment along the direction of the field per particle is

$$\begin{aligned}\bar{\mu} &= \mu(P_+ - P_-) = \mu \cdot \frac{e^{\beta\varepsilon} - e^{-\beta\varepsilon}}{z_1} \\ &= \frac{2\mu \sinh(\beta\varepsilon)}{2\cosh(\beta\varepsilon) + 1}.\end{aligned}$$

If n is the number of particles per unit volume of the gas, the paramagnetic magnetization is

$$M = n\bar{\mu} = \frac{2n\mu \sinh(\beta\varepsilon)}{2\cosh(\beta\varepsilon) + 1}.$$

In a weak magnetic field, $\mu B \ll k_B T$, or $\beta\varepsilon \ll 1$, $\sinh(\beta\varepsilon) \approx \beta\varepsilon$, $\cosh(\beta\varepsilon) \approx 1$, and so

$$M = \frac{2n\mu\beta\varepsilon}{3} = \frac{2n\mu^2 B}{3k_B T}.$$

Hence the paramagnetic susceptibility is

$$\chi = \frac{M}{H} = \frac{\mu_0 M}{B} = \frac{2n\mu_0\mu^2}{3k_B T}.$$

b. For a classical magnetic dipole $\boldsymbol{\mu}$ the potential energy in a magnetic field **B** is

$$\varepsilon = -\boldsymbol{\mu} \cdot \mathbf{B} = -\mu B \cos\theta,$$

where θ is the angle between $\boldsymbol{\mu}$ and **B**. The probability of finding it along the θ direction is given by the distribution function

$$e^{-\frac{\varepsilon}{k_B T}} = e^{\mu B\beta \cos\theta}.$$

Hence the mean magnetic moment along **B** is

$$\begin{aligned}\bar{\mu} &= \frac{\int_0^\pi e^{\mu B\beta\cos\theta} \mu \cos\theta \cdot 2\pi \sin\theta d\theta}{\int_0^\pi e^{\mu B\beta\cos\theta} \cdot 2\pi \sin\theta d\theta} \\ &= \frac{1}{B\beta}[\mu B\beta \coth(\mu B\beta) - 1] \\ &= \mu\left[\coth(\mu B\beta) - \frac{1}{\mu B\beta}\right].\end{aligned}$$

For a weak field

$$\mu B \ll \frac{1}{\beta}, \quad \text{or} \quad \mu B\beta \ll 1,$$

and

$$\coth(\mu B\beta) \approx \frac{1}{\mu B\beta} + \frac{\mu B\beta}{3}.$$

Thus the paramagnetic magnetization for a weak field is

$$M = n\bar{\mu} \approx \frac{n\mu^2 B\beta}{3},$$

and the paramagnetic susceptibility is

$$\chi = \frac{M}{H} = \frac{\mu_0 M}{B} = \frac{n\mu_0\mu^2}{3k_B T}.$$

1065

What is the typical saturation value for the density of magnetization of iron/steel (state units)? From this value and the values of other specified quantities, and stated assumptions about the nature of ferromagnetism, estimate the intrinsic-spin magnetic moment of the electron.

(Columbia)

Sol: In an applied magnetic field, the magnetization of iron increases with the intensity of the field. When the field increases to a certain value, the magnetization will not increase further but reach a saturation value M_s, at which time, according to the theory of ferromagnetism, all magnetic moments due to electron spin are aligned along the direction of the applied magnetic field. Let the number density of electrons in a ferromagnetic material be n and the magnetic moment due to the spin of an electron be μ, then

$$M_s = n\mu.$$

The saturation magnetization of iron is typically $M_s = 10^6$ Am^{-1}, and as $n = 3 \times 10^{28}$ m^{-3}, the intrinsic-spin magnetic moment of an electron is $\mu = \frac{M_s}{n} \approx 3 \times 10^{-23}\,\text{Am}^2$.

1066

2. (i) Determine the colors of the following semiconductors, as viewed by transmission, when they are in pure form.

Semiconductor	Direct Bandgap (eV)
GaP	2.3
TiO_2	3
ZnS	3.6
CdS	2.5

(ii) Assuming that GaP is doped with an impurity that creates an impurity level 0.7 eV above the valence band, what will be its color as observed by transmission?

(iii) Based on your results, what would be the colors of most oxides and alkali halides in their pure form? Why?

Sol:

(i) Assuming that no excitons are created, electrons in the valence band can absorb only those photons of energy $h\nu$ that result in the promotion of these electrons to the conduction band. The visible spectrum lies between 400 and 800 nm, viz., 3 eV > $h\nu$ > 1.5 eV. Hence, for materials in the table that have $E_g \geq$ 3eV, absorption of photons in the visible range is not possible. Hence, TiO_2 and ZnS will be transparent.

GaP absorbs light in the green, blue, and violet regions and transmits the yellow, orange, and red parts of the visible light, which will hence appear orange by transmission (see the following figure). CdS absorbs blue and violet and appears yellow by transmission.

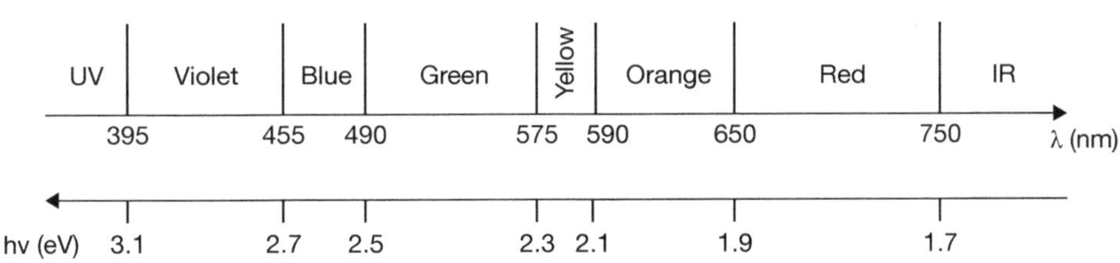

Fig. 1.38

(ii) The presence of an impurity level "n" in a material, results in transitions between this level and the valence band. Photons of energy $h\nu = E_g - E_D$ can be selectively absorbed. In GaP, the red will be absorbed and the transmission color will be yellow.

(iii) The bandgap gives an estimate of the bonding energy of the valence electrons, a higher value of the bandgap indicates that the electrons are more tightly bound. Most oxides (BeO, Al_2O_3, MgO) and alkali halides (LiF, NaCl) will have bandgaps >3 eV and will hence be transparent in their pure form.

1067

Figure 1.39 displays neutron diffraction data for scattering from a crystal of MnO, which is known to have the same structure as NaCl. Although the neutron data changes dramatically when the temperature is reduced below 120 K, X-ray diffraction patterns are nearly independent of temperature (exhibiting the same peaks as the 293 K neutron data) and indicate a lattice constant $a = 4.43$ Å.

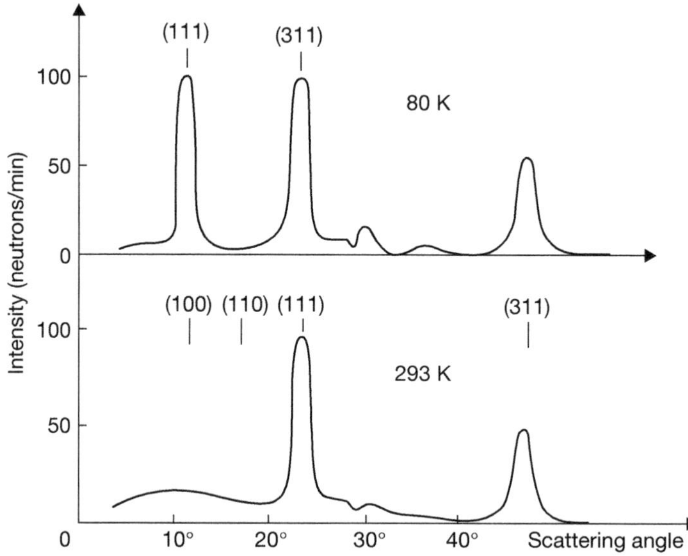

Fig. 1.39

a. Explain quantitatively why a crystal of the rock salt (NaCl) structure should have no Bragg reflections in the (100) and (110) directions (for either neutrons or X-rays).

b. From the figure, what is the approximate lattice constant indicated by the neutron data taken at 80 K?

c. How does the state of the MnO crystal at 80 K differ from its state at 293 K?

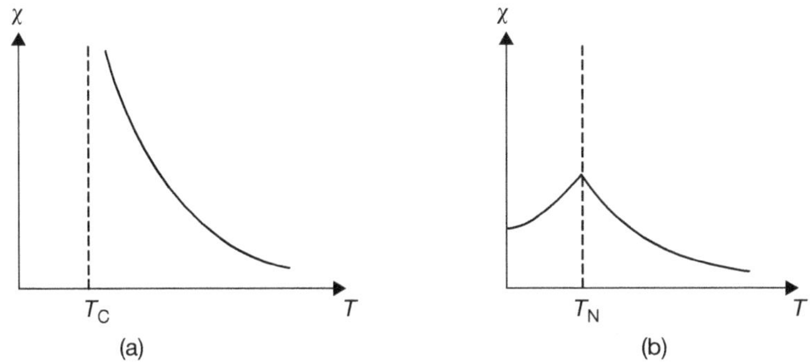

Fig. 1.40

d. The magnetic susceptibility χ is schematically plotted for two different materials in Fig. 1.40. For each plot, describe the magnetic nature of the material in the low temperature range. Which plot could belong to MnO?

(Princeton)

Sol:

a. MnO and NaCl have the same face-centered cubic (fcc) lattice structure, for which the geometrical structure factor vanishes if the indices of a diffraction plane are a mixture of odd and even numbers. Hence, crystals of such a structure have no Bragg reflection in directions such as (100) and (110).

b. Figure 1.39 shows that at $T = 293$ K the scattering angle is $2\theta \approx 24°$, and at $T = 80$ K the scattering angle is $2\theta' \approx 12°$. For fcc structure, the interplanar distance is $d = \frac{na}{3}$, where a is the lattice constant. Bragg's law $2d \sin\theta = n\lambda$ then gives

$$a \sin\theta = a' \sin\theta'.$$

Hence the lattice constant at $T = 80$ K is

$$a' = \frac{a \sin\theta}{\sin\theta'} = 4.43 \times \frac{\sin 12^\circ}{\sin 6^\circ} = 8.81\ \text{Å}.$$

c. The MnO crystal, which has fcc structure, has lattice constant 4.43 Å at $T = 293$ K and 8.81 Å at $T = 80$ K, and some new diffraction peaks at the lower temperature. The occurrence of these additional peaks is caused by the ordering of the magnetic moments at low temperatures. Specifically, the spins of Mn^{2+} have the same one direction on some (111) plane, but on an adjacent (111) plane the spins are in the opposite direction. Thus, considering one (111) plane the spins are in an ferromagnetic arrangement, but considering adjacent (111) planes the spins are in an antiferromagnetic arrangement. Because of this situation, only if the phase difference between adjacent (111) planes occupied by Mn^{2+} is π can a diffraction maximum be generated. This gives rise to the reflection at scattering angle $2\theta' = 12°$.

   By X-ray diffraction the lattice constant is determined to be 4.43 Å, which indicates that the lattice constant of a chemical cell is 4.43 Å. In order to get the same spin direction, it is necessary to enlarge each side of a chemical cell by a factor of 2, so as to form a magnetic cell.

d. Figure 1.40(a) refers to a ferromagnetic substance. When $T < T_C$, the Curie temperature, the substance is in an ordered ferromagnetic phase in which the molecular spins are arranged in parallel alignment. Above T_C the substance is paramagnetic. Figure 1.40(b) illustrates a case of antiferromagnetism. Below the Néel temperature, T_N, neighboring spins favor an antiparallel arrangement. Above T_N, the substance also becomes paramagnetic. Since at low temperatures the substance described by Fig. 1.40(b) is antiferromagnetic, it could be a plot of χ vs. T for MnO.

1068

a. You are given two pieces of transparent material which to the eye appear identical in all respects. One is a crystal and the other is a glass. Describe two independent experiments for distinguishing between the crystal and the glass.

b. You are given four identical looking objects:

1. a permanent magnet,
2. a diamagnetic material,
3. a paramagnetic material,
4. an unmagnetized ferromagnetic material.

Is it possible to identify all of the objects without the use of any externally produced electromagnetic field? If so, describe how it could be done. If not, give reasons.

(*Chicago*)

Sol:

a. Experiment one: Heating the material. A crystal would melt at a definite melting temperature which remains constant during melting. Glass, on the other hand, has no definite melting point.

Experiment two: Taking X-ray diffraction patterns. A crystal will give regular spots in the pattern, while glass only gives a diffused pattern.

b. It is possible to identify the objects by the following steps.

1. Bring the objects in turn near some iron powder. The permanent magnet will strongly attract the powder.
2. Bring the permanent magnet near each of the other three objects in turn. The unmagnetized ferromagnetic object will be strongly attracted to it.
3. Arrange the remaining two objects near the permanent magnet symmetrically as shown in Fig. 1.41. Measure the magnetic field at the points A_1 and B_1 between the two objects with a gaussmeter. The point nearer the paramagnetic material will show a greater magnetic intensity.

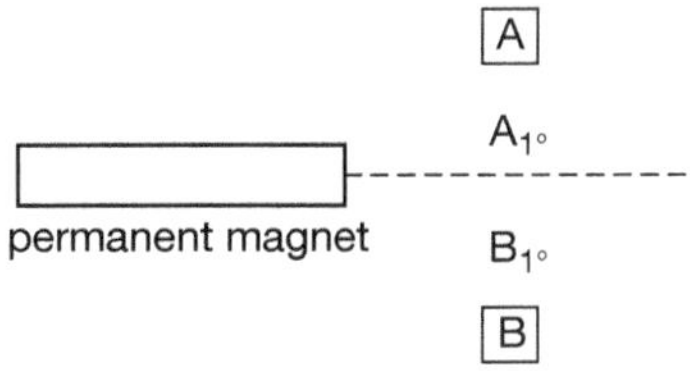

Fig. 1.41

1069

A long thin uniformly magnetized needle of hard steel, length 12 cm, is suspended by a torsionless suspension and makes free, small-angle oscillations about an axis through the center in a horizontal plane in the earth's magnetic field. The period of oscillation is 3–5 seconds. Estimate the density of magnetization of the steel.

(*Columbia*)

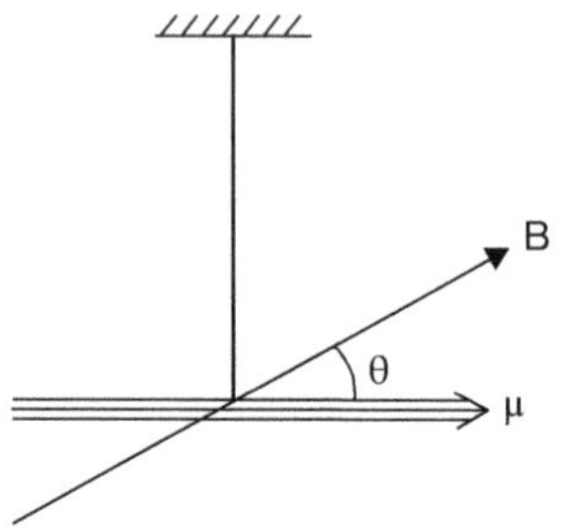

Fig. 1.42

Sol: As only small-angle oscillations are involved, the earth's magnetic field can be considered uniform.

Let μ be the magnetic moment of the steel needle. Its potential energy in the earth's magnetic field is $V = -\boldsymbol{\mu} \cdot \mathbf{B} = -\boldsymbol{\mu}\mathbf{B} \cos \theta$, where θ is the angle between $\boldsymbol{\mu}$ and $\mathbf{B}$ as shown in Fig. 1.41. The equation of motion is

$$I\ddot{\theta} = -\frac{dV}{d\theta} = -\mu B \sin\theta \approx -\mu B\theta,$$

I being the moment of inertia of the needle, for small angle oscillations. The angular frequency of oscillation is given by

$$\omega^2 = \frac{\mu B}{I}.$$

Let I, V, ρ, $\mathbf{M}$ be respectively the length, volume, density and magnetization of the steel needle. Then

$$I = \frac{\rho V l^2}{12}, \quad \boldsymbol{\mu} = V\mathbf{M},$$

and thus

$$\omega^2 = \frac{12BM}{\rho l^2},$$

on

$$M = \frac{\pi^2 \rho l^2}{3BT^2},$$

where T is the period of oscillation. With $l = 0.12$ m, $T = 4$ s, $\rho = 78 \times 10^3$ kgm^{-3}, $B \sim 5 \times 10^{-5}$ Wm^{-2}, we find

$$M = 4.6 \times 10^5 \,\mathrm{Am}^{-1}.$$

1070

Magnetism in condensed matter manifests itself in various forms for different materials and at different temperatures. Four types of magnetic behavior are:

1. Ferromagnetism
2. Diamagnetism
3. Antiferromagnetism
4. Paramagnetism.

   a. For an ideal example of a material exhibiting each of the above magnetic behavior, plot the magnetic susceptibility as a function of temperature. Pay attention to the relative scales in your diagrams so that the relative strengths of the magnetic phenomena are presented. Give one example of a real material that approximates each ideal form of magnetism.

   b. Indicate any critical behavior involved in any of these phenomena.

   c. Explain the "exchange interaction" and the part it plays in the above magnetic behaviors. Give a brief quantum mechanical explanation of the exchange interactions, the Hamiltonian for the interaction, and the connection between the exchange interaction and the Weiss molecular field.

   d. Explain briefly the difference between the magnetic behaviors of sodium metal and Cr^{2+}-doped (0.01%) Al_2O_3 single crystal.

(SUNY, Buffalo)

Sol:

a. Diamagnetism is a property of certain substances that in an external magnetic field the constituting atoms can generate induced currents whose magnetic field is in a direction opposite to that of the applied magnetic field, so that the magnetic susceptibility $\chi < 0$. Generally χ is the order 10^{-5} cm^{-3}, and is hardly affected by temperature, as shown by curve A of Fig. 1.43. A substance whose atoms have null permanent magnetic moment exhibits diamagnetism, e.g. lead.

In a paramagnetic material Langevin paramagnetism dominates. It is caused by the lining up of the permanent magnetic moments, with the resultant magnetic susceptibility $\chi \propto \frac{1}{T}$, as shown by curve B of Fig. 1.43. In addition, in a metal there is also Pauli paramagnetism due to conduction electrons, each of which carries a spin magnetic moment, with the magnetic susceptibility independent of temperature as shown by curve C. Generally speaking, the magnetic susceptibility of a paramagnetic substance is larger than that of a diamagnetic substance, but is of the same order of magnitude. It is necessary for a paramagnetic material to have atoms with nonzero permanent magnetic moments. Examples are metallic elements like sodium and aluminium, and chemical compounds of transition and rare-earth elements like Cr^{2+}-doped Al_2O_3.

A ferromagnetic substance typically has a magnetic susceptibility that is 10^3 times larger than that of a paramagnetic material and varies with temperature according to $\chi \propto \frac{1}{T - T_C}$, as shown by curve D. At T below the Curie temperature Tc, a ferromagnetic material is spontaneously magnetized; at T above Tc, it is paramagnetic. Typical examples are iron, cobalt and nickel.

An antiferromagnetic material has a magnetic susceptibility χ which is of the same order of magnitude as that of a paramagnetic material but varies with temperature as follows: At T higher than the Néel temperature T_N, $\chi \propto \frac{1}{T + \Theta}$, where Θ is a positive constant; at $T < T_N$, χ decreases with decreasing temperature, as shown by curve E. Antiferromagnetism is exhibited by many compounds involving transition metals, e.g. MnO.

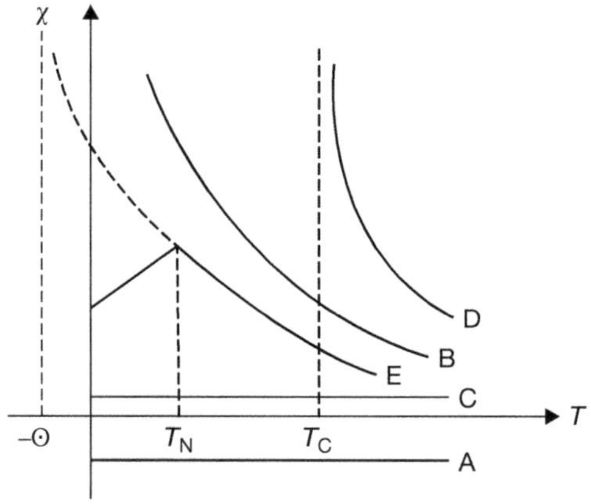

Fig. 1.43

b. For a ferromagnet, the phase transition ferromagnetism ↔ paramagnetism occurs at the Curie temperature T_C; for an antiferromagnetic substance, antiferromagnetism ↔ paramagnetism occurs at the Néel temperature T_N.

c. In the ferromagnetic region atomic or molecular magnetic moments are magnetized spontaneously, implying the existence of an internal molecular field. This indicates that neighboring moments interact with each other and that the interaction is spin-dependent. If the spin angular momenta of atoms i and j are $\mathbf{S}_i\hbar$ and $\mathbf{S}_j\hbar$ respectively, their interaction gives rise to an exchange Hamiltonian

$$-2J_e\mathbf{S}_i\cdot\mathbf{S}_j,$$

where J_e is the exchange integral. Suppose that the ferromagnetic substance has N atoms per unit volume and that each atom contributes one electron to ferromagnetism. Then the exchange Hamiltonian for the crystal is

$$H_{ex} = -\sum_{\mathbf{R}_i}\sum_{\mathbf{R}_j}{}'2J_e(\mathbf{R}_i-\mathbf{R}_j)\mathbf{S}(\mathbf{R}_i)\cdot\mathbf{S}(\mathbf{R}_j)\ ,$$

where $\mathbf{R}_i$, and $\mathbf{R}_j$ are two lattice vectors, and in the sum Σ', $\mathbf{R}_i\neq\mathbf{R}_j$. If we assume that only the exchange interactions of electrons of nearest-neighbor atoms are appreciable, then

$$H_{ex} = -\sum_{\mathbf{R}_i}\left[2\sum_{\substack{\text{neighboring}\\ \mathbf{R}_j}}J_e(\mathbf{R}_i-\mathbf{R}_j)\mathbf{S}(\mathbf{R}_j)\right]\cdot\mathbf{S}(\mathbf{R}_i).$$

This means that, if the exchange integral $J_e > 0$, parallel spins correspond to a state of lower energy. In a ferromagnetic domain the spin magnetic moments spontaneously tend to the parallel arrangement, indicating that the exchange integral is positive for interactions between neighboring atoms.

To find the connection between the exchange interaction and the Weiss molecular field, consider a crystal in which each atom has z nearest-neighbor atoms. H_{ex} can then be written as

$$-2J_e\mathbf{S}_0\cdot\sum_{j=1}^{z}\mathbf{S}_j,$$

where $\mathbf{S}_0\hbar$ is the spin angular momentum of the atom under consideration and $\mathbf{S}_j\hbar$ ($j = 1, 2, \ldots, z$) are those of its nearest neighbors. The corresponding magnetic moments $\boldsymbol{\mu}_0$, $\boldsymbol{\mu}_j$ are $g_s\mu_B\mathbf{S}_0$, $g_s\mu_B\mathbf{S}_j$, where g_s is a constant and μ_B is the Bohr magneton. Thus

$$H_{ex} = -\boldsymbol{\mu}_0\left[\frac{2J_e}{(g_s\mu_B)^2}\sum_{j=1}^{z}\boldsymbol{\mu}_i\right].$$

Comparing this with the expression for the energy of a magnetic dipole in a magnetic field, $-\boldsymbol{\mu}\cdot\mathbf{B}$, shows that the exchange interaction gives rise to an effective field

$$\frac{2J_e}{(g_s\mu_B)^2}\sum_{j}^{z}\boldsymbol{\mu}_j = \frac{2J_e z}{N(g_s\mu_B)^2}\mathbf{M},$$

where N is the number of atoms per unit volume and **M** is the magnetization. This has the form of the Weiss molecular field $\gamma\mathbf{M}$ with the field constant

$$\gamma = \frac{2J_e z}{N(g_s\mu_B)^2}.$$

d. In sodium metal, the sodium atom has a full-shell construction plus a valence electron which is free. Therefore its magnetism is due to the Pauli paramagnetism of free electrons and the diamagnetism of ions with complete shells. The sum of these two contributions results in a very weak paramagnetism. In a Cr^{2+}-doped Al_2O_3 single crystal, on the other hand, there are non-vanishing permanent magnetic moments which line up to give a stronger paramagnetism.

1071

When a metal enters the superconducting state it becomes more ordered and its free energy decreases.

What is the nature of the new ordering? Consider a superconducting ring carrying a circulating persistent current. Suppose that at some instant thermal fluctuations cause a small decrease in this current. In terms of your answer above, explain why the system will restore the current to its previous value. Explain why there is an energy gap when one tries to add energy to a superconductor by shining light on it, but not when one adds energy by applying a dc current.

(SUNY, Buffalo)

Sol: When a metal enters the superconducting state, electrons lying deep inside the Fermi sphere remain normal electrons, but electrons very near the Fermi surface couple to form Cooper pairs. These are the superconducting electrons and are condensed into the lowest energy state, which does not contribute to entropy and is a highly ordered state. Thus when a metal enters the superconducting state it becomes ordered and its free energy decreases.

At some instant, some Cooper pairs may be destroyed by thermal fluctuations. The density of superconducting electrons then decreases, which causes a small decrease in the current. At that time, the ordering of the system becomes less and the free energy increases. But the energy of the system must be conserved, so some other electrons near the Fermi surface inevitably couple to form superconducting electrons, and their density increases. The system will thus restore the current to its previous value.

There is an energy gap Δ between the superconducting ground state and the lowest excited state. When shining light on a superconductor, the latter absorbs light appreciably only if the energy of the photon is larger than or equal to Δ. This means that there is an energy gap when one tries to add energy to a superconductor by shining light on it. On the other hand, when a dc current is applied, a Cooper pair moves as a whole and does not have to overcome the energy gap.

1072

Briefly describe one of the Josephson tunneling effects.

(*Wisconsin*)

Sol: A dc current can flow through a junction formed by two pieces of superconductors S_1 and S_2 separated by a thin insulating film T of thickness about 10 Å, even though the potential difference across the junction may be zero. This indicates that the superconducting electrons can penetrate an insulating potential barrier between two superconductors resulting in a tunneling superconducting current. Such a current can be explained by a coupling between the wave functions in the two superconductors as briefly described below.

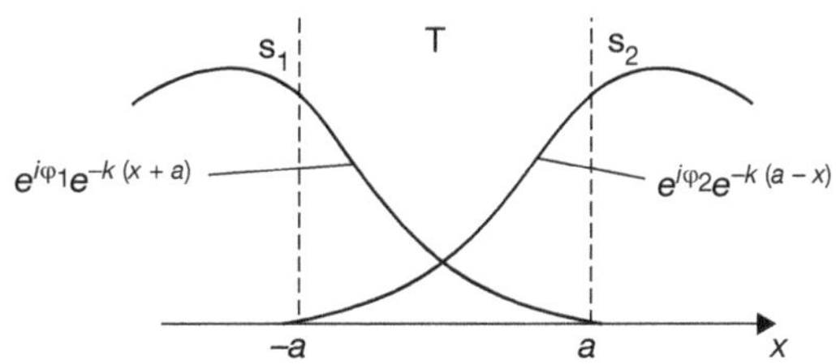

Fig. 1.44

The motion of superconducting electrons (a Cooper pair) can be described by wave functions in S_1 and S_2 which attenuate exponentially in the insulating layer, as shown in Fig. 1.44.

Let φ_1 and φ_2 be the phases of the waves in S_1 and S_2 respectively. The total wave function in the insulating layer is

$$\psi = e^{i\varphi_1}e^{-k(x+a)} + e^{i\varphi_2}e^{-k(a-x)}$$
$$= e^{-ka}\left(e^{i\varphi_1 - kx} + e^{i\varphi_2 + kx}\right).$$

The current density is

$$j = qJ,$$

where J is the probability current density

$$J = -\frac{i\hbar}{2m}(\psi^*\nabla\psi - \psi\nabla\psi^*).$$

Hence the current density crossing the junction is

$$\begin{aligned} j &= -\frac{i\hbar q}{2m}\left(\psi^*\frac{d\psi}{dx} - \psi\frac{d\psi^*}{dx}\right) \\ &= \frac{2\hbar qk}{m}e^{-2ka}\sin(\phi_2 - \phi_1) \\ &= j_0 \sin(\phi_2 - \phi_1) \end{aligned}$$

with $j_0 = \dfrac{2\hbar qk}{m}e^{-2ka}$. Therefore, as long as the two coupling superconducting wave functions have a nonzero phase difference, a superconducting current exists although the potential difference across the junction may be absent.

1073

1. In a two-fluid model, a superconductor is modeled to be consisting of two fluids: one with normal electrons (density = n_e) and the other formed by Cooper electron pairs obeying Bose–Einstein statistics (pair density = n_p).

(i) Determine an expression for the free electron density as a function of the Fermi energy, $E_F(N)$ at T = 0 K in the normal state.

(ii) Below the critical temperature, T_c, the superconducting Fermi energy is given by $E_F(S) = E_F(N)(1 - \delta)$, where $0 < \delta << 1$. Determine *(a)* the expression n_e in terms of n_0 and *(b)* the expression for the density of Cooper pairs, n_p as a function of δ and n_0.

(iii) An evaluation based on Bose statistics has found that $n_p = 2.6\left(\dfrac{m k_B T_c}{\pi\hbar^2}\right)^{3/2}$, where m is the mass of electrons. Find the normal electron density n_e, Cooper pair density, n_p, δ, and forbidden bandgap for Pb, whose T_c is 7.2 K and $E_F(N) = 6$ eV.

Sol:

(i) **In the normal state**, the Fermi energy is given by

$$E_F = \left(\frac{\hbar^2}{2m}\right)(3\pi^2 n)^{2/3}$$

Rearranging, we get the electron density as

$$n_0 = \frac{1}{3\pi^2}\left(\frac{2m E_F(N)}{\hbar^2}\right)^{3/2}$$

(ii) In the superconducting state, it is given that $E_F(S) = E_F(N)(1 - \delta)$

(a) This results in

$$n_e = \frac{1}{3\pi^2}\left(\frac{2m E_F(S)}{\hbar^2}\right)^{3/2} = \frac{1}{3\pi^2}\left(\frac{2m E_F(N)(1-\delta)}{\hbar^2}\right)^{3/2}$$

$$\frac{n_e}{n_0} = (1-\delta)^{3/2}$$

$$n_e = n_0(1-\delta)^{3/2} \approx n_0\left(1 - \left(\frac{3}{2}\right)\delta\right)$$

(Using binomial expansion)

(b) From the conservation of electric charge,

$$n_p = \frac{1}{2}(n_0 - n_e) = \left(\frac{3}{4}\right)\delta . n_0$$

(iii) Substituting the given values in the given expression, we get

$$n_0 = 6.72 \times 10^{28}\ e/m^3$$

$$n_p = 3.45 \times 10^{23}\ \text{e/m}^3$$

$$\delta = 6.8 \times 10^{-6}$$

Bandgap $= \delta \times E_F = 0.04$ meV

1074

Give a simple-minded argument for how two electrons in a lattice might experience an attractive interaction which can lead to the formation of a Cooper pair.

Consider such a pair of electrons in a spin-singlet state with the total momentum equal to zero. The pair is in a metal with Fermi energy ε_F, density of states N. The Schrödinger equation for such a pair can be written as $(H_0 + H_1)\,\psi(\mathbf{r}) = E\,\psi(\mathbf{r})$, where H_0 is the kinetic energy operator of the pair, H_1 is the two-electron interaction, and $\mathbf{r}$ is the relative coordinate.

(SUNY, Buffalo)

Sol: The electrons interact indirectly through an exchange of lattice wave phonons. For example, an electron with wave vector $\mathbf{k}_1$ emits a phonon with wave vector $\mathbf{q}$ and is scattered into a state of wave vector $\mathbf{k}_1 - \mathbf{q}$; another electron whose original wave vector is $\mathbf{k}_2$ absorbs the phonon emitted by the first electron and enters into a state of wave vector $\mathbf{k}_2 + \mathbf{q}$. An alternative process is that the first electron absorbs a phonon with wave vector $-\mathbf{q}$ emitted by the second electron and is scattered from a state $\mathbf{k}_1$ into $\mathbf{k}_1 - \mathbf{q}$; the second electron is scattered from a state $\mathbf{k}_2$ into $\mathbf{k}_2 + \mathbf{q}$. According to quantum theory, if the energy difference $|E(\mathbf{k}_1) - E(\mathbf{k}_1 - \mathbf{q})| < \hbar\,\omega_{\mathbf{q}}$, then the interaction energy is negative, which means that there is a net attractive interaction between the electrons. Obviously, only within the range $\pm\hbar\,\omega_D$, where ω_D is the Deyby frequency, near the Fermi surface can electrons satisfy this condition.

For a pair of independent electrons the resultant wave is

$$\varphi(\mathbf{r}_1, \mathbf{r}_2) = \frac{1}{V_c} e^{i(\mathbf{k}_1\cdot\mathbf{r}_1 + \mathbf{k}_2\cdot\mathbf{r}_2)} = \frac{1}{V_c} e^{i\mathbf{k}\cdot\mathbf{r}},$$

where V_c is the volume of the crystal, $\mathbf{r} = \mathbf{r}_1 - \mathbf{r}_2$, $\mathbf{k} = \mathbf{k}_1 = -\mathbf{k}_2$ as the total momentum is zero.

If the two electrons interact the resultant wave can be expressed as a linear combination of the independent-electron waves:

$$\psi(\mathbf{r}) = \sum_{\mathbf{k}} g(\mathbf{k})\,\varphi_{\mathbf{k}}(\mathbf{r}).$$

Substitution in the Schrödinger equation

$$(H_0 + H_1)\psi(\mathbf{r}) = E\psi(\mathbf{r})\ ,$$

where

$$H_0 = -\frac{\hbar^2}{2m}\left(\nabla^2_{\mathbf{r}_1} + \nabla^2_{\mathbf{r}_2}\right) = -\frac{\hbar^2}{m}\nabla^2_{\mathbf{r}}, \qquad H_1 = V(\mathbf{r})\ ,$$

gives

$$\frac{\hbar^2 k^2}{m} g(\mathbf{k}) + \sum_{\mathbf{k}'} g(\mathbf{k}')\,V_{\mathbf{k},\mathbf{k}'} = Eg(\mathbf{k}) \tag{1}$$

with

$$V_{\mathbf{k},\mathbf{k}'} = \frac{1}{V_c}\int V(\mathbf{r}) e^{i(\mathbf{k}-\mathbf{k}')\cdot\mathbf{r}}\,d\mathbf{r}.$$

To first approximation we can take

$$V_{\mathbf{k},\mathbf{k}'} = \begin{cases} -V, & \text{for } |\varepsilon(\mathbf{k})|,\ |\varepsilon(\mathbf{k}')| \le \hbar\,\omega_D, \\ 0, & \text{otherwise,} \end{cases}$$

Where $\varepsilon(\mathbf{k}) = \dfrac{\hbar^2\mathbf{k}^2}{2m} - \varepsilon_F$ and V is a positive constant. Equation (1) now becomes

$$g(\mathbf{k}) = \frac{V\sum_{\mathbf{k}'} g(\mathbf{k}')}{\frac{\hbar^2\mathbf{k}^2}{m} - E}.$$

Summing up each side of the above over $\mathbf{k}$, we have

$$\sum_{\mathbf{k}} g(\mathbf{k}) = V\sum_{\mathbf{k}'} g(\mathbf{k}') \sum_{\mathbf{k}} \frac{1}{\left(\frac{\hbar^2\mathbf{k}^2}{m} - E\right)},$$

or

$$1 = V\sum_k \frac{1}{\frac{\hbar^2k^2}{m} - E} \approx V\int_0^{k_D} \frac{N(k)dk}{\frac{\hbar^2k^2}{m} - E} = V\int_0^{\hbar\omega_D} \frac{N(\varepsilon)d\varepsilon}{2(\varepsilon + \varepsilon_F) - E}.$$

As $\hbar\,\omega_D \ll E_F$, we can take $N(\varepsilon) \approx N(0)$. Hence

$$1 \approx N(0)V\int_0^{\hbar\omega_D} \frac{d\epsilon}{2(\varepsilon + \varepsilon_F) - E} = \frac{N(0)V}{2}\ln\left(\frac{E - 2\,\varepsilon_F - 2\hbar\,\omega_D}{E - 2\,\varepsilon_F}\right),$$

or

$$E = 2\,\varepsilon_F - \frac{2\hbar\,\omega_D}{\exp(2/N(0)V)-1} < 2\,\varepsilon_F$$

as $N(0)V$ is positive.

Therefore two electrons near the Fermi surface will combine into a bound state forming a Cooper pair if a small net attractive interaction $-V$ exists between them.

1075

1. Below the superconducting transition temperature T_c, a finite fraction n_s/n of the total number of conduction electrons can participate in a supercurrent; n_s is known as the number density of superconducting electrons. The London equation for simple superconductors is a phenomenological equation relating the supercurrent $\mathbf{j}_s$ to the magnetic vector potential $\mathbf{A}$:

$$\mathbf{j}_s = \frac{-n_s e^2}{m_e c}\mathbf{A} \qquad (1)$$

where m_e is the electron mass. Using the appropriate Maxwell equation, show how Eq. (1) leads to the Meissner effect. What is the London penetration depth Λ?

2. Another typical feature of most superconductors is the appearance of an energy gap of width Δ centered about the Fermi energy in the set of excited one-electron levels. Briefly discuss how any two of the following individual experimental measurements can discern the existence and magnitude of Δ.

a. Specific heat measurements at $T \ll T_C$.

b. Tunneling of single electrons across a thin insulator separating two superconductors.

c. Microwave (where $\hbar\omega \leq \Delta$) absorption in thin films.

d. Infrared ($\hbar\omega \leq \Delta$) absorption in bulk samples.

e. Acoustic attenuation at $T \ll T_C$.

(*Princeton*)

Sol:

1. The Meissner effect refers to the fact that in the superconducting state magnetic induction vanishes and materials become strongly diamagnetic. The London equation

$$\mathbf{j}_s = -\frac{n_s e^2}{m_e c}\mathbf{A} = -\frac{1}{c\lambda^2}\mathbf{A},$$

where $\lambda^2 = \frac{m_e}{n_s e^2}$, gives

$$\nabla\times \mathbf{j}_s = -\frac{1}{c\lambda^2}\,\nabla\times\mathbf{A} = -\frac{1}{c\lambda^2}\mathbf{B}.$$

Inside a superconductor the electric field vanishes and we have the Maxwell equation

$$\nabla\times \mathbf{B} = \frac{4\pi}{c}\mathbf{j}_s.$$

Hence

$$\mathbf{B} = -c\lambda^2\,\nabla\times\mathbf{j}_s = -\frac{c^2\lambda^2}{4\pi}[\nabla(\nabla\cdot\mathbf{B}) - \nabla^2\mathbf{B}],$$

or, using Maxwell's equation $\nabla\cdot\mathbf{B} = 0$,

$$\nabla^2\mathbf{B} = \frac{1}{\Lambda^2}\mathbf{B},$$

where

$$\Lambda = \left(\frac{c^2}{4\pi}\lambda^2\right)^{1/2} = \left(\frac{m_e c^2}{4\pi n_s e^2}\right)^{1/2}.$$

Let x denote depth from the surface of the superconductor, then

$$\frac{d^2\mathbf{B}}{dx^2} = \frac{\mathbf{B}}{\Lambda^2},$$

giving

$$\mathbf{B} = \mathbf{B}_0 e^{-\frac{x}{\Lambda}}.$$

This shows that $\mathbf{B}$ decays exponentially such that $\mathbf{B} = \frac{1}{e}\mathbf{B}_0$ at $x = \Lambda$. Λ is known as the London penetration depth. For a great majority of metals showing superconductivity, the penetration depths are in the range 10^{-8}–10^{-7} m. For $x \gg \Lambda$, $\mathbf{B} \to 0$, indicating that the magnetic field exists only in a thin layer of thickness $\approx \Lambda$ beneath the surface of the superconductor. Thus the magnetic field inside a superconductor is zero. This is the Meissner effect.

2. The exponential decrease of specific heat with temperature for a superconductor at low temperatures implies that there is an energy gap Δ, centered at about the Fermi energy, between the ground and lowest excited states in the energy spectrum of the electrons, as shown in Fig. 1.45. Two of the experiments mentioned may be used to determine the magnitude of Δ. These are the following.

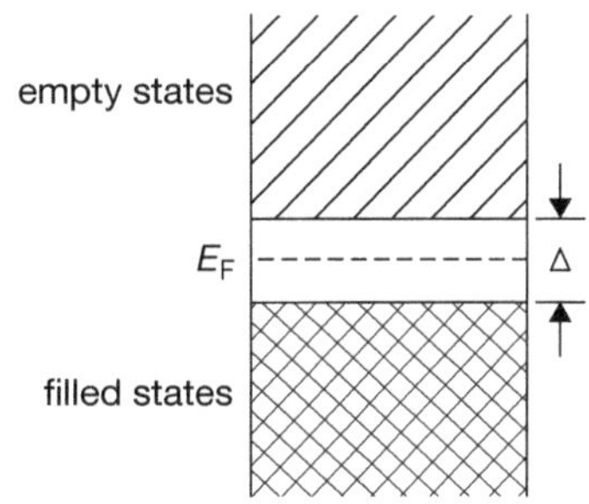

Fig. 1.45

a. Specific heat measurements at $T \ll T_C$. For $T \ll T_c$, the specific heat of the system is expressible as

$$C_S \approx \frac{A}{T}\exp\left(-\frac{\Delta}{k_B T}\right),$$

where A is a constant. For $T \approx T_c$ we find experimentally

$$C_S = a\exp\left(-\frac{b}{t}\right),$$

Where $t = \frac{T}{T_c}$ and a is a constant. Comparing the above, we obtain

$$\frac{b}{t} = \frac{\Delta}{k_B T},$$

or

$$\Delta = b\,k_B T_c = 1.52\,k_B T_c,$$

using the experimentally determined value of b.

b. Acoustic attenuation of ultrasonic waves in the superconducting state is different from that in the normal state. Theory gives

$$\frac{\alpha_S}{\alpha_N} = \frac{2}{\exp\left(\frac{\Delta}{k_B T}\right) + 1},$$

where α_S and α_N are the attenuation coefficients for the superconducting state and the normal state respectively. Then using the experimentally obtained ratio of $\frac{\alpha_s}{\alpha_N}$, Δ can be determined

---

1076

A ring is made of 1 mm diameter lead wire formed into a circle of 10 cm diameter as shown in Fig. 1.46. The ring is in the superconducting state and has a 100 A current flowing in it. It is observed that there is no detectable change in current for a period of 1 year. If the detector is sensitive to a change in current of a little as 1 μA, calculate the experimental upper limit for the resistivity of lead in the superconducting state.

Note: An answer within a factor of three will suffice.

(*Chicago*)

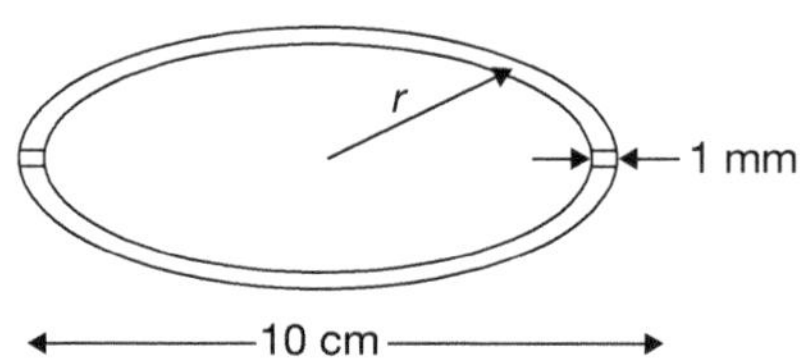

Fig. 1.46

Sol:

Let ρ be the resistivity of lead. Then the resistance of the ring is $R = \rho\frac{l}{S}$, where $l = 2\pi r$, S is the cross-sectional area of the lead wire. The ring may be considered as an RL circuit, so that

$$-L\frac{dI}{dt} = IR,$$

or

$$I(t) = I(0)e^{-Rt/L},$$

i.e.

$$R = \frac{L}{t} \ln \frac{I(0)}{I(t)}.$$

The self-inductance L is given by the flux linkage $\mathbf{\Phi}$ and the current I by $L = \mathbf{\Phi}/I$, thus

$$I(0) = 100\,\text{A}, \quad I(t) = 100 - 10^{-6}\,\text{A}.$$

The magnetic field distribution in the ring is somewhat complicated, but for a rough estimate we can approximate the magnetic field inside the ring by a uniform field equal in magnitude to that at the center of the ring. Thus

$$\mathbf{\Phi} = \pi r^2 B = \pi r^2 \cdot \frac{\mu_0 I}{2r} = \frac{1}{2}\pi \mu_0 r I$$

and

$$L = \frac{1}{2}\pi \mu_0 r.$$

As $I(t) \approx I(0), \frac{Rt}{L} \ll 1$.Then

$$\begin{aligned}
\Delta I &= I(0) - I(t) \approx \frac{I_0 Rt}{L}, \\
\ln \frac{I(0)}{I(t)} &= \ln\left[1 + \frac{\Delta I}{I(t)}\right] \approx \frac{\Delta I}{I(0)} \approx 10^{-8}, \\
R &\approx \frac{\pi \mu_0 r}{2t}\left[\frac{\Delta I}{I(0)}\right] = \frac{\pi}{2} \times \frac{4\pi \times 10^{-7} \times 5 \times 10^{-2}}{365 \times 24 \times 3600} \times 10^{-8} \\
&= 3.13 \times 10^{-23}\,\Omega.
\end{aligned}$$

This gives

$$\rho = \frac{SR}{2\pi r} = \frac{\pi \times (0.5 \times 10^{-3})^2 \times 3.13 \times 10^{-23}}{2\pi \times 5 \times 10^{-2}} = 7.8 \times 10^{-29}\,\Omega\text{m}.$$

1.4

OTHER TOPICS (1077–1081)

1077

A currently important technique for precisely measuring the lifetimes of states of multiply ionized atoms consists of exciting a beam of the desired ions with a laser tuned to a resonance wavelength of the ion under study, and measuring the emission intensity from the ion beam as a function of down-stream distance. When a particular ion beam with a resonance wavelength at 4885 Å is excited with a 4880 Å argon-ion laser output, the intensity is found to drop by a factor of two, 10 mm down-stream from the excitation point. If ion beam velocity is $v/c = 5 \times 10^{-3}$, calculate the angle the laser beam must make with the normal to the ion beam to achieve resonance and the lifetime of the excited state of the ion.

(*Wisconsin*)

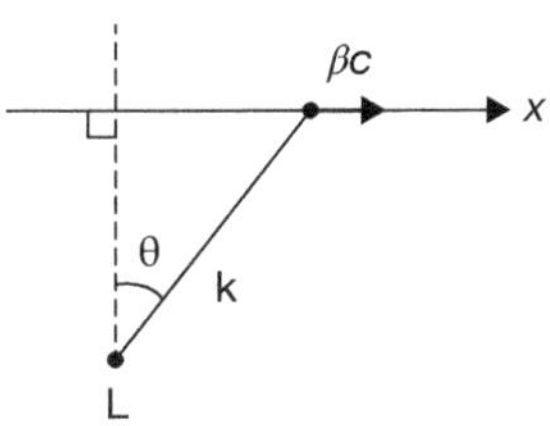

Fig. 1.46

Sol: Take two coordinate frames Σ, Σ'. Σ is the laboratory frame in which the laser source L is at rest and the ions move with velocity βc along the x-axis as shown in Fig. 1.46. Σ' is the rest frame of an ion in the beam and moves with velocity βc relative to Σ along the x direction. Consider the transformation of the 4-vector $\left(\mathbf{k}, \frac{\omega}{c}\right)$. We have

$$\frac{\omega'}{c} = \gamma\left(\frac{\omega}{c} - \beta k \sin\theta\right),$$

where $\gamma = \left(1-\beta^2\right)^{-\frac{1}{2}}$, $k = \frac{\omega}{c}$. Hence

$$\begin{aligned}\sin\theta &= \frac{1}{\beta}\left(1 - \frac{\lambda}{\lambda'}\sqrt{1-\beta^2}\right) \\ &\approx \frac{1}{\beta}\left[1 - \frac{\lambda}{\lambda'}\left(1 - \frac{\beta^2}{2}\right)\right].\end{aligned}$$

With $\lambda = 4885$ Å, $\lambda' = 4880$ Å, $\beta = 5 \times 10^{-3}$, we find

$$\sin\theta = 0.2074, \quad \text{or} \quad \theta = 11.97^\circ.$$

As the excited ions move with velocity βc, on account of time dilation the laboratory lifetime of an ion is $\gamma\tau$, where τ is the proper lifetime. Hence the excited ion beam will decay according to

$$I = I_0 \exp\left(-\frac{t}{\gamma\tau}\right),$$

giving

$$\begin{aligned}\tau &= \frac{t}{\gamma \ln\left(\frac{I_0}{I}\right)} = \frac{x}{\gamma\beta c \ln\left(\frac{I_0}{I}\right)} \\ &\approx \left(1 - \frac{\beta^2}{2}\right)\frac{x}{\beta c \ln\left(\frac{I_0}{I}\right)} = 9.6 \times 10^{-9}\,\text{s}.\end{aligned}$$

1078

Tiny balls of a reflecting material are floating on the surface of a transparent material. Begin by assuming that they form a perfect square lattice with lattice spacing of 10,000 Å as shown in Fig. 1.47. A coherent beam of light going straight down hits the surface. The light has a wavelength of 5000 Å.

a. At what angles (θ, φ) will we observe light going out if we stand back from the surface?

b. If the incoming beam were infinitely wide and infinitely well collimated and if the lattice were infinitely large, the outgoing beam would of course also move in perfectly precise directions. If we have a horizontal screen 10 meters above the surface, perfectly sharp spots would appear on it. If, however, the lattice were a square containing "only" 10,000 balls, what spot width would you expect to see on the screen? (An order of magnitude estimate is good enough, but do give an argument for it.)

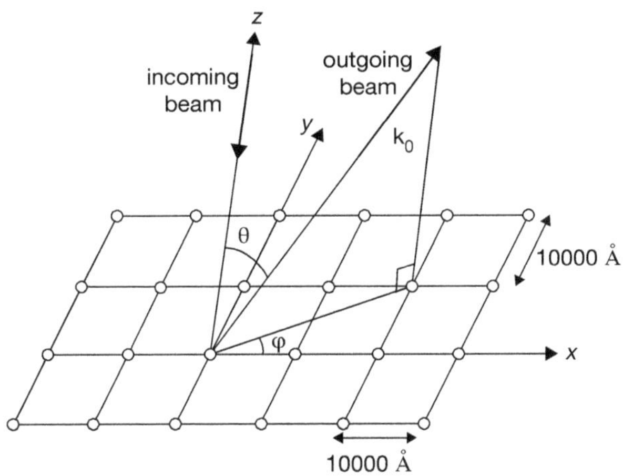

Fig. 1.47

c. Now assume that each ball were held in its place in the lattice by forces which behaved like tiny springs. Each ball independently jiggles up and back about its lattice position in a random pattern of motion with a mean squared displacement of about 2000 Å. In what way will this motion change the pattern of the outgoing light in comparison to the case in which the balls were at their lattice sites? A comparison of the "in motion" intensity pattern to the "at rest" pattern can tell us whether the motion of the balls is horizontal or vertical. How can one distinguish between these two cases?

(*Chicago*)

Sol:

a. The direction of the outgoing beam is that of an interference maximum of the reflected light. Let the unit vector of the interference-maximum direction be

$$\mathbf{k}_0 = \sin\theta\cos\varphi\,\mathbf{e}_x + \sin\theta\sin\varphi\,\mathbf{e}_y + \cos\theta\,\mathbf{e}_z.$$

A plane through the origin with unit normal $\mathbf{k}_0$ is given by

$$\mathbf{k}_0 \cdot \mathbf{r} = 0,$$

$\mathbf{r}$ being any vector lying on the plane. If $\mathbf{R}$ is the radius vector of a point on the xy plane, its distance from this plane is $d_A = \mathbf{k}_0 \cdot \mathbf{R}$. Then for the lattice points A(d, 0, 0), B(0, d, 0), C(d, d, 0), shown in Fig. 1.48, d being the edge of a unit cell, we have

$$\begin{aligned} d_A &= d\sin\theta\cos\varphi, \\ d_B &= d\sin\theta\sin\varphi, \\ d_C &= d\sin\theta(\cos\varphi + \sin\varphi). \end{aligned}$$

The conditions for $\mathbf{k}_0$ to be a direction of interference maximum are

$$d_A = n_1\lambda_0, \quad d_B = n_2\lambda_0, \quad d_C = n_3\lambda_0,$$

where n_1, n_2, n_3 are integers, and λ_0 is the wavelength of the incident light. Note that as $d_C = d_A + d_B$, the last condition is already included in the first two.

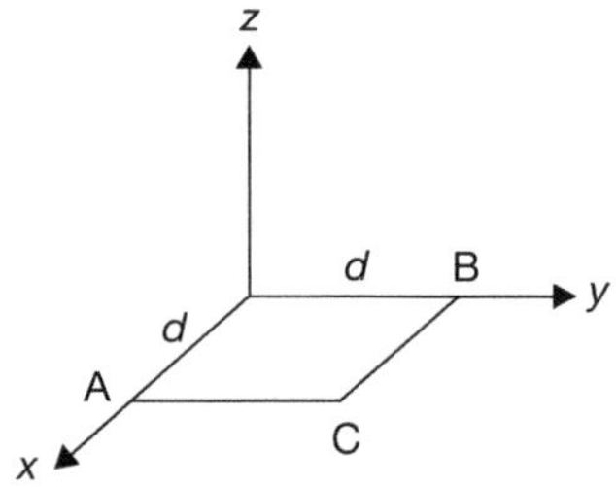

Fig. 1.48

On account of translational symmetry of the lattice, the distances from the plane to other lattice points are also integral multiples of the wavelength. As

$$\tan\varphi = \frac{n_2}{n_1}, \quad d^2\sin^2\theta = (n_1^2 + n_2^2)\lambda_0^2$$

we have

$$\theta = \arcsin\left(\sqrt{n_1^2 + n_2^2}\,\frac{\lambda_0}{d}\right), \quad \text{or} \quad \theta = \pi - \arcsin\left(\sqrt{n_1^2 + n_2^2}\,\frac{\lambda_0}{d}\right),$$

$$\varphi = \arctan\frac{n_2}{n_1}, \quad \text{or} \quad \varphi = \pi + \arctan\frac{n_2}{n_1}.$$

b. The edge of a square containing 10,000 balls is 100 d. Now the beam passing through the square will produce diffraction and the width of a spot on the screen is

$$\Delta x = \Delta y \approx \frac{\lambda}{100d} \times D = \frac{5000}{100 \times 10000} \times 10 = 0.05\,\text{m},$$

D being the distance between the screen and the square lattice.

c. The oscillatory motion of the balls will make the original sharp spots diffuse into larger spots as well as produce some background light where there are no spots.

If the motion of the balls is horizontal, the phase of the light along the z-direction is less affected. Then the interference pattern is such that central spots are brighter and smaller and are on a darker background, while spots near the edges are fainter and larger, and are on a brighter background. If the motion of the balls is vertical, the phase of the light about the z direction will be affected more, with the result that the effects on the spots and background light will be the reverse of the above.

1079

Describe briefly the basic parts and principle of operation of a hot-cathode ionization gauge for measuring pressure in a vacuum system. State approximately the high and low pressure limits of the gauge and the factors which determine these limits.

(*Wisconsin*)

Sol: A hot-cathode ionization gauge consists of a cathode F, an accelerator A and a collector C, as shown in Fig. 1.49 together with a typical circuit for its operation.

The cathode filament, heated by an electric current, emits electrons which move toward the accelerator A maintained at a positive potential. Some of the electrons are absorbed by A while others pass through the gaps between the grids of the accelerator toward the collector C, kept at a negative potential, which repulses the electrons as they approach C. Thus the electrons oscillate to and fro between C and A until they are absorbed by A. During such oscillations the electron collide frequently with the gas molecules in the tube, giving rise to positive ions and electrons. The positive ions are collected by the collector C to give an ionic current i_+, and the electrons are collected by the accelerator to give an electric current i_e. Theory shows that the pressure p in the tube can be expressed as

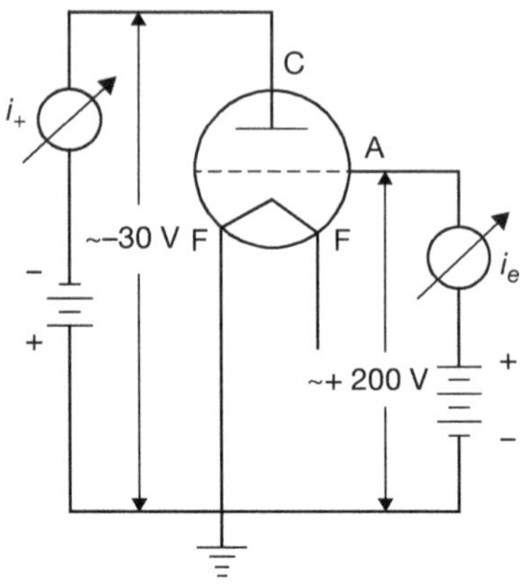

Fig. 1.49

$$p = K\frac{i_+}{i_e},$$

where K is a constant. If i_e is kept constant, then

$$p = K_1 i_+,$$

which enables one to determine the pressure in a vacuum system by measuring the ionic current i_+.

The high pressure limit of the ionization gauge is about 10^{-3} torr. When $p > 10^{-3}$ torr, the filament is easily burnt out; furthermore, i_+ does not increase with p, as due to the chemical cleaning of the hot filament it behaves as an absorber of gas. The low pressure limit is about 10^{-7} torr, below which the current collected by the collector is mainly the photoemission current; only a small fraction of it is the ionic current with the result that p does not vary linearly with i_+.

1080

A physicist designs a thermionic converter for a concentrated solar power plant. He chooses a cathode material with a work function of 1.8 eV, the surface area of 0.2 cm^2, and Richardson–Dushman constant of 160 A/cm^{2}K^2. He fixes the distance between the anode and cathode as 0.5 cm and the potential across the cathode and anode is found to be 0.5 V. The operating temperature of the concentrated solar power plant can reach up to 1,200 K.

a. Calculate the thermal efficiency of the converter.

b. What should be the work function to get an increase of 25% in emission?

c. Does this increase affect the efficiency of the converter? If so, how?

d. Discuss the ways to enhance the efficiency of a thermionic converter.

Sol:

a. **Thermal efficiency of the converter:**

To determine the efficiency of the converter, the thermionic current density of emission, total current emitted, output power, and thermal power input must be determined.

Thermionic current density:

According to the Richardson–Dushman equation:

$$J_T = AT^2 \exp\left(-\frac{\phi}{kT}\right)$$

where, A is a constant = 160 A/cm²K², T is the operation temperature = 1,200 K, ϕ is the work function = 1.8 eV, k is the Boltzmann constant = 1.38×10^{-23} J/K

$$J_T = 160 \times 1{,}200^2 \times \exp\left(-\frac{1.8 \times 1.6 \times 10^{-19}}{1.38 \times 10^{-23} \times 1{,}200}\right)$$

$$J_T = 6.45 \text{ A/cm}^2$$

Total current:

The total current is given by

$$I_T = J_T \times \text{Area}_{\text{cathode}} = 6.45 \times 0.2$$

$$I_T = 1.29 \text{ A}$$

Output power:

$$P_{\text{out}} = I_T V = 1.29 \times 0.5 = 0.645 \text{ W}$$

Thermal power input:

$$P_{\text{in}} = \text{Area}_{\text{cathode}} \times \sigma T^4$$

where σ is the Stefan–Boltzmann constant

$$P_{\text{in}} = 2 \times 10^{-5} \times 5.67 \times 10^{-8} \times 1{,}200^4 = 2.35 \text{ W}$$

Efficiency:

The efficiency of the thermionic energy converters can be determined using the formula

$$\eta = \frac{P_{\text{out}}}{P_{\text{in}}} \times 100\,\% = \frac{0.645}{2.35} \times 100\%$$

$$\eta = 27.4\,\%$$

So, in this scenario, the thermionic converter has an efficiency of approximately 27.4% and generates a power output of approximately 0.645 W. This demonstrates the conversion of thermal energy into electrical energy using thermionic emission.

b. Work function to get an increase of 25% in emission:

Given that J_T is exponentially dependent on $-\Phi/kT$, a 25% increase in J_T corresponds to a reduction in the exponent.

We can set up the relationship as follows:

$$J_{T(\text{new})} = 1.25\,J_T$$

$$AT^2\exp\left(-\frac{\phi_{\text{new}}}{kT}\right) = 1.25\,AT^2\exp\left(-\frac{\phi_{\text{old}}}{kT}\right)$$

$$\exp\left(-\frac{\phi_{\text{new}}}{kT}\right) = 1.25\,\exp\left(-\frac{\phi_{\text{old}}}{kT}\right)$$

$$-\frac{\phi_{\text{new}}}{kT} = \ln(1.25) - \frac{\phi_{\text{old}}}{kT}$$

$$\phi_{\text{new}} = \phi_{\text{old}} - kT\ln(1.25)$$

$$= 1.8\text{ eV} - \left(\frac{1.38\times10^{-23}\times1{,}200\times\ln(1.25)\text{ eV}}{1.6\times10^{-19}}\right) = 1.776\text{ eV}$$

So, to achieve a 25% increase in electron emission at a temperature of 1,200 K, the new work function should be approximately 1.776 eV.

c. New efficiency of the converter:

The new thermionic current is given by

$$J_T = 160\times1{,}200^2\times\exp\left(-\frac{1.776\times1.6\times10^{-19}}{1.38\times10^{-23}\times1{,}200}\right) = 8.13\text{ A/cm}^2$$

The new total current is given by

$$I_T = J_T\times\text{Area}_{\text{cathode}} = 8.13\times0.2$$
$$I_T = 1.63\text{ A}$$

The new output power is

$$P_{\text{out}} = I_T V = 1.63\times0.5 = 0.815\text{ W}$$

The input power will not change based on the work function, so the new efficiency is

$$\eta = \frac{P_{\text{out}}}{P_{\text{in}}}\times100\% = \frac{0.815}{2.35}\times100\%$$
$$\eta = 34.6\%$$

So, in this scenario, the thermionic converter has an efficiency of approximately 34.6% and generates a power output of approximately 0.815 W.

The change in work function from 1.8 eV to approximately 1.776 eV results in a considerable improvement in efficiency, about 10%.

d. Ways to enhance the efficiency of a thermionic converter:

The efficiency can be enhanced by optimizing various aspects of the converter's design and operation. Here are some methods to improve efficiency:

1. Choosing a cathode material with a lower work function (Φ) can increase the emission of electrons at a given temperature. Materials with work functions close to the thermal energy level (kT) will improve efficiency. Surface treatments or coatings on the cathode can improve electron emission and reduce the work function.
2. Increasing the operating temperature (T) of the cathode can significantly enhance efficiency. Higher temperatures lead to greater electron emission and, in turn, increased thermionic current. However, this must be balanced with material limitations and structural integrity.
3. Effective cooling of the cathode while maintaining the high temperature at the emission surface is important. Proper heat management can increase efficiency and extend the lifetime of the device.
4. The separation distance (d) between the cathode and anode should be minimized without causing arcing or breakdown. This reduces the electric field strength and enhances electron transfer.
5. Maintaining a high-quality vacuum environment is critical to reduce electron scattering and increase thermionic emission.
6. Increasing the emission area (A) of the cathode or using multiple cathodes can boost thermionic current and, subsequently, power output.
7. Combining thermionic converters with other energy conversion technologies, such as thermophotovoltaics, can enhance overall efficiency by utilizing a broader range of the electromagnetic spectrum.
8. Fine-tuning operating conditions, such as the voltage applied between the cathode and anode, can impact efficiency. The voltage should be optimized for maximum power output.

It's important to note that improving efficiency often involves trade-offs between different parameters. For instance, higher operating temperatures can increase efficiency but may reduce the lifetime of the device. Engineering and optimizing these factors require a balanced approach to meet the specific needs and constraints of the application.

1081

For each of the following describe in one sentence each the phenomenon observed (not the explanation).

a. Raman effect

b. Ramsauer effect

c. Kerr effect

d. Faraday effect

e. Paschen-Back effect.

(*Wisconsin*)

Sol:

a. A phonon is scattered inelastically by a crystal with the creation or annihilation of a phonon or magnon.

b. The mean free path of low energy electrons is increased in an inert gas.

c. Certain transparent substances become doubly refracting when an electric field is applied at right angles to the direction of the incident light.

d. The plane of polarization of linearly polarized light rotates when it travels in ceratin transparent substances placed in a magnetic field which is parallel to the direction of the light.

e. When a light source is placed in a strong magnetic field, a spectral line is seen to split into two lines when observed along the direction of the field, and into three lines when observed perpendicular to the field.

Part II
Relativity

2.1

SPECIAL RELATIVITY (2001–2007)

2001

In this problem you are asked to derive some aspects of Lorentz transformations, using some thought experiments devised by Einstein. Einstein's thought experiments involve an idealized clock (Fig. 2.1(a)) in which a light wave (or a massive particle) is bouncing back and forth between two mirrors. The clock "ticks" when the light wave makes a round trip from mirror A to mirror B and back.

a. Derive the time dilation formula of special relativity as follows. Assuming that light has velocity c in every frame, how much slower does the clock tick in a frame (Fig. 2.1(b)) in which it has velocity V in a direction perpendicular to the separation between the mirrors than it ticks in its rest frame.

b. According to special relativity, the time dilation factor is the same for all clocks of a given velocity, independent of how they operate. Suppose that in Einstein's clock a massive particle rather than a photon is bouncing between the mirrors. In the rest frame of the clock, the massive particle has velocity $v < c$. Derive the Lorentz transformation law for velocities as follows. What must be the velocity of the massive particle in the frame considered in exercise (a) if it is required that the clock, when operated with a massive particle, ticks more slowly in this frame than in its rest frame by the same time dilation factor found in (a)?

c. Now derive the Lorentz contraction law for length. Suppose again that the clock is operated with a photon of velocity c, but suppose that it is rotated by 90° before being set in motion, so that now (Fig. 2.1(c)) it has velocity V parallel to the separation between the mirrors. How much shorter must the clock be in the moving frame than in its rest frame, so that the ticking is slowed down in the moving frame relative to the rest frame by the same time dilation factor found in (a)?

(Princeton)

Sol:

a. Suppose the mirrors A and B are separated by a distance L in the rest frame. In that frame a light wave will take a time $t_0 = \frac{2L}{c}$ for the round trip A → B → A. This is the proper time interval between two consecutive ticks of the clock. Let t_0' be the time interval between two consecutive ticks

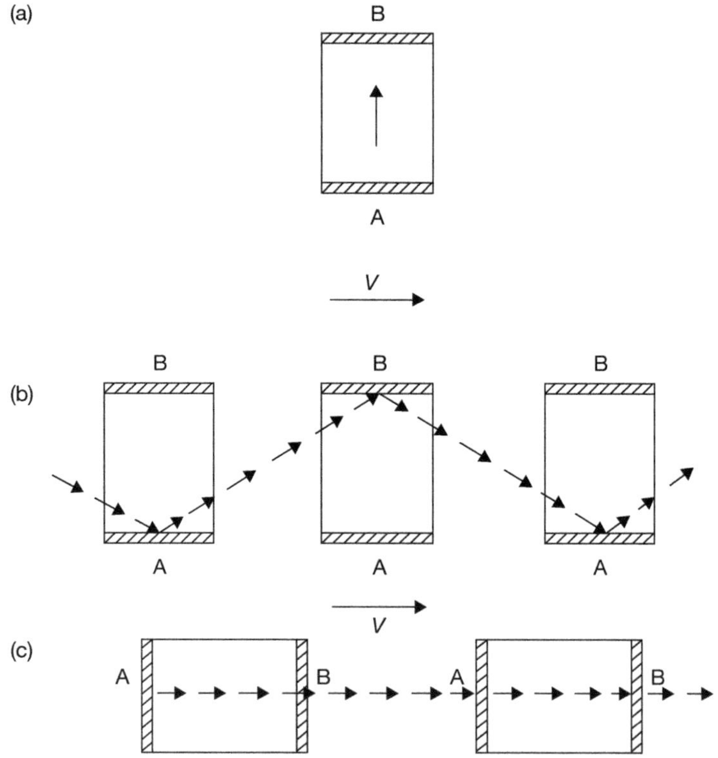

Fig. 2.1

of the clock in a frame in which the mirrors move with velocity V as shown in Fig 2.1(b). It is noted that when the light wave is bounced back at the mirror B, the latter has already moved a distance $\frac{t_0'}{2}V$. As shown in Fig. 2.2, we have, since light has velocity c in all directions

$$L^2 + \left(\frac{t_0' V}{2}\right)^2 = \left(\frac{t_0'}{2}c\right)^2,$$

or

$$t_0' = \frac{2L}{\sqrt{c^2 - V^2}} = \left(1 - \frac{V^2}{c^2}\right)^{-\frac{1}{2}} t_0 .$$

Hence the ticking of the clock in a frame in which it moves with velocity V in a direction perpendicular to the separation of the mirrors is slower by a factor $\left(1 - \frac{V^2}{c^2}\right)^{-\frac{1}{2}}$.

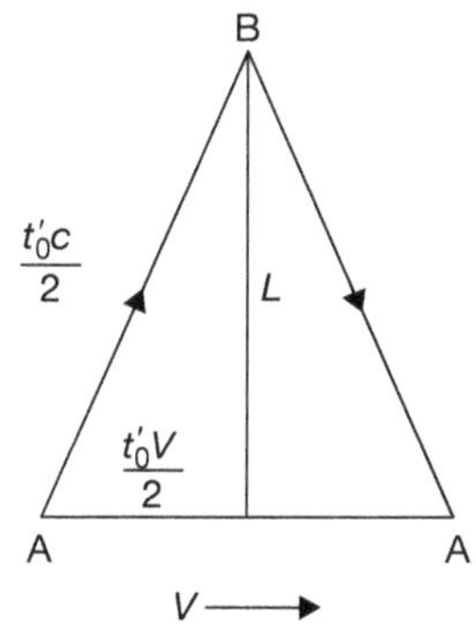

Fig. 2.2

b. The light wave is now replaced by a massive particle of velocity v in the rest frame of the clock. Then the proper time interval between two consecutive ticks is $t = \frac{2L}{v}$. Suppose the velocity of the massive particle in the frame in which the clock has velocity V, as in (a), is v' in the direction that it bounces at B. We have

$$L^2 + \left(\frac{t'V}{2}\right)^2 = \left(\frac{t'v'}{2}\right)^2,$$

or

$$t' = \frac{2L}{\sqrt{v'^2 - V^2}} = \frac{t}{\sqrt{(\frac{v'}{v})^2 - (\frac{V}{v})^2}}$$

where t' is the time interval between two consecutive ticks in the moving frame of the clock. We required that

$$\frac{1}{\sqrt{(\frac{v'}{v})^2 - (\frac{V}{v})^2}} = \frac{1}{\sqrt{1 - (\frac{V}{c})^2}},$$

which gives

$$v' = \sqrt{v^2 + \left(1 - \frac{v^2}{c^2}\right) V^2}.$$

c. Suppose the mirrors A and B are separated by L' in the moving frame of the clock. After leaving A, suppose the light wave reaches B at time t_1'. As B has moved a distance Vt_1', we have

$$L' + Vt_1' = ct_1',$$

or

$$t_1' = \frac{L'}{c - V}.$$

Supposing the light wave, after bouncing at B, takes time t_2' to reach A again, we have by the same reasoning

$$L' + Vt_1' - V(t_1' + t_2') = ct_2',$$

or

$$t_2' = \frac{L'}{c + V}.$$

Hence the interval between two consecutive ticks in the moving frame is

$$t_0' = t_1' + t_2' = \frac{2L'}{c\left(1 - \frac{V^2}{c^2}\right)} = \left(\frac{L'}{L}\right) \frac{t_0}{1 - \frac{V^2}{c^2}}.$$

We require that

$$\left(\frac{L'}{L}\right) \frac{1}{1 - \frac{V^2}{c^2}} = \frac{1}{\sqrt{1 - \frac{V^2}{c^2}}},$$

which gives

$$L' = \left(1 - \frac{V^2}{c^2}\right)^{1/2} L.$$

2002

a. A muon at rest lives 10^{-6} sec, and its mass is 100 MeV/c^2. How energetic must a muon be to reach the Earth's surface if it is produced high in the atmosphere (say $\sim 10^4$ m up)?

b. Suppose to a zeroth approximation that the Earth has a 1 gauss magnetic field pointing in the direction of its axis, extending out to 10^4 cm.

How much, and in what direction, is a muon of energy E normally incident at the equator deflected by the field?

c. Very high energy protons in cosmic rays can lose energy through collision with 3 K radiation (cosmological background) in the process $p + \lambda \to p + \pi$. How energetic need a proton be to be above threshold for this reaction?

(*Princeton*)

Sol:

a. A clock moving at velocity v appears to go slow by a factor $\gamma = \left(1 - \frac{v^2}{c^2}\right)^{-\frac{1}{2}}$. Thus a muon of velocity v will have a lifetime $\gamma\tau_0$ in the laboratory frame, where τ_0 is its rest lifetime. For a muon produced at a height h with velocity v to reach the Earth's surface, it must have a lifetime

$$\gamma\tau_0 = \frac{h}{v},$$

which gives

$$\gamma v = c\sqrt{\gamma^2 - 1} = \frac{h}{\tau_0}.$$

Hence the muon of rest mass m must have a total energy

$$\begin{aligned} m\gamma c^2 &= mc^2\sqrt{\left(\frac{h}{\tau_0 c}\right)^2 + 1} \approx \frac{mc^2 h}{\tau_0 c} \\ &= \frac{100 \times 10^4}{10^{-6} \times 3 \times 10^8} = 3.3 \times 10^3\,\text{MeV}. \end{aligned}$$

b. Suppose the muon has rest mass m, charge q and velocity $\mathbf{v}$. It suffers gravity, which is negligible, and magnetic force $\mathbf{F} = q\mathbf{v} \times \mathbf{B}$. As the magnetic force does no work on the muon, for $\mathbf{F} \cdot \mathbf{v} = 0$, the muon energy $E = m\gamma c^2$ is constant. It follows that the magnitudes of its velocity $\mathbf{v}$ and momentum $\mathbf{p} = m\gamma\mathbf{v}$, where γ is the Lorentz factor $\left(1 - \frac{v^2}{c^2}\right)^{-\frac{1}{2}}$, are constant.

Take Cartesian coordinates as shown in Fig. 2.3 with the origin on the Earth's surface, y-axis vertically upward, z-axis pointing north, and x-axis pointing west. With $\mathbf{v} = (\dot{x}, \dot{y}, \dot{z})$, $\mathbf{B} = (0, 0, B)$, the equation of motion in SI units

$$\frac{d\mathbf{p}}{dt} = m\gamma\frac{d\mathbf{v}}{dt} = q\mathbf{v} \times \mathbf{B}$$

can be written as

$$\ddot{x} = \omega\dot{y}, \qquad \ddot{y} = -\omega\dot{x}, \qquad \ddot{z} = 0$$

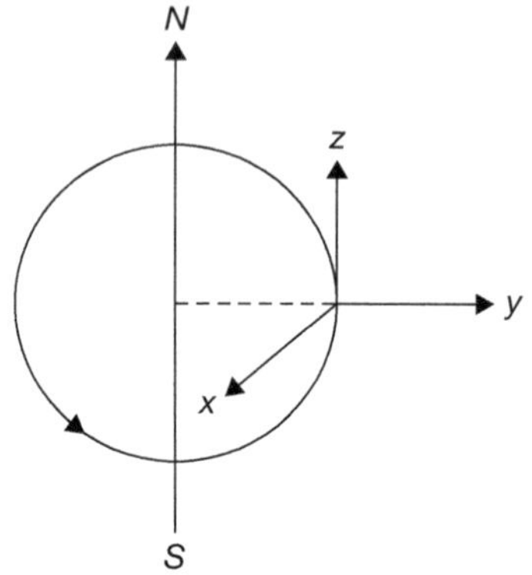

Fig. 2.3

with $\omega = \frac{qB}{m\gamma}$. With the initial condition that at $t = 0$, when the muon is just within the range of the Earth's magnetic field, $x = z = 0$, $y = y_0$, $\dot{x} = \dot{z} = 0$, $\dot{y} = -v$, the solution is

$$x = \frac{v}{\omega}\cos(\omega t) - \frac{v}{\omega}, \qquad y = y_0 - \frac{v}{\omega}\sin(\omega t), \qquad z = 0.$$

Hence

$$x = \frac{v}{\omega}\sqrt{1 - \frac{(y - y_0)^2\omega^2}{v^2}} - \frac{v}{\omega}.$$

On the Earth's surface, $y = 0$, $x = x_e$ given by

$$x_e = \frac{v}{\omega}\sqrt{1 - \left(\frac{y_0\,\omega}{v}\right)^2} - \frac{v}{\omega}.$$

We have $B = 1$ gauss $= 10^{-4}$ Wb m^{-2}, $|q| = e = 1.6 \times 10^{-19}$ C,

$$\frac{v}{\omega} = \frac{m\gamma v}{qB} = \frac{p}{qB},$$

so that if p is 1 MeV/c,

$$\left|\frac{v}{\omega}\right| = \frac{1.6 \times 10^{-19} \times 10^6}{1.6 \times 10^{-23} \times 3 \times 10^8} = \frac{1}{3} \times 10^2\,\text{m}.$$

Thus if p is greater than, say, 10 MeV/c, we can take $\left|\frac{v}{\omega}\right| \gg y_0$ which is ~ 10^2 m, and approximate

$$x_e \approx \frac{v}{\omega}\left[1 - \frac{1}{2}\left(\frac{y_0\,\omega}{v}\right)^2\right] - \frac{v}{\omega} \approx \frac{\omega y_0^2}{2v}$$
$$= \frac{qBy_0^2}{2p} = \frac{cqBy_0^2}{2\sqrt{E^2 - m^2c^4}}.$$

Hence the angle of deflection is

$$\alpha = \frac{x_e}{y_0} = \frac{cqBy_0}{2\sqrt{E^2 - m^2c^4}}.$$

If E is that given by (a), 3.3×10^3 MeV,

$$\begin{aligned}\alpha &= \frac{cqBy_0}{2E} = \frac{3\times 10^8 \times 1.6\times 10^{-19}\times 10^{-4}\times 10^2}{2\times 3.3\times 10^9\times 1.6\times 10^{-19}} \\ &= 5\times 10^{-4} \quad \text{rad} = 1.7'.\end{aligned}$$

The deflection is to the west for positive muons and to the east for negative muons.

c. The line element of the energy-momentum four-vector, $E^2 - P^2c^2$ is invariant under Lorentz transformation and, if the total energy E and total momentum P are conserved, remains constant in a reaction. Let E_p, E_λ, $\mathbf{p}_p$, $\mathbf{p}_\lambda$ be respectively the energy and momentum of the proton and a background photon. Before the collision,

$$\begin{aligned}E^2 - P^2c^2 &= (E_p + E_\lambda)^2 - (\mathbf{p}_p + \mathbf{p}_\lambda)^2c^2 \\ &= \left(E_p^2 - p_p^2c^2\right) + 2E_pE_\lambda + \left(E_\lambda^2 - p_\lambda^2c^2\right) - 2\mathbf{p}_p\cdot\mathbf{p}_\lambda c^2 \\ &= m_p^2c^4 + 2E_pE_\lambda - 2p_pp_\lambda c^2\cos\theta,\end{aligned}$$

where θ is the angle between p_p and p_λ, as $E_p^2 = m^2c^4 + p_p^2c^2$, $E_\lambda = p_\lambda c$. At threshold, the proton and pion are produced with zero velocity in the center of mass frame of the system. Thus

$$\bar{E}^2 - \bar{P}^2c^2 = \left(m_pc^2 + m_\pi c^2\right)^2.$$

Equating the above two expressions and making the approximation $E_p \approx p_pc$ for energetic protons, we find

$$E_p \approx \frac{m_\pi(m_\pi + 2m_p)c^4}{2(1-\cos\theta)E_\lambda}.$$

The lowest threshold proton energy is given by $\cos\theta = -1$, or $\theta = \pi$, and is

$$E_p \approx \frac{m_\pi(m_\pi + 2m_p)c^4}{4E_\lambda}.$$

The energy of a background photon can be taken to be ~ kT, where k is the Boltzman constant and T = 3 K. As $k = 1.38 \times 10^{-23}$ J deg^{-1} = 8.63×10^{-11} MeV deg^{-1}, $m_\pi = 140$ MeV, $m_p = 938$ MeV, the lowest threshold is

$$E_p \approx \frac{140\times(140 + 2\times 938)}{4\times 8.63\times 10^{-11}\times 3} = 2.7\times 10^{14}\,\text{MeV}.$$

2003

Consider a 2-dimensional plane in Minkowski space (spacetime) defined by the equation $ax + by \pm - dt = 0$, where (x, y, t) are the coordinates in spacetime, "a,""b," and "d" are constants. This plane is observed from two inertial reference frames S and S', where S' is moving relative to S with a velocity v in the positive x direction.

a. Find the equation of the transformed plane in the S' frame.

b. Is the transformed plane still a plane in the S' frame?

c. Calculate the apparent angle of this plane when observed from the S' frame.

Sol:

a. **Equation of the transformed plane in the S' frame:**

The coordinate transformation between the frames S and S' is given by the Lorentz transformation:

$$x' = \frac{x - vt}{\sqrt{1 - \frac{v^2}{c^2}}} = \gamma(x - vt)\ ;$$
$$y' = y$$
$$t' = \frac{t - \frac{vx}{c^2}}{\sqrt{1 - \frac{v^2}{c^2}}} = \gamma\left(t - \frac{vx}{c^2}\right)$$

where (x', y', t') are the coordinates in S' frame.

Taking the inverse Lorentz transformation, we get

$$x = \gamma(x' + vt')\ ;$$
$$y = y'$$
$$t = \gamma\left(t' + \frac{vx'}{c^2}\right)$$

Substituting these values into the equation of the plane $ax + by - dt = 0$, we get:

$$a\gamma\left(x' + vt'\right) + by' - d\gamma\left(t' + \frac{vx'}{c^2}\right) = 0$$

Simplifying, we get:

$$a\gamma x' + a\gamma vt' + by' - d\gamma t' - d\gamma vx'/c^2 = 0$$

Factoring out the common terms,

$$\left(a\gamma - \frac{dv\gamma}{c^2}\right)x' + by' - (d\gamma - a\gamma v)t' = 0$$

Let's define new constants *A*, *B*, and *D*:

$$A = \left(a\gamma - \frac{dv\gamma}{c^2}\right)$$
$$B = b$$
$$D = d\gamma - a\gamma v$$

Now, we have the equation of the transformed plane in the *S*' frame:

$$Ax' + By' - Dt' = 0$$

b. Transformed plane is still a plane in the *S*' frame

Yes, the transformed plane is still a plane in the *S*' frame.

The equation $Ax' + By' - Dt' = 0$ represents a plane in the *S*' frame, just like the original equation $ax + by - dt = 0$ represented a plane in the *S* frame. This demonstrates the theorem "a plane is a plane under linear transformation" in special relativity, as the Lorentz transformation is a linear transformation between the coordinates of the two frames.

c. Apparent angle of this plane

The apparent angle θ' of the plane when observed from the *S*' frame can be calculated as:

$$\tan(\theta') = \frac{y'}{x'}$$

Substituting the values for x', y'

$$\tan(\theta') = \frac{y}{\gamma(x - vt)}$$

Express *t* in terms of *x* and *y*, using the plane equation $ax + by - dt = 0$

$$t = \frac{ax + by}{d}$$

Substituting this in the equation and simplifying

$$\tan(\theta') = \frac{y}{\gamma\left[x - v\left(\frac{ax+by}{d}\right)\right]} = \frac{y}{\gamma\left(x - \frac{vax}{d} - \frac{vby}{d}\right)}$$

$$\text{tant}\,(\theta') = \frac{y}{\gamma x\left(1 - \frac{va}{d} - \frac{vby}{d}\right)}$$

This is the expression for the apparent angle θ' of the plane when observed from the *S*' frame. The presence of the velocity *v* and the constants *a*, *b*, and *c* in the expression reflects how the angle changes due to the relative motion of the frames and the properties of the plane.

2004

This problem is to be analyzed according to the doctrines of special relativity. An observer A at rest relative to fixed distant stars sees an isotropic distribution of stars in a finite galaxy: the number of stars seen within an element of solid angle $d\Omega$ is $Pd\Omega = \frac{N}{4\pi}d\Omega$, where N is the total number of stars that A can see.

Another observer B moves uniformly along the z-axis relative to A with a large velocity $v = \beta c$. Let θ and φ be respectively the polar (with respect to the z-axis) and azimuthal angles in the inertial frame of B, and let $P'(\theta', \varphi')d\Omega'$ be the number of stars in the element $d\Omega' = \sin\theta' d\theta' d\varphi'$ seen by B.

a. Compute P' as a function of θ', φ', assuming that every star seen by A can be seen by B.

b. Also in particular discuss for the limiting case $\beta \to 1$ what B sees in (1) the forward direction, (2) the backward direction.

(Princeton)

Sol:

a. Consider transformation of the propagation four-vector $k^\alpha = \left(\mathbf{k}, \frac{\omega}{c}\right)$ of the light wave from a star, where $|\mathbf{k}| \equiv k = \frac{\omega}{c}$:

$$k'\cos\theta' = \gamma\left(k\cos\theta - \tfrac{\omega}{c}\beta\right) = \gamma k(\cos\theta - \beta),$$
$$k' = \gamma\left(\tfrac{\omega}{c} - k\beta\cos\theta\right) = \gamma k(1 - \beta\cos\theta).$$

These combine to give

$$\cos\theta' = \frac{\cos\theta - \beta}{1 - \beta\cos\theta},$$

or

$$1 + \beta\cos\theta' = \frac{1 - \beta^2}{1 - \beta\cos\theta}.$$

Differentiating we have

$$\frac{d\cos\theta'}{d\cos\theta} = \frac{1 - \beta^2}{(1 - \beta\cos\theta)^2}.$$

In spherical coordinates

$$d\Omega = \frac{rd\theta \cdot r\sin\theta d\varphi}{r^2} = -d\cos\theta \cdot d\varphi.$$

Thus

$$d\Omega' = \frac{d\cos\theta'}{d\cos\theta}d\Omega = \frac{(1-\beta^2)}{(1-\beta\cos\theta)^2}d\Omega,$$

or, for the inverse transformation,

$$d\Omega = \frac{(1-\beta^2)}{(1+\beta\cos\theta')^2}d\Omega'.$$

As the same stars are seen by both A and B,

$$P'(\theta', \varphi')d\Omega' = P(\theta, \varphi)d\Omega,$$

and so

$$P'(\theta', \varphi') = \frac{N}{4\pi}\frac{d\Omega}{d\Omega'} = \frac{N}{4\pi}\frac{1-\beta^2}{(1+\beta\cos\theta')^2}.$$

b. **1.** For the light from a star in the forward direction of the observer to reach him, it must propagate in the –z direction. Then $\theta' = \pi$, and

$$P'(\theta', \varphi') = \frac{N}{4\pi}\frac{1-\beta^2}{(1-\beta)^2} = \frac{N}{4\pi}\frac{1+\beta}{1-\beta} \to \infty$$

as $\beta \to 1$.

2. For stars in the backward direction of the observer, $\theta' = 0$, and

$$P'(\theta', \varphi') = \frac{N}{4\pi}\frac{1-\beta^2}{(1+\beta)^2} = \frac{N}{4\pi}\frac{1-\beta}{1+\beta} \to 0$$

as $\beta \to 1$.

Hence the stars appear to be highly concentrated in the forward direction and none can be seen in the backward direction.

2005

An observer at rest with respect to fixed distant stars observes an isotropic distribution of stars, $dN = \frac{N}{4\pi}d\Omega$, in any solid angle $d\Omega$. Consider another observer who accelerates with a constant (relative to his instantaneous inertial frame) acceleration a. If he starts at rest at time $t = 0$ with respect to the fixed observer, determine the distribution of stars, $dN' = N'(\theta', \varphi')d\Omega'$, that he observes at time t', where t' is the time measured in his frame.

(Princeton)

Sol: Let Σ, Σ' respectively be the frame of the observer at rest with respect to distant stars and the instantaneous rest frame of the accelerated observer, with the *x-axis* along the direction of relative motion. At $t = t' = 0$, the latter observer starts at rest with respect to the former. Consider the instant when Σ' moves with velocity v relative to Σ.

The acceleration four-vector of a point in an inertial frame is defined as

$$a^\alpha = \frac{\gamma_0}{c}\frac{du^\alpha}{dt}$$

with u^α being the velocity four-vector $\left(\frac{\gamma_0}{c}\mathbf{u},\ \gamma_0\right)$, where $\mathbf{u}$ is the velocity of the point, $\gamma_0 = \left(1 - \frac{u^2}{c^2}\right)^{-\frac{1}{2}}$. For the accelerated observer, $\mathbf{u}' = 0, \gamma_0' = 1, \mathbf{a}' = (a, 0, 0)$, $a^{\alpha'} = \left(\frac{a}{c^2},\ 0,\ 0,\ 0\right)$ in Σ' and $\mathbf{u} = (v, 0, 0), \gamma_0 = \gamma, a^\alpha = \left(\frac{\gamma}{c^2}\frac{d}{dt}(\gamma v), 0, 0, \frac{\gamma}{c}\frac{d\gamma}{dt}\right)$ in Σ. Four-vector transformation then gives

$$\frac{\gamma}{c^2}\frac{d}{dt}(\gamma v) = \gamma\frac{a}{c^2},$$

or

$$\frac{d}{dt}(\gamma v) = a,$$

giving

$$\gamma v = at$$

since a is constant and $v = 0$ at $t = 0$. It follows that

$$v = \frac{at}{\sqrt{1 + (\frac{at}{c})^2}}.$$

We still have to express t in terms of t'. Lorentz transformation gives

$$dt = \gamma\left(dt' + \frac{v}{c^2}dx'\right) = \gamma dt'.$$

Hence

$$t' = \int_0^{t'} dt' = \int_0^t \sqrt{1 - \left(\frac{v}{c}\right)^2}\, dt = \int_0^t \frac{dt}{\sqrt{1 + (\frac{at}{c})^2}} = \frac{c}{a}\sinh^{-1}\left(\frac{at}{c}\right),$$

or

$$t = \frac{c}{a}\sinh\left(\frac{at'}{c}\right),$$

and

$$v = \frac{c\sinh(at'/c)}{\sqrt{1 + \sinh^2(at'/c)}} = \frac{c\sinh(at'/c)}{\cosh(at'/c)} = c\tanh(at'/c).$$

Using spherical coordinates with the polar axis along the x'-axis and the solution of **Problem 2004**, we have

$$N(\theta', \varphi') = \frac{N}{4\pi}\frac{d\Omega}{d\Omega'} = \frac{N}{4\pi}\frac{1-\beta^2}{(1+\beta\cos\theta')^2}$$

with $\beta = \tanh\left(\frac{at'}{c}\right)$, or

$$dN' = \frac{N}{4\pi}\frac{\left[1-\tanh\left(\frac{at'}{c}\right)\right]}{\left[1+\tanh\left(\frac{at'}{c}\right)\cos\theta'\right]^2}\sin\theta' d\theta' d\varphi'.$$

2006

In flat space-time, a spherical star has total luminosity L. Thus, at a distance r from its center, an observer at rest relative to the star sees an energy flux of $L/4\pi r^2$.

a. Using spherical polar coordinates centered on the star, write down the energy-momentum tensor for the radiation.

b. An observer at radius r is in motion with velocity $\mathbf{v}$; $v_{\parallel}$ is the component of $\mathbf{v}$ along the line joining the observer to the center of the star, and $v_{\perp}$ is the component perpendicular to that line. What will the observer measure the energy flux to be in terms of L, r, $v_{\parallel}$ and $v_{\perp}$?

c. A photon of frequency ν_c in the rest frame of the star is emitted from its surface. What frequency will the observer in part (b) measure for this photon?

(Princeton)

Sol:

a. The total luminosity of the star is

$$L = \oint \mathbf{N}\cdot d\mathbf{S} = \int \mathbf{N}\cdot \hat{\mathbf{r}} r^2 d\Omega = 4\pi r^2 N,$$

where $\mathbf{N} = \mathbf{E}\times\mathbf{H}$ is the Poynting vector, which gives the energy flux, of the electromagnetic field of the star. The above relation suggests that

$$\mathbf{N} = N(r)\hat{\mathbf{r}},$$

i.e., $\mathbf{N}$ is radial and independent of the angles θ, φ of a spherical coordinate system with origin at the center. At large distances from the source the electromagnetic waves can be considered plane. Hence we can take

$$E = E_\theta, \qquad H = H_\varphi, \qquad \sqrt{\epsilon_0}\,E = \sqrt{\mu_0}\,H.$$

The contravariant energy-momentum tensor $T^{\alpha\beta}$ has components

$$\begin{aligned}
T^{ij} &= T^{ji} = -\epsilon_0 E_i E_j - \mu_0 H_i H_j + U\delta_{ij}, \\
T^{4j} &= T^{j4} = \frac{N_j}{c}, \\
T^{44} &= U,
\end{aligned}$$

where $i, j = 1, 2, 3$, $\mathbf{N} = \mathbf{E} \times \mathbf{H}$ is the Poynting vector and $U = \frac{1}{2}(\epsilon_0 E^2 + \mu_0 H^2)$ is the energy density, the two being related by $N = Uc$. Thus

$$\begin{aligned}
T^{11} &= -\epsilon_0 E_r^2 - \mu_0 H_r^2 + U = U = \frac{L}{4\pi r^2 c}, \\
T^{22} &= -\epsilon_0 E_\theta^2 - \mu_0 H_\theta^2 + \frac{1}{2}(\epsilon_0 E_\theta^2 + \mu_0 H_\varphi^2) \\
&= -\frac{1}{2}\epsilon_0 E_\theta^2 + \frac{1}{2}\mu_0 H_\varphi^2 = 0, \\
T^{33} &= 0, \\
T^{44} &= U = \frac{L}{4\pi r^2 c}, \\
T^{12} &= T^{21} = -\epsilon_0 E_r E_\theta - \mu_0 H_r H_\theta = 0, \\
T^{13} &= T^{31} = T^{23} = T^{32} = 0, \\
T^{41} &= T^{14} = \frac{N_r}{c} = \frac{L}{4\pi r^2 c}, \\
T^{42} &= T^{24} = \frac{N_\theta}{c} = 0, \\
T^{43} &= T^{34} = \frac{N_\varphi}{c} = 0,
\end{aligned}$$

since $E_r = E_\varphi = H_r = H_\theta = 0$, $N_\theta = N_\varphi = 0$.

b. Decompose **E**, **B** into components parallel and perpendicular to the velocity **v** of Σ' relative to Σ:

$$\mathbf{E} = \mathbf{E}_\parallel + \mathbf{E}_\perp, \quad \mathbf{B} = \mathbf{B}_\parallel + \mathbf{B}_\perp.$$

Lorentz transformation gives

$$\mathbf{E}'_\parallel = \mathbf{E}_\parallel, \;\; \mathbf{B}'_\parallel = \mathbf{B}_\parallel,$$

$$\mathbf{E}'_\perp = \gamma(\mathbf{E}_\perp + \mathbf{v} \times \mathbf{B}_\perp), \;\; \mathbf{B}'_\perp = \gamma\left(\mathbf{B}_\perp - \frac{\mathbf{v} \times \mathbf{E}_\perp}{c^2}\right).$$

These can be combined as follows:

$$\begin{aligned}
\mathbf{E}' &= \gamma[(\mathbf{E} - \mathbf{E}_\parallel) + \mathbf{v} \times (\mathbf{B} - \mathbf{B}_\parallel)] + \mathbf{E}_\parallel \\
&= \gamma(\mathbf{E} + \mathbf{v} \times \mathbf{B}) - (\gamma - 1)\mathbf{E}_\parallel - \gamma \mathbf{v} \times \mathbf{B}_\parallel \\
&= \gamma(\mathbf{E} + \mathbf{v} \times \mathbf{B}) - (\gamma - 1)\left(\frac{\mathbf{E} \cdot \mathbf{v}}{v^2}\right)\mathbf{v}, \\
\mathbf{B}' &= \gamma\left(\mathbf{B} - \frac{\mathbf{v} \times \mathbf{E}}{c^2}\right) - (\gamma - 1)\left(\frac{\mathbf{B} \cdot \mathbf{v}}{v^2}\right)\mathbf{v},
\end{aligned}$$

as $\mathbf{v} \times \mathbf{B}_\parallel = 0$, $\mathbf{v} \times \mathbf{E}_\parallel = 0$.

The quantity $\mathbf{E}\cdot\mathbf{B}$ is invariant under Lorentz transformation. In Σ the waves are plane so that $\mathbf{E}\cdot\mathbf{B}=0$. Then $\mathbf{E}'\cdot\mathbf{B}'=0$, i.e., the waves are plane in all other inertial frames. Hence

$$\mu_0^2N^2=(\mathbf{E}\times\mathbf{B})^2=E^2B^2-(\mathbf{E}\cdot\mathbf{B})^2=E^2B^2.$$

Consider

$$\begin{aligned}E'^2&=\gamma^2[E^2+(\mathbf{v}\times\mathbf{B})^2+2\mathbf{E}\cdot\mathbf{v}\times\mathbf{B}]+(\gamma-1)^2\left(\frac{\mathbf{E}\cdot\mathbf{v}}{v^2}\right)^2v^2\\&\quad-2\gamma(\gamma-1)\left(\frac{\mathbf{E}\cdot\mathbf{v}}{v^2}\right)\mathbf{v}\cdot(\mathbf{E}+\mathbf{v}\times\mathbf{B})\\&=\gamma^2\left[E^2+v^2B^2-(\mathbf{v}\cdot\mathbf{B})^2-2v_{\parallel}EB-\left(\frac{\mathbf{v}\cdot\mathbf{E}}{c}\right)^2\right]\\&=\gamma^2\left[E^2-\frac{2v_{\parallel}E^2}{c}+\frac{v_{\parallel}^2E^2}{c^2}+\left(\frac{v_{\perp}E}{c}\right)^2-\left(\frac{\mathbf{v}_{\perp}\cdot\mathbf{E}}{c}\right)^2-(\mathbf{v}_{\perp}\cdot\mathbf{B})^2\right]\\&=\gamma^2\left[E^2\left(1-\frac{v_{\parallel}}{c}\right)^2+\left(\frac{\mathbf{v}_{\perp}\times\mathbf{E}}{c}\right)^2-(v_{\perp}\cdot\mathbf{B})^2\right].\end{aligned}$$

Using

$$\begin{gathered}(\mathbf{v}\times\mathbf{B})^2=v^2B^2-(\mathbf{v}\cdot\mathbf{B})^2,\\ \mathbf{E}\cdot\mathbf{v}\times\mathbf{B}=-\mathbf{v}\cdot\mathbf{E}\times\mathbf{B}=-v_{\parallel}|\mathbf{E}\times\mathbf{B}|=-v_{\parallel}EB,\\ \mathbf{v}\cdot\mathbf{v}\times\mathbf{B}=0,\\ \mathbf{v}\cdot\mathbf{E}=(v_{\perp}+v_{\parallel})\cdot\mathbf{E}=\mathbf{v}_{\perp}\cdot\mathbf{E},\\ \mathbf{v}\cdot\mathbf{B}=\mathbf{v}_{\perp}\cdot\mathbf{B},\\ v^2=v_{\parallel}^2+v_{\perp}^2,\quad\gamma^2-1=\frac{\gamma^2v^2}{c^2}.\end{gathered}$$

As $\mathbf{E}$, $\mathbf{B}$, $\hat{\mathbf{r}}$ are mutually perpendicular and $E=cB$, we have

$$(\mathbf{v}_{\perp}\cdot\mathbf{B})^2=\left(\frac{\mathbf{v}_{\perp}\cdot\mathbf{E}\times\hat{\mathbf{r}}}{c}\right)^2=\left(\frac{\mathbf{v}_{\perp}\times\mathbf{E}\cdot\hat{\mathbf{r}}}{c}\right)^2=\left(\frac{\mathbf{v}_{\perp}\times\mathbf{E}}{c}\right)^2.$$

Hence

$$E'^2=\gamma^2\left(1-\frac{v_{\parallel}}{c}\right)^2E^2,$$

and similarly

$$B'^2=\gamma^2\left(1-\frac{v_{\parallel}}{c}\right)^2B^2.$$

Therefore

$$N'^2=\frac{E'^2B'^2}{\mu_0^2}=\gamma^4\left(1-\frac{v_{\parallel}}{c}\right)^4\frac{E^2B^2}{\mu_0^2},$$

or

$$N' = \left(1 - \frac{v_\parallel^2 + v_\perp^2}{c^2}\right)^{-1}\left(1 - \frac{v_\parallel}{c}\right)^2 \frac{L}{4\pi r^2}.$$

c. The last component of the propagation four-vector $k^\alpha = \left(\mathbf{k}, \frac{\omega}{c}\right)$, where $k = \frac{\omega}{c}$, transforms according to

$$\frac{\omega'}{c} = \gamma\left(\frac{\omega}{c} - \frac{v}{c}k\cos\theta\right)$$

or

$$\omega' = \omega\gamma\left(1 - \frac{v_\parallel}{c}\right).$$

If a photon is emitted from the surface of the star with frequency ν_0, the observed frequency is thus

$$\nu' = \nu_0\left(1 - \frac{v_\perp^2 + v_\parallel^2}{c^2}\right)^{-\frac{1}{2}}\left(1 - \frac{v_\parallel}{c}\right).$$

However ν_0 is not the true frequency of the photon because of gravitational shift given by

$$\nu_0 = \nu_c\left(1 - \frac{2GM}{rc^2}\right)^{\frac{1}{2}},$$

where M is the mass and r the radius of the star and G is the gravitational constant. Hence

$$\nu' = \nu_c\left(1 - \frac{2GM}{rc^2}\right)^{\frac{1}{2}}\left(1 - \frac{v_\perp^2 + v_\parallel^2}{c^2}\right)^{-\frac{1}{2}}\left(1 - \frac{v_\parallel}{c}\right).$$

2007

Consider the collision of n bodies in a Lorentz frame S. The bodies are small enough to be adequately characterized by their rest masses and velocities, and they collide simultaneously at the same point. Assume the velocities and rest masses to be $\mathbf{v}_\alpha$ and m_α ($\alpha = 1, ..., n$) before collision, and $\mathbf{v}'_\alpha$ and m'_α ($\alpha = 1, ..., n$) after collision. (All m_α and m'_α are positive.)

For given $\mathbf{v}_\alpha$ and m_α call a collision maximally inelastic in S if after the collision the sum of kinetic energies of all bodies has the smallest value compatible with energy and momentum conservation. (The kinetic energy of a body is

$m_\alpha c^2\left[\left(1 - \frac{v_\alpha^2}{c^2}\right)^{-1/2} - 1\right].$)

a. Find v'_α and m'_α — to the extent that they are determinate — for a maximally inelastic collision.

b. Is the notion of a maximally inelastic collision Lorentz invariant? Explain.

(*Princeton*)

Sol:

a. The total kinetic energy after the collision is

$$T = \sum (m'_\alpha \gamma'_\alpha c^2 - m'_\alpha c^2)$$

with $\gamma'_\alpha = \left(1 - \frac{v'^2_\alpha}{c^2}\right)^{-\frac{1}{2}}$. As there is no external force acting on the system, energy is conserved:

$$\sum m'_\alpha \gamma'_\alpha c^2 = \sum m_\alpha \gamma_\alpha c^2 .$$

Thus

$$T = \sum (m_\alpha \gamma_\alpha c^2 - m'_\alpha c^2).$$

For T to be minimum, we require $\sum m'_\alpha$ to be a maximum compatible with conservation of energy and momentum.

Consider the collision in the center of mass frame S_{cm}, which is given by its velocity relative to S:

$$\mathbf{u} = \frac{\sum m_\alpha \gamma_\alpha \mathbf{v}_\alpha}{\sum m_\alpha \gamma_\alpha} .$$

Denoting all quantities in this frame by a bar we have

$$\begin{aligned} \overline{T} &= \sum (m_\alpha \overline{\gamma}'_\alpha c^2 - m'_\alpha c^2) \\ &= \sum (m_\alpha \overline{\gamma}_\alpha c^2 - m'_\alpha c^2). \end{aligned}$$

As $\overline{T} \geq 0$, $\sum m'_\alpha \leq \sum m_\alpha \overline{\gamma}_\alpha$. Hence $\sum m'_\alpha$ has the maximum value $\sum m_\alpha \overline{\gamma}_\alpha$.

For a system the quantity $E^2 - P^2c^2$ is Lorentz invariant. Thus

$$\left(\sum m_\alpha \gamma_\alpha c^2\right)^2 - \left(\sum m_\alpha \gamma_\alpha \mathbf{v}_\alpha\right)^2 c^2 = \left(\sum m_\alpha \overline{\gamma}_\alpha c^2\right)^2,$$

since $\overline{P} = 0$ in S_{cm}, i.e.,

$$\left(\sum m_\alpha \gamma_\alpha\right)^2 - \left(\sum m_\alpha \gamma_\alpha\right)^2 \frac{u^2}{c^2} = \left(\sum m_\alpha \overline{\gamma}_\alpha\right)^2,$$

or

$$\sum m_\alpha \overline{\gamma}_\alpha = \left(1 - \frac{u^2}{c^2}\right)^{\frac{1}{2}} \sum m_\alpha \gamma_\alpha .$$

Therefore $\sum m'_\alpha$ has the maximum value

$$\left(1-\frac{u^2}{c^2}\right)^{\frac{1}{2}}\sum m_\alpha\gamma_\alpha,$$

which determines $\sum m'_\alpha$ for a maximally inelastic collision. Momentum conservation

$$\sum m'_\alpha\gamma'_\alpha\mathbf{v}'_\alpha=\sum m_\alpha\gamma_\alpha\mathbf{v}_\alpha=\mathbf{u}\sum m_\alpha\gamma_\alpha$$

determines the collective velocity **u**, which is the same before and after the collision.

b. The notion of a maximally inelastic collision is Lorentz invariant as a Lorentz transformation has been used to transform to the center of mass frame and this does not affect the internal energy arising from the relative motion of the particles of the system.

2.2

GENERAL RELATIVITY (2008–2023)

2008

a. What is the form of the "global extension" of Newton's first law of motion in general relativity?
Identify all terms in the generalized equation of motion.

b. Show the correspondence between the equation of motion in (a) and Newton's law of gravity in a linear approximation to first order in *v/c*.
The following expression may be useful in answering part (b):

$$\Gamma^{\rho}_{\mu\nu} = \frac{1}{2} g^{\rho\lambda}\left(\partial_\nu g_{\lambda\mu} + \partial_\mu g_{\lambda\nu} - \partial_\lambda g_{\mu\nu}\right).$$

(*SUNY, Buffalo*)

Sol:

a. In special relativity, a particle subject to no force will have a constant four-velocity. That is, $\frac{du^\alpha}{d\tau} = 0$. This is exactly Newton's first law or law of inertia.
In the general covariant form this becomes $\frac{Du^\alpha}{D\tau} = 0$, or

$$\frac{du^\alpha}{d\tau} + \Gamma^\alpha_{\mu\nu}\frac{dx^\mu}{d\tau}\frac{dx^\nu}{d\tau} = 0. \qquad (1)$$

On the left-hand side, the first term denotes the four-acceleration of the particle, the second term represents the action of a gravitational field of intensity $\Gamma^\alpha_{\mu\nu}$.

b. In the absence of gravitational field, the space is flat or Euclidean. Use rectangular Cartesian coordinates and let

$$x^0 = ct, \quad x^1 = x, \quad x^2 = y, \quad x^3 = z.$$

The space-time line element is

$$ds^2 = \eta_{\mu\nu} dx^\mu dx^\nu,$$

with $\eta_{\mu\nu} = -\delta_{\mu\nu}$, except $\eta_{00} = 1$.

When gravitational field is present, the space is curved or Riemannian with line element

$$ds^2 = g_{\mu\nu} dx^\mu dx^\nu,$$

where $g_{\mu\nu}$ is symmetric and diagonal when rectangular Cartesian coordinates are used. If the field is weak,

$$g_{\mu\nu} = \eta_{\mu\nu} + h_{\mu\nu},$$

where $h_{\mu\nu} \ll 1$ are functions of the space coordinates: For small velocities $v \ll c$,

$$\begin{aligned}\left(\frac{ds}{d\tau}\right)^2 &= (\eta_{\mu\nu} + h_{\mu\nu}) \frac{dx^\mu}{d\tau}\frac{dx^\nu}{d\tau} \\ &\approx \left(\frac{dx^0}{d\tau}\right)^2 - \left(\frac{dx^j}{dt}\right)^2 \left(\frac{dt}{d\tau}\right)^2 \\ &= (c^2 - v^2)\left(\frac{dt}{d\tau}\right)^2 \approx c^2 \left(\frac{dt}{d\tau}\right)^2,\end{aligned}$$

or

$$ds \approx cdt,$$

and

$$\frac{dx^j}{ds} \simeq \frac{1}{c}\frac{dx^j}{dt} \approx 0, \quad \frac{dx^0}{ds} \approx 1, \quad (j = 1, 2, 3).$$

The motion of the particle is given by the geodesic equation

$$\frac{d^2 x^\alpha}{ds^2} + \Gamma^\alpha_{\mu\nu} \frac{dx^\mu}{ds}\frac{dx^\nu}{ds} = 0.$$

For a weak field and low velocities this gives approximately

$$\begin{aligned}\frac{d^2 x^\alpha}{ds^2} &\approx -\Gamma^\alpha_{00}\left(\frac{dx^0}{ds}\right)^2 \\ &\approx -\Gamma^\alpha_{00} \\ &= -\frac{1}{2} g^{\alpha\lambda}\left(\frac{\partial g_{\lambda 0}}{\partial x^0} + \frac{\partial g_{0\lambda}}{\partial x^0} - \frac{\partial g_{00}}{\partial x^\lambda}\right) \\ &= \frac{1}{2} g^{\alpha\alpha}\left(2\frac{\partial g_{\alpha 0}}{\partial x^0} - \frac{\partial g_{00}}{\partial x^\alpha}\right).\end{aligned}$$

As $g^{\alpha\alpha} = \frac{1}{g_{\alpha\alpha}}$ for diagonal $g_{\mu\nu}$, we have for $i = 1, 2, 3$

$$\frac{1}{c^2}\frac{d^2x^i}{dt^2} = \frac{1}{2(1+h_{ii})}\frac{\partial g_{00}}{\partial x^i}$$
$$\approx \frac{1}{2}(1-h_{ii})\frac{\partial h_{00}}{\partial x^i},$$

or

$$\frac{d^2x^i}{dt^2} \approx \frac{c^2}{2}\frac{\partial h_{00}}{\partial x^i},$$

retaining only terms of first order in $h_{\mu\nu}$ and their derivatives. In vector form this is

$$\frac{d^2\mathbf{r}}{dt^2} \approx \frac{c^2}{2}\nabla h_{00},$$

which is the same as Newton's equation

$$\frac{d^2\mathbf{r}}{dt^2} = -\nabla\Phi$$

if we set

$$h_{00} \approx -\frac{2\Phi}{c^2} + \text{constant}.$$

2009

Imagine a futuristic scenario in a distant part of the universe where a massive neutron star, known as NS-X1, with a mass of $2.0 \times 10^{30}\ kg$, is located. Astronomers have placed a probe in a stable circular orbit at a radius of $r = 10^6\ m$ from NS-X1 to study its gravitational effects. The system is accurately described by the Schwarzschild metric,

$$ds^2 = -\left(1 - \frac{2GM}{c^2 r}\right)c^2dt^2 + \left(1 - \frac{2GM}{c^2 r}\right)^{-1}dr^2 + r^2(d\theta^2 + \sin^2\theta\, d\varphi^2)$$

a. Determine the time dilation experienced by the probe as it orbits the neutron star.

b. Calculate the annual precession of the orbit.

c. Explain the implications of the abovementioned results.

Sol:

a. Time dilation: To calculate the time dilation experienced by the probe as it orbits the neutron star, we need to consider the Schwarzschild metric's impact on the probe's proper time. The Schwarzschild metric's temporal component, g_{tt}, affects the proper time experienced by the probe. The proper time, dτ, experienced by the probe is related to the coordinate time, dt, by $d\tau = \sqrt{g_{tt}}\, dt$, where $g_{tt} = 1 - \frac{2GM}{c^2 r}$.

$$d\tau = \sqrt{\left(1 - \frac{2GM}{c^2 r}\right)}\, dt$$

Now, let's calculate the time dilation factor D.

Rearranging the formula gives:

$$D = \frac{d\tau}{dt} = \sqrt{\left(1 - \frac{2GM}{c^2 r}\right)}$$

Substituting the values,

$$D = \sqrt{\left(1 - \frac{2 \times 6.6743 \times 10^{-11} \times 2 \times 10^{30}}{(3 \times 10^{8})^2 \times 10^{6}}\right)} = \sqrt{1 - (2.97 \times 10^{-3})}$$

$$D = 1 - \frac{2.97 \times 10^{-3}}{2} = 1 - (1.485 \times 10^{-3}) = 0.998515$$

The probe's time is dilated by a factor of $D = 0.998515$ compared to a distant observer, as it orbits the massive neutron star NS-X1.

b. **Annual precession of the orbit:** To calculate the precession of the orbit using the Schwarzschild metric, we can use the following formula considering one complete revolution of the probe:

$$\Delta\varphi = 6\pi \frac{GM}{ac^2(1 - e^2)}$$

Where:

- Δϕ is the advance in radians per orbit.
- *G* is the gravitational constant.
- *M* is the mass of the central object (NS-X1 in this case).
- *c* is the speed of light.
- *a* is the semimajor axis of the probe's orbit.
- *e* is the eccentricity of the probe's orbit.

We need to determine the number of revolutions the probe completes in a year and then multiply $\Delta\phi$ by that number to find the annual advance. Since the probe is in a circular orbit, $e = 0$, and we can use the formula for the circumference of a circle to find the number of revolutions per year.

The circumference of the circular orbit is given by: $C = 2\pi a$

The time it takes to complete one revolution is the orbital period (T), and it can be calculated using the formula: $T = \frac{C}{v} = \frac{2\pi a}{v}$.

Where v is the orbital speed, which is given by: $v = \sqrt{\frac{GM}{a}}$.

On substitution, $T = \frac{2\pi a}{\sqrt{\frac{GM}{a}}} = 2\pi\sqrt{\frac{a^3}{GM}}$

Now, let's calculate the number of revolutions per year ($N_{rev/y}$):

$$N_{rev/y} = \frac{1}{2\pi\sqrt{\frac{a^3}{GM}}}$$

Now that we know the number of orbits per year, we can calculate the annual precession:

$$\Delta\varphi_{annual} = N_{rev/y} \times \Delta\varphi$$

$$\Delta\varphi_{annual} = \frac{1}{2\pi\sqrt{\frac{a^3}{GM}}} \times 6\pi\frac{GM}{ac^2(1-e^2)} = \frac{3\sqrt{(\frac{GM}{a})^3}}{c^2 a} = 3\sqrt{\frac{(GM)^3}{c^4 a^5}}$$

On substitution of the known values:

$$\Delta\varphi_{annual} = 3\sqrt{\frac{(GM)^3}{c^4 a^5}} = 3\sqrt{\frac{(6.6743 \times 10^{-11} \times 2 \times 10^{30})^3}{(3 \times 10^8)^4 (10^6)^5}}$$

$$\Delta\varphi_{annual} = 0.05\ rad/year = 2.86\ deg/year$$

c. **Implications of the above mentioned results:** The result obtained in part (a) means that for every second experienced by a distant observer, the probe in orbit around NS-X1 experiences about 0.998515 seconds due to the intense gravitational field near the neutron star. This time dilation effect is a direct consequence of the Schwarzschild metric and general relativity's predictions, and it has been observed in various astrophysical contexts. Time dilation effects can impact the synchronization of events, and these effects must be accounted for when conducting experiments or observations in such environments.

The magnitude of the annual precession of the orbit indicates the presence of a very strong gravitational field near the neutron star NS-X1. Neutron stars are known for their extreme density, and this observation highlights the profound influence they have on the surrounding spacetime. In summary, the annual advance of the orbit around the massive neutron star NS-X1 is a valuable observation that contributes to our understanding of extreme gravitational environments, confirming the predictions of general relativity, and has practical implications for scientific research and education. It serves as a reminder of the remarkable precision and consistency of physical laws even in the most extreme cosmic settings.

2010

The Schwarzschild metric describing the space-time outside a spherical mass M is given by

$$ds^2 = -\left(1 - \frac{2GM}{c^2 r}\right)c^2 dt^2 + \left(1 - \frac{2GM}{c^2 r}\right)^{-1} dr^2 + r^2(d\theta^2 + \sin^2\theta d\varphi^2).$$

a. Find the radius at which a photon can describe a circular orbit.

b. What would a fixed observer at this radius measure as the period of the photon orbit?

c. If this observer flashes a signal each time the photon passes him, at what interval would an observer at infinity who is stationary measure these flashes?

(*Princeton*)

Sol:

a. Use a coordinate frame with origin at the center of the spherical mass and let

$$x^0 = ct, \quad x^1 = r, \quad x^2 = \theta, \quad x^3 = \varphi.$$

Then the Schwarzschild metric tensor has components

$$\begin{aligned} g_{00} &= -\left(1 - \tfrac{k}{r}\right), \quad g_{11} = \left(1 - \tfrac{k}{r}\right)^{-1}, \\ g_{22} &= r^2, \quad g_{33} = r^2 \sin^2\theta, \\ g_{\mu\nu} &= 0 \text{ for } \mu \neq \nu, \end{aligned}$$

where $k = \frac{2GM}{c^2}$. The world-line of the photon is a null geodesic given by

$$\begin{aligned} \frac{d}{ds}\left(g_{\alpha\beta}\frac{dx^\beta}{ds}\right) &= \frac{1}{2}\frac{\partial g_{\mu\lambda}}{\partial x^\alpha}\frac{dx^\mu}{ds}\frac{dx^\lambda}{ds}, \\ g_{\mu\nu}\frac{dx^\mu}{ds}\frac{dx^\nu}{ds} &= 0, \end{aligned}$$

where s is some independent variable. As the mass is spherically symmetric, we can choose coordinates such that the orbit of the photon is in the plane $\theta = \frac{\pi}{2}$.

Since $g_{\mu\nu}$ do not depend on x^3 explicitly, we have

$$g_{3\beta}\frac{dx^\beta}{ds} = g_{33}\frac{dx^3}{ds} = r^2\frac{d\varphi}{ds} = h, \quad \text{a constant.}$$

They also do not depend on x^0 explicitly, so

$$g_{0\beta}\frac{dx^\beta}{ds} = g_{00}\frac{dx^0}{ds} = -c\left(1-\frac{k}{r}\right)\frac{dt}{ds} = b, \quad \text{a constant.}$$

We also have

$$\begin{aligned} g_{\mu\nu}\frac{dx^\mu}{ds}\frac{dx^\nu}{ds} &= g_{00}\left(\frac{dx^0}{ds}\right)^2 + g_{11}\left(\frac{dx^1}{ds}\right)^2 + g_{33}\left(\frac{dx^3}{ds}\right)^2 \\ &= -c^2\left(1-\frac{k}{r}\right)\left(\frac{dt}{ds}\right)^2 + \frac{1}{\left(1-\frac{k}{r}\right)}\left(\frac{dr}{ds}\right)^2 + r^2\left(\frac{d\varphi}{ds}\right)^2 = 0. \end{aligned}$$

Hence

$$\left(\frac{dr}{ds}\right)^2 + \left(1-\frac{k}{r}\right)\frac{h^2}{r^2} - b^2 = 0.$$

Letting $\frac{1}{r} = u$ we have

$$\begin{aligned} \frac{du}{d\varphi} &= -\frac{1}{r^2}\frac{dr}{d\varphi}, \\ \frac{dr}{ds} &= \frac{dr}{d\varphi}\frac{d\varphi}{ds} = \frac{h}{r^2}\frac{dr}{d\varphi} = -h\frac{du}{d\varphi}, \end{aligned}$$

and Eq. (4) becomes

$$h^2\left(\frac{du}{d\varphi}\right)^2 + (1-ku)h^2u^2 - b^2 = 0.$$

Differentiating both sides with respect to φ, we obtain

$$\frac{d^2u}{d\varphi^2} + u - \frac{3ku^2}{2} = 0,$$

i.e.,

$$\frac{d^2u}{d\varphi^2} + u - \frac{3GM}{c^2}u^2 = 0.$$

For the orbit to be circular, $u = \frac{1}{r} = \text{constant}$. The last equation then gives

$$r = \frac{3GM}{c^2}.$$

for the radius of the circular orbit.

b. For the photon moving in a circular orbit $r = \frac{3GM}{c^2}$,

$$(ds)^2 = -\left(1 - \frac{2GM}{c^2 r}\right)c^2(dt)^2 + r^2(d\varphi)^2 = 0,$$

giving

$$dt = \frac{\sqrt{3}\, r}{c} d\varphi$$

and hence the period in the coordinate time

$$\Delta t = \frac{\sqrt{3}\, r}{c} \int_0^{2\pi} d\varphi = \frac{2\pi\sqrt{3}}{c} \frac{3GM}{c^2} = \frac{6\sqrt{3}\,\pi GM}{c^3}.$$

For the observer at rest at $r = \frac{3GM}{c^2}$, a stationary standard clock gives

$$-(cd\tau)^2 = -\left(1 - \frac{2GM}{c^2 r}\right)(cdt)^2,$$

or

$$d\tau = \frac{dt}{\sqrt{3}}.$$

Hence the observer will measure the period, according to the standard clock, as

$$\Delta\tau = \frac{\Delta t}{\sqrt{3}} = \frac{6\pi GM}{c^3}.$$

c. For the stationary observer at infinity, a standard clock at rest in his neighborhood gives

$$-\left(cd\,\tau_\infty^2\right) = -(cdt)^2,$$

as $r = \infty$, thus

$$d\tau_\infty = dt,$$

where t is the coordinate time. Hence according to this observer,

$$\Delta\,\tau_\infty = \Delta t = \frac{6\sqrt{3}\,\pi GM}{c^3}.$$

2011

a. Write down the geodesic equation.

b. What is the physical significance of

(1) time-like geodesics? (2) null geodesics?

c. With a certain choice of coordinate variables ρ, θ, φ, ψ a solution of Einstein's equations has the metric

$$ds^2 = -\frac{1}{U(p)}d\rho^2 + 4U(\rho)l^2(d\psi + \cos\theta d\varphi)^2 + (\rho^2 + l^2)(d\theta^2 + \sin^2\theta d\varphi^2),$$

where

$$\begin{aligned} -\infty &\le \rho \le \infty, \\ 0 &\le \theta \le \pi, \\ 0 &\le \varphi \le 2\pi, \\ 0 &\le \psi \le 4\pi, \end{aligned}$$

θ, φ, and ψ are cyclic coordinates, and

$$U(p) = -1 + \frac{2(m\rho + l^2)}{\rho^2 + l^2},$$

where m and l are constant parameters.

1. Find the equations which describe geodesics in the plane $\varphi = 0$, $\theta = \frac{\pi}{2}$.
2. Solving for null geodesic equations in the plane of case (1), show that there are null geodesics which lie in the surface $\varphi = 0$, $\theta = \frac{\pi}{2}$, $\rho = m + \sqrt{m^2 + l^2}$. In a few words, what is it that is physically interesting about these null geodesics?

   Information: The Christoffel symbols are defined by $\Gamma^{g}_{\beta v} = \frac{1}{2}g^{\alpha\delta}(\partial_\beta g_{\delta v} - \partial_\delta g_{\beta v} + \partial_v g_{\delta\beta})$.

(*Princeton*)

Sol:

a. The geodesic equation is

$$\frac{d^2x^\mu}{ds^2} + \Gamma^\mu_{\alpha\beta}\frac{dx^\alpha}{ds}\frac{dx^\beta}{ds} = 0.$$

b. **1.** Time-like geodesics are those along which the time order of two events cannot be changed by a coordinate transformation.

2. Null geodesics are paths of light rays or photon in a space-time and are characterized by $ds = 0$, or

$$g_{\mu v}\frac{dx^\mu}{ds}\frac{dx^v}{ds} = 0.$$

The parameter s in the equation is to be considered as an unspecified independent variable.

c. 1. Let

$$x^0 = \rho, \quad x^1 = \psi, \quad x^2 = \theta, \quad x^3 = \varphi.$$

The metric tensor has nonzero components

$$\begin{aligned} g_{00} &= -\frac{1}{U} \quad g_{11} = 4Ul^2, \quad g_{22} = \rho^2 + l^2, \\ g_{33} &= 4Ul^2 \cos^2\theta + (\rho^2 + l^2)\sin^2\theta, \\ g_{23} &= g_{32} = 4Ul^2 \cos\theta, \end{aligned}$$

with

$$U = -1 + \frac{2(mp + l^2)}{\rho^2 + l^2}.$$

It is more convenient to use an alternative form of the geodesic equation:

$$\frac{d}{ds}\left(g_{\mu\nu}\frac{dx^\nu}{ds}\right) = \frac{1}{2}\frac{\partial g_{\alpha\beta}}{\partial x^\mu}\frac{dx^\alpha}{ds}\frac{dx^\beta}{ds}.$$

As all the $g_{\mu\nu}$ do not depend on x^1 explicitly and $g_{1\nu} = 0$ for $\nu \neq 1$, the above gives the first integral

$$g_{11}\frac{d\psi}{ds} = k, \qquad \text{a constant.}$$

Introduce the local proper time τ defined by $-d\tau^2 = ds^2$ and interpret ψ $(-\infty \le \psi \le \infty)$ as the coordinate time τ. The metric gives

$$-d\tau^2 = ds^2 = g_{11}\, d\psi^2.$$

ρ is interpreted as radial distance $(0 \le \rho \le \infty)$. Let $\rho \to \infty$, then as $U \to -1$, $g_{11} \to -4l^2$, we find $k^2 = \left(g_{11}\frac{d\psi}{ds}\right)^2 = g_{11}(\infty) = -4l^2$. If the geodesic is confined to the plane $\varphi = 0$, $\theta = \frac{\pi}{2}$, then $d\varphi = d\theta = 0$ and the metric gives

$$g_{00}\left(\frac{dp}{ds}\right)^2 + g_{11}\left(\frac{d\psi}{ds}\right)^2 = 1,$$

or

$$\frac{d\rho}{ds} = \sqrt{\frac{1}{g_{00}}\left(1 - \frac{k^2}{g_{11}}\right)} = \sqrt{-U\left(1 + \frac{4l^2}{4Ul^2}\right)} = \sqrt{-1 - U(\rho)}$$

for the geodesic.

2. For a null geodesic,

$$g_{\mu\nu}\frac{dx^\mu}{ds}\frac{dx^\nu}{ds} = 0,$$

where s is some unspecified independent variable. For a geodesic confined to the plane $\varphi = 0$, $\theta = \frac{\pi}{2}$, we have

$$\frac{d\varphi}{ds} = \frac{d\theta}{ds} = 0,$$

and thus

$$g_{00}\left(\frac{d\rho}{ds}\right)^2 + g_{11}\left(\frac{d\psi}{ds}\right)^2 = 0,$$

or

$$\frac{d\rho}{d\psi} = \sqrt{-\frac{g_{11}}{g_{00}}} = 2Ul.$$

$\rho =$ constant, $U = 0$ is obviously a solution of this equation. This corresponds to

$$\rho^2 + l^2 = 2m\rho + 2l^2,$$

or, taking the positive root,

$$\rho = m + \sqrt{m^2 + l^2}.$$

Hence there are null geodesics lying in the surface $\varphi = 0$, $\theta = \frac{\pi}{2}$, $\rho = m + \sqrt{m^2 + l^2}$.

This result is physically interesting as it indicates that light can propagate along the surface of a sphere of radius $\rho = m + \sqrt{m^2 + l^2}$, but not outside this sphere.

2012

To first order in $1/r$, the metric near a spherical star of mass m ($c = c_1$) is

$$ds^2 = -\left(1 \pm \frac{2m}{r}\right)dt^2 + \left(1 \pm \frac{2m}{r}\right)dl^2,$$

where t and l are time and distance coordinates and r is the distance from the star.

a. Choose the sign in the coefficient of dt^2 and give a physical reason for your choice.

b. In t, l coordinates, define a coordinate velocity of light c'. Sketch c' as a function of r for both choices of sign in the coefficient of dl^2. Over what range of r is your graph meaningful? What happens to an electromagnetic plane wave as it passes near a massive body? On the basis of the above discussion, choose the correct sign in the coefficient of dl^2.

(*Princeton*)

Sol:

a. Let $x^0 = t$, $x^1 = x$, $x^2 = y$, $x^3 = z$. The metric tensor has zero components except

$$g_{00} = -(1 + h_{00}), \quad g_{11} = g_{22} = g_{33} = 1 + h_{ii},$$

where $h_{00} = \pm\frac{2m}{r}$, $h_{ii} = \pm\frac{2m}{r}$. Introduce the local proper time interval $d\tau$ by $d\tau^2 = -ds^2$. To determine the sign of h, consider the motion of a particle of velocity $v \ll 1$, the velocity of light, at a large distance from the star $r \to \infty$. As

$$d\tau^2 = -ds^2 = (1 + h_{00})dt^2 - (1 + h_{ii})dl^2,$$

i.e.,

$$\left(\frac{d\tau}{dt}\right)^2 \approx 1 - v^2 \approx 1,$$

we can make the approximation

$$\frac{d\tau}{dt} \approx 1, \qquad \frac{dx^i}{d\tau} = \frac{dx^i}{dt}\frac{dt}{d\tau} \approx 0 \quad (i = 1, 2, 3).$$

The acceleration is given by the geodesic equation

$$\begin{aligned} \frac{d^2x^i}{d\tau^2} &= -\Gamma^i_{\alpha\beta}\frac{dx^\alpha}{d\tau}\frac{dx^\beta}{d\tau} \\ &= -\Gamma^i_{00}. \end{aligned}$$

From the expression for the Christoffel symbols (**Problem 2011**), we have

$$\begin{aligned} \Gamma^i_{00} &= -\frac{1}{2}g^{ii}\frac{\partial g_{00}}{\partial x^i}. \\ &\approx \frac{1}{2}(1 - h_{ii})\frac{\partial h_{00}}{\partial x^i} \approx \frac{1}{2}\frac{\partial h_{00}}{\partial x^i}, \end{aligned}$$

as $g^{ii} = \frac{1}{g_{ii}}$ with g being diagonal. Hence we have, in vector form,

$$\frac{d^2\mathbf{1}}{dt^2} = -\frac{1}{2}\nabla h_{00}.$$

However, Newton's equation, which applies to the case of $v \ll c, r \to \infty$, is

$$\frac{d^2\mathbf{1}}{dt^2} = -\nabla\Phi$$

with $\Phi = -\frac{m}{r}$ (note $G = 1$ and gravitatnioal force is attractive). A comparison of these two equations gives

$$h_{00} = 2\Phi = -\frac{2m}{r}.$$

Hence the negative sign is to be adopted for h_{00}.

b. For light propagation, $ds^2 = 0$. The coordinate velocity of light is therefore

$$c' = \frac{dl}{dt} = \sqrt{\frac{1 + h_{00}}{1 + h_{ii}}} = \sqrt{\frac{1 - \frac{2m}{r}}{1 \pm \frac{2m}{r}}}$$

$$= \begin{cases} 1 & \text{for negative sign in } h_{ii}, \\ \sqrt{\frac{r - 2\text{m}}{r + 2\text{m}}} & \text{for positive sign in } h_{ii}. \end{cases}$$

c' as a function of r is shown in Fig. 2.4 for the two signs. Note for c' to be physically meaningful we require $r \geq 2m$ in the positive sign case. When a plane electromagnetic wave (light) passes near a massive body, its velocity

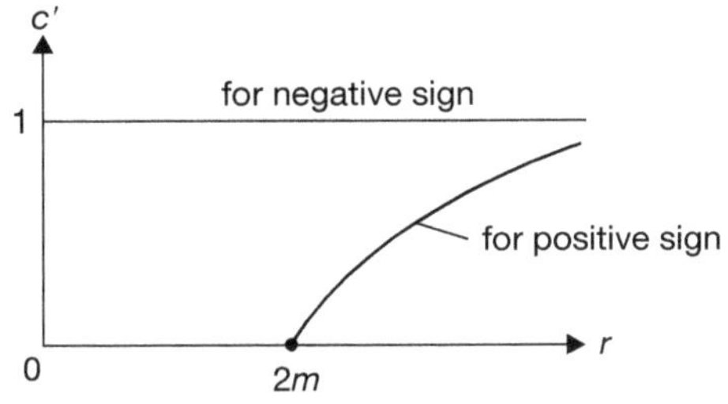

Fig. 2.4

c' will decrease. This effect is not accounted for if the negative sign is used for h_{ii}. Hence the positive sign is to be used in the coefficient of dl^2.

2013

Consider a massive object described by the Schwarzschild metric, such as a nonrotating spherically symmetric star. A light source is situated at a distance r_s (Schwarzschild radius) above the event horizon, and an observer is located at a distance $r_o > r_s$ from the center of the object. The light source emits light of frequency ν_s, and the observer measures the frequency ν_o of the light received.

a. Derive an expression for the redshift z experienced by the light as it travels from the source to the observer in terms of the Schwarzschild radii of the source and observer, r_s and r_o, and the speed of light c.

b. Assume that the light source is emitting light at a frequency of $\nu_s = 6 \times 10^{14}$ Hz and is located at a distance of 2.5 times the Schwarzschild radius (r_s) from the center of the star. If the observer is located very far away from the star, calculate the observed frequency ν_o and the redshift z.

Sol:

a. **Expression for redshift (z) in terms of Schwarzschild radii**

In the Schwarzschild metric, the line element is given by:

$$ds^2 = -\left(1 - \frac{2GM}{c^2 r}\right) c^2 dt^2 + \left(1 - \frac{2GM}{c^2 r}\right)^{-1} dr^2 + r^2(d\theta^2 + \sin^2\theta \, d\varphi^2).$$

Now, let's consider a photon emitted at the source (at $r = r_s$) and received by the observer (at $r = r_o$). The photon's trajectory can be described by $ds^2 = 0$ because it travels at the speed of light.

For the photon's emission and reception events, we have: At the source (emission event): $ds^2 = 0$

We can rewrite this as:

$$0 = -\left(1 - \frac{2GM}{c^2 r_s}\right) c^2 dt_s^2,$$

which simplifies to:

$$1 - \frac{2GM}{c^2 r_s} = c^2 dt_s^2 \tag{1}$$

At the observer (reception event): $ds^2 = 0$. Similarly, we have:

$$0 = -\left(1 - \frac{2GM}{c^2 r_o}\right) c^2 dt_o^2,$$

which simplifies to:

$$1 - \frac{2GM}{c^2 r_o} = c^2 dt_o^2 \tag{2}$$

Now, we can find the time dilation factor between the source and observer, that is, the ratio of proper time intervals (Δt_s (1) and Δt_o (2)) for the photon:

$$\sqrt{1-\frac{2GM}{c^2 r_s}}\,\Delta t_0 = \Delta t_s\sqrt{1-\frac{2GM}{c^2 r_o}}$$

The ratio of the frequencies is inversely proportional to the time intervals:

$$\frac{\nu_o}{\nu_s} = \frac{\Delta t_s}{\Delta t_o} = \frac{\sqrt{1-\frac{2GM}{c^2 r_s}}}{\sqrt{1-\frac{2GM}{c^2 r_0}}} = \sqrt{\frac{1-\frac{2GM}{c^2 r_s}}{1-\frac{2GM}{c^2 r_0}}}$$

Recall the formula for gravitational redshift:

$$z = \frac{(\nu_s - \nu_o)}{\nu_o} = \left(\frac{\nu_s}{\nu_o} - 1\right)$$

Now, substitute the formula for $\frac{\nu_o}{\nu_s}$

$$z = \sqrt{\frac{1-\frac{2GM}{c^2 r_0}}{1-\frac{2GM}{c^2 r_s}}} - 1$$

This is the formula for redshift (z) expressed in terms of Schwarzschild radii and the speed of light.

b. Given the values of ν_s, r_s, and the fact that r_o is very large (approaching infinity), we can calculate ν_o and z. We know:

$$G = 6.674\times10^{-11}\,m^3\,kg^{-1}\,s^{-2},\quad M = 10^{30}\,kg\ \&\ c = 3x\,10^8\,m/s.$$

$$\nu_o = \nu_s\sqrt{\frac{1-\frac{2GM}{c^2 r_s}}{1-\frac{2GM}{c^2 r_0}}}$$

Since r_o is very large, we can approximate $1-\frac{2GM}{c^2 r_o}$ to be 1. Therefore:

$$\nu_o \approx \nu_s\sqrt{1-\frac{2GM}{c^2 r_s}}$$

Substituting the given values: $\nu_s = 6\times10^{14}\,Hz\ \&\ r_s = 2.5\times\frac{2GM}{c^2}$

$$\nu_o \approx 6\times10^{14}\sqrt{1-\frac{2GM}{c^2\left(2.5\times\frac{2GM}{c^2}\right)}}$$

Calculating the above mentioned expression, we get:

$$\nu_o \approx 3.6\times10^{14}\,Hz$$

Now, we can calculate the redshift z:

$$z = \frac{(\nu_s - \nu_o)}{\nu_o} = \frac{(6 \times 10^{14} - 3.6 \times 10^{14})}{3.6 \times 10^{14}} \approx 0.66$$

So, the observed frequency is approximately 3.6×10^{14} Hz and the redshift is approximately 0.66.

2014

The line element for a static spherically symmetric relativistic star can be written as

$$ds^2 = e^{2\Phi(r)} dt^2 - \frac{dr^2}{1 - r^2 Y(r)} - r^2 d\theta^2 - r^2 \sin^2\theta d\varphi^2,$$

where Φ and Y are functions of r. The proper mass density is $\rho(r)$, the pressure is $p(r)$, and the proper number density is $n(r)$.

a. In terms of the two given functions in the line element, what are the expressions for the mass of the star and the red shift at the star center as measured by an observer a great distance away.

b. In terms of the given functions Φ, Y and n, what is the total number of baryons in the star?

c. In terms of the given functions, what is the condition of static equilibrium?

(*Princeton*)

Sol:

a. The volume element of the space is

$$dV = \sqrt{-g}\, dx^1 dx^2 dx^3,$$

where

$$g = \det g_{\mu\nu} = -\frac{e^{2\Phi} r^4 \sin^2\theta}{1 - r^2 Y(r)}.$$

Thus the mass of the star is

$$M = \int_0^R \int_0^\pi \int_0^{2\pi} \sqrt{-g}\rho(r) dr d\theta d\varphi$$

$$= 4\pi \int_0^R \frac{\rho(r) e^{\Phi} r^2}{\sqrt{1 - r^2 Y(r)}} dr.$$

In the neighborhood of the star center, the local proper time τ_0 is related to the coordinate time t by

$$d\tau_0 = \sqrt{g_{00}(0)}\, dt.$$

Similarly for an observer at a great distance r the local proper time interval is

$$d\tau = \sqrt{g_{00}(r)}\, dt,$$

Hence

$$\frac{\lambda}{\lambda_0} = \frac{d\tau}{d\tau_0} = \sqrt{\frac{g_{00}(r)}{g_{00}(0)}} = \exp[\Phi(r) - \Phi(0)],$$

and the red shift is

$$z = \frac{\lambda - \lambda_0}{\lambda_0} = \exp[\Phi(r) - \Phi(0)] - 1.$$

b. The baryon number-current density 4-vector is

$$J^\alpha = n(r)u^\alpha,$$

where u^α is the velocity 4-vector $\frac{dx^\alpha}{d\tau}$, and the total number of baryons in the star is

$$N = \int_0^R \sqrt{-g}J^0\, drd\theta d\varphi.$$

As $J^0 = n(r)\frac{dx^0}{d\tau} = \frac{n(r)}{\sqrt{g_{00}}} = n(r)e^{-\Phi}$, we have

$$N = 4\pi \int_0^R \frac{n(r)r^2}{\sqrt{1 - r^2 Y(r)}} dr.$$

c. The equation of hydrostatic equilibrium is

$$-\frac{\partial p}{\partial x^\mu} = (p + \rho)\frac{\partial}{\partial x^\mu} \ln \sqrt{g_{00}}.$$

As p, ρ, g_{00} all depend on r alone, let $\mu = 1$ in the above and obtain

$$\begin{aligned} -\frac{dp(r)}{dr} &= [p(r) + \rho(r)]\frac{d}{dr}\ln e^{\Phi(r)} \\ &= [p(r) + \rho(r)]\frac{d\Phi(r)}{dr}. \end{aligned}$$

2015

The Solar System is accurately described by the Schwarzschild metric

$$d\tau^2 = -\left(1 - \frac{2GM}{rc^2}\right)c^2 dt^2 + \frac{1}{1 - \frac{2GM}{rc^2}}dr^2 + r^2(d\theta^2 + \sin^2\theta d\varphi^2),$$

where M is the mass of the Sun, t the time coordinate, r the radial coordinate, and θ and φ are polar angles.

A radio pulse is sent from the Earth, reflected off a satellite of Jupiter (the satellite is a point), and received on Earth. Jupiter is a distance r_2 from the Sun, the Earth a distance r_1. Assume that Jupiter is on the other side of the Sun relative to us. Let r_0 be the distance of closest approach of the radio pulse to the Sun. Calculate the gravitational delay in the round-trip time of the radio pulse as a function of r_0, to the lowest order in G. Estimate very roughly the magnitude of the effect, given that

$$\begin{aligned}
\text{mass of the Sun} &\approx 2 \times 10^{33}\,\text{gm}, \\
\text{radius of the Sun} &\approx 7 \times 10^{10}\,\text{cm}, \\
\text{Sun-Earth distance} &\approx 1.5 \times 10^{13}\,\text{cm}, \\
\text{Sun-Jupiter distance} &\approx 8 \times 10^{13}\,\text{cm}, \\
G &\approx 6.67 \times 10^{-8}\,\text{cm}^3/\text{gm}\,\text{sec}^2.
\end{aligned}$$

(*Princeton*)

Sol: Let

$$x^0 = t, \quad x^1 = r, \quad x^2 = \theta, \quad x^3 = \varphi.$$

Then

$$g_{00} = -c^2\left(1 - \frac{k}{r}\right), \quad g_{11} = \left(1 - \frac{k}{r}\right)^{-1}, \quad g_{22} = r^2,$$
$$g_{33} = r^2 \sin^2\theta, \quad g_{\mu\nu} = 0 \text{ for } \mu \neq \nu,$$

where $k = \frac{2GM}{c^2}$. For convenience choose coordinates such that the Earth (E), Sun (S) and Jupiter (J) are in the $\theta = \frac{\pi}{2}$ plane. The equations of the geodesic along which the radio pulse propagates,

$$\frac{d}{ds}\left(g_{\mu\nu}\frac{dx^\nu}{ds}\right) = \frac{1}{2}g_{\alpha\beta,\mu}\frac{dx^\alpha}{ds}\frac{dx^\beta}{ds},$$

where s is some orbital parameter, give two first integrals,

$$g_{3\nu}\frac{dx^\nu}{ds} = g_{33}\frac{dx^3}{ds} = r^2\sin^2\theta\frac{d\varphi}{ds} = r^2\frac{d\varphi}{ds} = \text{constant},$$

$$g_{0\nu}\frac{dx^\nu}{ds} = g_{00}\frac{dx^0}{ds} = -c^2\left(1-\frac{k}{r}\right)\frac{dt}{ds} = \text{constant},$$

since

$$g_{\alpha\beta,3} = \frac{\partial g_{\alpha\beta}}{\partial\varphi} = 0, \quad g_{\alpha\beta,0} = \frac{\partial g_{\alpha\beta}}{\partial t} = 0.$$

Taking ratio of the integrals gives

$$\frac{r^2}{1-\frac{k}{r}}\frac{d\varphi}{dt} = F, \quad \text{a constant.}$$

For radio waves, $ds^2 = 0$ and the Schwarzschild metric reduces to

$$\left(1-\frac{k}{r}\right) - \frac{1}{c^2}\left[\left(1-\frac{k}{r}\right)^{-1}\left(\frac{dr}{dt}\right)^2 + \frac{F^2}{r^2}\left(1-\frac{k}{r}\right)^2\right] = 0.$$

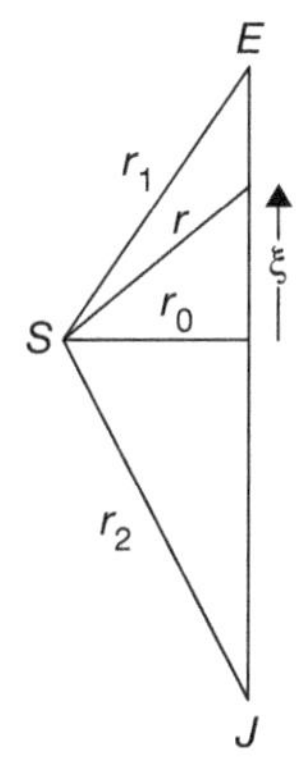

Fig. 2.5

The relative positions of E, S, J, are shown in Fig. 2.5, where $r_0 \ll r_1, r_2$. At r_0, the closest approach of the radio pulse to the Sun, $\frac{dr}{dt} = 0$ and the last equation above gives

$$F = cr_0\left(1-\frac{k}{r_0}\right)^{-\frac{1}{2}}.$$

Then the coordinate speed of the pulse, c', is given by

$$\begin{aligned} c'^2 &= \left(\frac{dr}{dt}\right)^2 + r^2\left(\frac{d\varphi}{dt}\right)^2 \\ &= \left(1-\frac{k}{r}\right)^2 c^2 + kr\left(\frac{d\varphi}{dt}\right)^2 \\ &= \left(1-\frac{k}{r}\right)^2 c^2\left[1+\left(\frac{r_0}{r}\right)^2\left(1-\frac{k}{r_0}\right)^{-1}\frac{k}{r}\right], \end{aligned}$$

or

$$c' \approx \left(1 - \frac{k}{r}\right)c,$$

to first order in $\frac{k}{r}$ and $\frac{r_0}{r}$.

The time taken for one round trip of the pulse if the Sun did not exert any effect on the speed of light is

$$T_0 = \frac{2}{c}\left(\sqrt{r_1^2 - r_0^2} + \sqrt{r_2^2 - r_0^2}\right),$$

while the actual time is

$$T = 2\int_0^{\sqrt{r_1^2 - r_0^2}} \frac{d\xi}{c'} + 2\int_0^{\sqrt{r_2^2 - r_0^2}} \frac{d\xi}{c'},$$

where $\xi = \sqrt{r^2 - r_0^2}$. As $d\xi = \frac{rdr}{\xi}$,

$$\begin{aligned} T &\approx \frac{2}{c}\left[\int_{r_0}^{r_1} \left(1 + \frac{k}{r}\right)\frac{rdr}{\sqrt{r^2 - r_0^2}} + \int_{r_1}^{r_2} \left(1 + \frac{k}{r}\right)\frac{rdr}{\sqrt{r^2 - r_0^2}}\right] \\ &= \frac{2}{c}\left[\sqrt{r_1^2 - r_0^2} + \sqrt{r_2^2 - r_0^2} + k\ln\left(\frac{r_1 + \sqrt{r_1^2 - r_0^2}}{r_0}\right)\right. \\ &\quad \left. + k\ln\left(\frac{r_2 + \sqrt{r_2^2 - r_0^2}}{r_0}\right)\right] \\ &\approx \frac{2}{c}\left[\sqrt{r_1^2 - r_0^2} + \sqrt{r_2^2 - r_0^2} + k\ln\left(\frac{4r_1 r_2}{r_0^2}\right)\right]. \end{aligned}$$

Hence the delay time is

$$\Delta T = T - T_0 \approx \frac{4GM}{c^3}\ln\left(\frac{4r_1 r_2}{r_0^2}\right).$$

Taking $r_0 \simeq$ the Sun's radius and using the given data we find

$$\Delta T = 2.7 \times 10^{-4}\,\text{s}.$$

2016

A conformal transformation of a space-time is one where the metric g_{ab} of an original space-time is transformed into the metric $\tilde{g}_{ab}$ of a new space-time such that $\tilde{g}_{ab}$ is related to g_{ab} by

$$\tilde{g}_{ab} = \Omega^2 g_{ab},$$

where Ω is a function of the space-time coordinates x^a.

a. Suppose in the old space-time one has a solution to the source-free Maxwell's equations

$$\nabla_a F^{ab} = 0,\ \nabla_{[a} F_{bc]} = 0,$$

F being the antisymmetric field strength tensor. Show that F_{ab} is also a solution to these equations in the new space-time with metric $\tilde{g}_{ab}$. (You may wish to recall that $\Gamma^a_{ac} = \frac{1}{\sqrt{g}}\partial_c(\sqrt{g})$ where $g = -\det(g_{ab})$.)

b. The metric of a $\kappa = 0$ Robertson–Walker space-time is sometimes written as

$$ds^2 = -c^2 dt^2 + \left(\frac{t}{t_0}\right)^{2/3}(dx^2 + dy^2 + dz^2)\ .$$

Show that this space-time is conformal to Minkowski space-time.

c. By constructing solutions to Maxwell's equations in Minkowski space, find a formula for the cosmological red shift of light emitted by a galaxy at $t = t_1$, and seen by observer at $t = t_2$. Assume both source and observer are at rest with respect to the co-moving time coordinate t.

(*Princeton*)

Sol:

a. As $\tilde{g}_{ab} = \Omega^2 g_{ab}$, we have $\tilde{g}^{ab} = \Omega^{-2} g^{ab}$ and $\tilde{g} = -\det(\tilde{g}_{ab}) = -\det(\Omega^2 g_{ab}) = -\Omega^8 \det(g_{ab}) = \Omega^8 g$. Also, for an antisymmetric tensor $T^{\alpha\beta}$ we have

$$\nabla_\alpha T^{\alpha\beta} = \frac{1}{\sqrt{g}}\frac{\partial}{\partial x^\alpha}\left(\sqrt{g}\,T^{\alpha\beta}\right).$$

Then in the new space-time we have

$$\begin{aligned}
\tilde{\nabla}_a F^{ab} &= \tilde{\nabla}_a\left(\tilde{g}^{al}\tilde{g}^{bm} F_{lm}\right)\\
&= \frac{l}{\sqrt{\tilde{g}}}\frac{\partial}{\partial x^a}\left(\sqrt{\tilde{g}}\,\Omega^{-4} g^{al} g^{bm} F_{lm}\right)\\
&= \frac{1}{\sqrt{\tilde{g}}}\frac{\partial}{\partial x^a}\left(\sqrt{g}F^{ab}\right)\\
&= \frac{1}{\Omega^4}\frac{1}{\sqrt{g}}\frac{\partial}{\partial x^a}\left(\sqrt{g}F^{ab}\right) = \frac{1}{\Omega^4}\nabla_a F^{ab} = 0
\end{aligned}$$

and

$$\begin{aligned}\tilde{\nabla}_{[a}F_{bc]} &= \frac{\partial F_{bc}}{\partial x^a} + \frac{\partial F_{ca}}{\partial x^b} + \frac{\partial F_{ab}}{\partial x^c} \\ &= \nabla_{[a}F_{bc]} \\ &= 0.\end{aligned}$$

Hence, F_{ab} is also a solution to the source-free Maxwell's equation in the new space-time.

b. Rewriting the metric of the Robertson-Walker space-time as

$$\begin{aligned}ds^2 &= \left(\frac{t}{t_0}\right)^{\frac{2}{3}}\left[-c^2\left(\frac{t_0}{t}\right)^{\frac{2}{3}} dt^2 + dx^2 + dy^2 + dz^2\right] \\ &= a^2(t)\left[-d\tau^2 + dx^2 + dy^2 + dz^2\right] \\ &= \eta_{\mu\nu} dx^\mu dx^\nu,\end{aligned}$$

where

$$a(t) = \left(\frac{t}{t_0}\right)^{\frac{1}{3}}, \quad a d\tau = c dt,$$

we see that the metric becomes Minkowskian if

$$g_{\mu\nu} = \Omega^2 \eta_{\mu\nu}$$

with.$\Omega^2 = a^2(t)$. *As*

$$\tau = c\int \frac{dt}{a(t)} = c\int\left(\frac{t_0}{t}\right)^{\frac{1}{3}} dt = \frac{3c}{2} t_0^{\frac{1}{3}} t^{\frac{2}{3}}$$

putting any constant of integration to zero by resetting the time origins, we have

$$\Omega^2 = \left(\frac{t}{t_0}\right)^{\frac{2}{3}} = \frac{2\tau}{3c\, t_0}.$$

Hence the Robertson-Walker space-time, whose metric tensor is $g_{\mu\nu}$, is conformal to the Minkowski space-time, whose metric tensor is $\eta_{\mu\nu}$ the transformation being

$$g_{\mu\nu} = \frac{2\tau}{3c\, t_0}\eta_{\mu\nu}.$$

c. For light, $ds = 0$ and thus

$$a^2(d\tau^2 - dr^2) = 0,$$

where $dr^2 = dx^2 + dy^2 + dz^2$, or

$$dr = dr,$$

i.e.,

$$r = \tau + \text{constant}.$$

Take the origin at the light source and consider an observer at position r, both being at rest with respect to the co-moving frame of time t. The time interval between two consecutive waves is

$$dt_1 = \frac{a(t_1)}{c}d\tau$$

at the source and

$$dt_2 = \frac{a(t_2)}{c}d\tau$$

at the observer. Hence

$$\frac{\lambda_2}{\lambda_1} = \frac{cdt_2}{cdt_1} = \frac{a(t_2)}{a(t_1)} = \left(\frac{t_2}{t_1}\right)^{\frac{1}{3}},$$

and the red shift is

$$z = \frac{\Delta\lambda}{\lambda} = \frac{\lambda_2 - \lambda_1}{\lambda_1} = \left(\frac{t_2}{t_1}\right)^{\frac{1}{3}} - 1.$$

2017

Recently, a "double" quasar was discovered consisting of two indistinguishable images separated by an angle of 6 arc seconds. One interpretation is that we are seeing one quasar imaged as two by an intervening gravitational lens. To analyse this, suppose that the observer, the intervening galaxy, and the quasar are exactly lined up as indicated in Fig. 2.6.

Light from the quasar is gravitationally bent by the galaxy before reaching us. The angle ϕ, as measured in the rest frame of the galaxy, is given by $\phi = 4M/r$, where M is the mass of the galaxy and r the impact parameter of the light ray. Thus, as seen by us, the quasar is a ring whose diameter subtends an angle 2θ. The universe has a geometry described by the metric

$$ds^2 = -dt^2 + a^2(t)\left(d\rho^2 + \rho^2 d\theta^2 + \rho^2 \sin^2\theta d\phi^2\right).$$

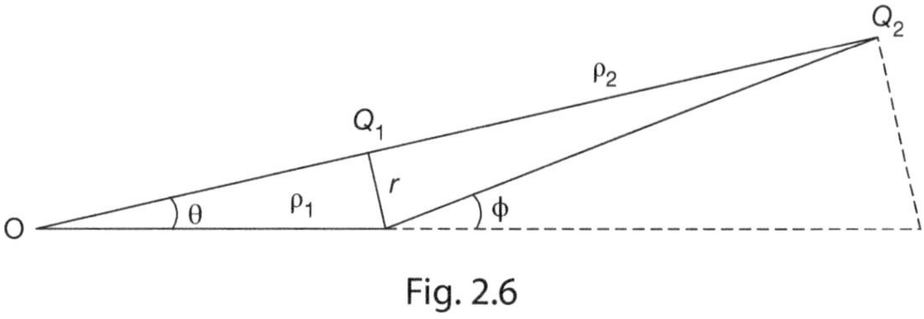

Fig. 2.6

The scale factor is $a(t) \propto t^{2/3}$. The quasar emits radiation at $t = t_2$, which passes the galaxy at time t_1 and reaches us at t_0.

a. Find an expression which relates θ to M, t_0, t_1 and t_2.

b. Re-express your results in terms of the physical variables M, the red shift of the galaxy, z_2, the red shift of the quasar, z_1, and the Hubble time H. $\left(\text{Recall that } H = \left[\frac{d}{dt}\ln a\right]_{t=t_0}\right)$

(Princeton)

Sol: For the rest frame of the galaxy take origin at the position O of the observer and let the radial coordinate of the galaxy Q_1 be ρ_1 and that of the quasar Q_2 be ρ_2. The effect of the expansion of the universe is accounted for by the scale factor $a(t)$ so that ρ_1 and ρ_2 do not change with time. As

$$\theta = \frac{r}{a(t_1)\rho_1}, \quad \phi - \theta = \frac{r}{a(t_1)(\rho_2 - \rho_1)},$$

we have

$$\theta = \frac{\rho_2 - \rho_1}{\rho_2}\phi.$$

With

$$\phi = \frac{4M}{r} = \frac{4M}{a(t_1)\rho_1\theta},$$

$$\theta = \left[\frac{\rho_2 - \rho_1}{\rho_1\rho_2}\cdot\frac{4M}{a(t_1)}\right]^{1/2}.$$

For light $ds^2 = 0$. As the light propagates radially, the metric becomes

$$-dt^2 + a^2(t)d\rho^2 = 0.$$

Thus, with $a(t) \propto t^{\frac{2}{3}}$,

$$\rho_1 = \int_0^{\rho_1} d\rho = \int_{t_2}^{t_0} \frac{dt}{a(t)} = A\int_{t_1}^{t_0} t^{-\frac{2}{3}}dt = 3A\left(t_0^{\frac{1}{3}} - t_1^{\frac{1}{3}}\right),$$

$$\rho_2 = \int_{t_2}^{t_0} \frac{dt}{a(t)} = 3A\left(t_0^{\frac{1}{3}} - t_2^{\frac{1}{3}}\right),$$

where A is a constant. Hence

$$\theta = \left\{\frac{4M}{3}\frac{\left[(t_1/t_2)^{\frac{1}{3}} - 1\right]}{\left[t_1\left(\frac{t_0}{t_1}\right)^{1/3} - 1\right]\left[\left(\frac{t_0}{t_1}\right)^{1/3} - 1\right]}\right\}^{1/2}.$$

b. Consider successive crests of a light wave emitted at a point P on the line OQ_2 at coordinate times t, $t + dt$ which are received at O at times t_0, $t_0 + dt_0$ respectively. Then

$$d\rho = \frac{dt}{a(t)},$$

and thus

$$\rho = \int_0^\rho d\rho = \int_t^{t_0} \frac{dt}{a(t)} = \int_{t+dt}^{t_0+dt_0} \frac{dt}{a(t)}.$$

Since dt, dt_0 are small quantities, this implies that

$$\frac{dt_0}{a(t_0)} = \frac{dt}{a(t)},$$

i.e., $\frac{\lambda_0}{\lambda} = \frac{dt_0}{dt} = \frac{a(t_0)}{a(t)}$. Hence the red shift z is given by

$$1 + z = \frac{\lambda_{\text{observed}}}{\lambda_{\text{emitted}}} = \frac{a(t_0)}{a(t)} = \left(\frac{t_0}{t}\right)^{\frac{2}{3}}.$$

Particularly,

$$z_1 = \left(\frac{t_0}{t_1}\right)^{\frac{2}{3}} - 1, \quad z_2 = \left(\frac{t_0}{t_2}\right)^{\frac{2}{3}} - 1.$$

Then with

$$H = \left(\frac{d}{dt}\ln a\right)_{t=t_0} = \left(\frac{1}{a}\frac{da}{dt}\right)_{t=t_0} = \frac{2}{3t_0},$$

we have

$$\theta = \left\{ \frac{2MH(1+z_1)\left[(1+z_2)^{\frac{1}{2}} - (1+z_1)^{\frac{1}{2}}\right]}{\left[(1+z_2)^{\frac{1}{2}} - 1\right]\left[(1+z_1)^{\frac{1}{2}} - 1\right]} \right\}^{\frac{1}{2}}.$$

2018

The metric outside a black hole is given by

$$ds^2 = -\left(1 - \frac{2m}{r}\right)dt^2 + \left(1 - \frac{2m}{r}\right)^{-1} dr^2 + r^2(d\theta^2 + \sin^2\theta d\varphi^2)$$

where θ, φ are spherical polar coordinates, r is a radial coordinate and t is a time coordinate.

a. Show that a massless particle moving in the plane $\theta = \pi/2$ moves along a trajectory such that

$$\frac{d^2u}{d\varphi^2} + u = 3u^2$$

where $u = m/r$.

b. Find the circular orbit for photons in this plane. Is this orbit stable to small perturbations?

The Christoffel symbol is given by

$$\Gamma^a_{bc} = \frac{1}{2} g^{ad} (\partial_b g_{dc} - \partial_d g_{cb} + \partial_c g_{bd}).$$

(*Princeton*)

Sol:

a. Letting $x^0 = t$, $x^1 = r$, $x^2 = \theta$, $x^3 = \varphi$, we have

$$g_{00} = -\left(1 - \frac{2m}{r}\right), \quad g_{11} = \left(1 - \frac{2m}{r}\right)^{-1}, \quad g_{22} = r^2,$$
$$g_{33} = r^2 \sin^2\theta, \; g_{\mu\nu} = 0 \quad \text{for}\, \mu \neq \nu.$$

The motion of the particle is confined to the plane $\theta = \frac{\pi}{2}$, so $\sin\theta = 1$, $d\theta = 0$. As $g_{\mu\nu}$ do not depend on φ and t explicitly, the geodesic equations

$$\frac{d}{ds}\left(g_{\mu\nu}\frac{dx^\nu}{ds}\right) = \frac{1}{2}\frac{\partial g_{\alpha\beta}}{\partial x^\mu}\frac{dx^\alpha}{ds}\frac{dx^\beta}{ds}$$

yield two first integrals:

$$\frac{d}{ds}\left(g_{33}\frac{dx^3}{ds}\right) = 0, \qquad \text{or} \qquad r^2\frac{d\varphi}{ds} = h, \text{ a constant},$$

$$\frac{d}{ds}\left(g_{00}\frac{dx^0}{ds}\right) = 0, \qquad \text{or} \qquad \left(1 - \frac{2m}{r}\right)\frac{dt}{ds} = l, \text{ a constant.}$$

As the particle is massless, it moves with the velocity of light and its path is a null geodesic given by

$$g_{\mu\nu}\frac{dx^\mu}{ds}\frac{dx^\nu}{ds} = 0,$$

i.e.,

$$l^2 - \left(\frac{dr}{ds}\right)^2 - \frac{h^2}{r^2}\left(1 - \frac{2m}{r}\right) = 0.$$

Here s is to be considered an arbitrary independent variable. Let $u = \frac{m}{r}$.

Then

$$\frac{dr}{ds} = -\frac{m}{u^2}\frac{du}{ds}$$

and the above becomes

$$l^2 - \frac{m^2}{u^4}\left(\frac{du}{ds}\right)^2 - \frac{h^2u^2}{m^2}(1-2u) = 0,$$

or

$$l^2 - \frac{h^2}{m^2}\left(\frac{du}{d\varphi}\right)^2 - \frac{h^2u^2}{m^2}(1-2u) = 0,$$

as

$$\frac{du}{ds} = \frac{du}{d\varphi}\frac{d\varphi}{ds} = \frac{hu^2}{m^2}\frac{du}{d\varphi}.$$

Differentiating with respect to φ gives

$$\frac{d^2u}{d\varphi^2} + u = 3u^2.$$

b. For the circular orbit, $u =$ constant and the orbit equation gives

$$u_0 = \frac{1}{3}, \text{ or } r_0 = 3m.$$

For small perturbations on this orbit, let $u = u_0 + u'$, where $u' \ll u_0$, and the orbit equation gives

$$\frac{d^2u'}{d\varphi^2} + u_0 + u' \approx 3\,u_0^3 + 6\,u_0u',$$

or

$$\frac{d^2u'}{d\varphi^2} - u' = 0,$$

This has the general solution

$$u' = Ae^{\varphi} + Be^{-\varphi},$$

A, B being constants. As u' is not oscillatory the orbit is unstable to small perturbations.

2019

a. Explain briefly why in Einstein's theory of general relativity it is impossible to have monopole or dipole gravitational radiation.

b. Suppose two compact stars, each of one solar mass, are in a circular orbit around each other with a radius of one solar radius. What is the approximate rate of energy loss due to gravitational radiation from this system? What is the time scale for decay of this orbit? Take

$$\begin{aligned}\text{solar mass} &= 2\times 10^{33}\,\text{gm},\\ \text{solar radius} &= 7\times 10^{10}\,\text{cm},\\ G &= 6.7\times 10^{-8}\,\text{cm}^3\text{gm}^{-1}\text{sec}^{-2},\\ c &= 3\times 10^{10}\,\text{cm sec}^{-1}.\end{aligned}$$

(*Princeton*)

Sol:

a. The total mass-energy of an isolated system is conserved, so it is not possible to radiate monopole gravitational radiation. Also as the total momentum of an isolated system is conserved, which means that the second time derivative of the mass dipole moment $\Sigma m\mathbf{r}$ is zero, it is not possible to radiate dipole gravitational radiation. The lowest multipole gravitational radiation is quadrupole.

b. For a system of two stars in a circular orbit around their common center of mass, the rate of energy loss by radiation of gravitational waves is

$$-\frac{dE}{dt}=\frac{32G^4}{5c^5r^5}(m_1m_2)^2(m_1+m_2)$$

where r is their mutual distance, which is constant for motion in a circular orbit. With $m_1=m_2=m$ and the data given we have

$$-\frac{dE}{dt}=\frac{64G^4}{5c^5r^5}m^5=1.57\times 10^{31}\,\text{ergs}^{-1}.$$

With

$$E=-\frac{Gm^2}{2r},$$

$$\frac{dr}{dt}=\frac{2r^2}{Gm^2}\frac{dE}{dt}=-\frac{128Gm^3}{5c^5r^3}.$$

As $\frac{dr}{dt}$ is the rate at which the two stars approach each other, the time taken for the complete collapse of the orbit is

$$\begin{aligned}\tau=\int_r^0\frac{dr}{\left(\frac{dr}{dt}\right)}&=-\frac{5c^5}{128G^3m^3}\int_r^0 r^3dr\\ &=\frac{5c^5}{512G^3m^3}r^4=2.4\times 10^{15}\,\text{s}.\end{aligned}$$

2020

a. In general relativity theory, can a source emit spherically symmetric monopole gravitational radiation? Explain your answer in a few sentences.

b. Can a source emit dipole gravitational radiation in general relativity theory? Explain in a few sentences.

c. Can a source emit quadrupole radiation in general relativity theory? Explain in a few sentences.

d. In general relativity theory a closed cosmological model cannot have a net charge different from zero. Explain why.

e. Energy conservation and charge conservation are on very different footings in general relativity theory, for in general one cannot write down a globally conserved energy but there is always a global conserved charge integral. Explain briefly.

f. If there exists a vector field ξ^i such that $\xi_{i;j} + \xi_{j;i} = 0$, then one can combine it with the stress-energy tensor to define a global energy integral. Explain.

(Princeton)

Sol:

a. It is not possible to emit spherically symmetric monopole gravitational radiation because the total mass-energy of an isolated system does not vary with time.

b. It is not possible to emit dipole gravitational radiation because the total momentum of an isolated system does not vary with time. The second derivative of the mass dipole moment with respect to time is zero. This is a principal aspect in which gravitational radiation differs from electromagnetic radiation.

c. It can emit quadrupole radiation, which is the lowest nonzero multipole gravitational radiation. As the volume integral $\int \mu(x) x^\alpha x^\beta \, dV$ of the mass density μ is not generally equal to zero, its third time-derivative, which appears in the energy loss formula, is also nonzero.

d. In a finite universe, each side of a closed curved surface encloses a finite region. The total electric flux passing through the surface in the outward direction equals $\frac{Q}{\epsilon_0}$, where Q is the net charge enclosed by the surface and ϵ_0 the permittivity of free space. This total flux can also be considered as passing through the surface inward into the outer region so that the net charge in the outer region enclosed by the surface is $-Q$. Hence the total charge in a finite universe is zero

e. In the absence of a gravitational field and in Galilean coordinates the conservation of energy-momentum is expressed by

$$\frac{\partial T_{ik}}{\partial x^k} = 0.$$

In the presence of a gravitational field, this is generalized to

$$T^{\mu\nu}_{;\nu} = \frac{1}{\sqrt{-g}}\frac{\partial}{\partial x^\nu}\left(\sqrt{-g}\,T^{\mu\nu}\right) + \Gamma^{\mu}_{\rho\nu}T^{\rho\nu} = 0,$$

where $T^{\mu\nu}$ is the energy-momentum tensor, which has rank 2 and is symmetric, $-g$ is the determinant of the metric tensor $g_{\mu\nu}$, $\Gamma^{\mu}_{\rho\nu}$ is a Christoffel symbol of the second kind. Owing to the presence of the term outside the differentiation sign, it is generally not possible to reduce the above equation to a form showing a time rate of variation of a volume integral being equal to a surface integral of a flux crossing the closed surface. Hence in general one cannot write down a globally conserved energy integral.

On the other hand, the charge-current density is represented by a vector j^α whose conservation is given by

$$j^{\alpha}_{;\alpha} = \frac{1}{\sqrt{-g}}\frac{\partial}{\partial x^\alpha}(\sqrt{-g}j^\alpha) = 0.$$

This can be readily integrated to give a globally conserved charge integral.

f. Any vector field ξ_i that satisfies the condition

$$\xi_{i;j} + \xi_{j;i} = 0$$

is called a Killing vector field. By the rule of covariant differentiation we have

$$(\xi_i T^{ik})_{;k} = \xi_{i;k}T^{ik} + \xi_i T^{ik}{}_{;k}.$$

T^{ik} is the symmetric energy-momentum tensor, for which $T^{ik}; k = 0$. The property of ξ_i gives

$$\xi_{i;k}T^{ik} = -\xi_{k;i}T^{ik},$$

$$\xi_{i;k}T^{ik} = \xi_{k;i}T^{ki} = \xi_{k;i}T^{ik}.$$

Hence

$$\xi_{i;k}T^{ik} = 0,$$

and so

$$\left(\xi_i T^{ik}\right)_{;k} = 0.$$

As $\xi_i T^{ik}$ is a vector,

$$\left(\xi_i T^{ik}\right)_{;k} = \frac{1}{\sqrt{-g}}\frac{\partial}{\partial x^k}\left(\sqrt{-g}\,\xi_i T^{ik}\right) = 0.$$

Integrating the above over a 3-dimensional volume on whose boundaries T^{ik} vanish we have

$$\frac{d}{dx^0}\int\sqrt{-g}\,\xi_i T^{i0}\,dx^1\,dx^2\,dx^3 = \text{constant}.$$

If the Killing vector is time-like, this is the conservative global energy integral.

2021

In the weak field approximation $g_{\mu\nu} = \eta_{\mu\nu} + h_{\mu\nu}$, Einstein's vacuum equations can be written in the linearized form

$$\Box h_{\mu\nu} = \theta_{\mu,\nu} + \theta_{\nu,\mu}$$

(up to first order in h), where $\theta_\mu = h^{\alpha}_{\mu,\alpha} - \frac{1}{2}h^{\alpha}_{\alpha,\mu}$ (indices are raised and lowered with the Minkowski metric $\eta_{\mu\nu}$, and a comma denotes partial differentiation).

a. Consider the coordinate transformation $x_\mu \to x'_\mu = x_\mu + \varepsilon_\mu(x)(\varepsilon$ small) and find $h_{\mu\nu}$ in the new coordinates. Determine whether $h'_{\mu\nu}$ is also a solution of the Einstein equation, i.e., are the linearized vacuum equations gauge invariant?

b. Use (a) to determine the number of degrees of freedom of a plane wave solution to the homogeneous equation $\Box h_{\mu\nu} = 0$ (since there is a gauge for which $\theta_\mu = 0$).

(Princeton)

Sol:

a. As $\dfrac{\partial x'^\mu}{\partial x^\alpha} = \delta^\mu_\alpha + \varepsilon^\mu_{,\alpha}$,

$$\begin{aligned} g'_{\mu\nu}(x') &= \frac{\partial x'^\alpha}{\partial x^\mu}\frac{\partial x'^\beta}{\partial x^\nu}g_{\alpha\beta}(x) \\ &= \left(\delta^\alpha_\mu + \varepsilon^\alpha_{,\mu}\right)\left(\delta^\beta_\nu + \varepsilon^\beta_{,\nu}\right)g_{\alpha\beta}(x) \\ &\approx g_{\mu\nu}(x) + g_{\alpha\nu}(x)\,\varepsilon^\alpha_{,\mu} + g_{\mu\beta}(x)\,\varepsilon^\beta_{,\nu}. \end{aligned}$$

On the other hand, Taylor's expansion gives

$$g'_{\mu\nu}(x') = g'_{\mu\nu}(x) + g_{\mu\nu,\alpha}(x)\varepsilon^\alpha .$$

Thus

$$g'_{\mu\nu}(x) \approx g_{\mu\nu}(x) - g_{\mu\nu,\alpha}(x)\varepsilon^\alpha + g_{\alpha\nu}(x)\,\varepsilon^\alpha_{,\mu} + g_{\mu\beta}(x)\,\varepsilon^\beta_{,\nu} .$$

Retaining only terms of first order in the small quantities $h_{\mu\nu}$, ε^μ, etc, and their derivatives, we have

$$\begin{aligned} g_{\mu\nu,\alpha}\varepsilon^\alpha &\approx h_{\mu\nu,\alpha}\varepsilon^\alpha \approx 0, \\ g_{\alpha\nu}\,\varepsilon^\alpha_{,\mu} &\approx \eta_{\alpha\nu}\,\varepsilon^\alpha_{,\mu} \approx \varepsilon_{\nu,\mu} . \end{aligned}$$

Hence

$$h'_{\mu\nu} \approx h_{\mu\nu} + \varepsilon_{\nu,\mu} + \varepsilon_{\mu,\nu}$$

or, raising the index μ,

$$h'^{\mu}_{\nu} \approx h^{\mu}_{\nu} + \varepsilon_{\nu,}{}^{\mu} + \varepsilon^{\mu}{}_{,\nu} .$$

Consider

$$\begin{aligned} \Box h'_{\mu\nu} &= \Box h_{\mu\nu} + \Box \varepsilon_{\mu,\nu} + \Box \varepsilon_{\nu,\mu} \\ &= \theta_{\mu,\nu} + \theta_{\nu,\mu} + \Box \varepsilon_{\mu,\nu} + \Box \varepsilon_{\nu,\mu} . \end{aligned}$$

As

$$\begin{aligned} \theta'_\mu &= h'^{\alpha}_{\mu,\alpha} - \frac{1}{2} h'^{\alpha}_{\alpha,\mu} \\ &= h^\alpha_{\mu,\alpha} + \varepsilon_{\mu,}{}^{\alpha}{}_{\alpha} + \varepsilon^\alpha{}_{,\mu\alpha} - \frac{1}{2}\left(h^\alpha_{\alpha,\mu} + \varepsilon^\alpha_{\alpha,\mu} + \varepsilon^\alpha{}_{,\alpha\mu}\right) \\ &= \theta_\mu + \Box \varepsilon_\mu \end{aligned}$$

since

$$\begin{aligned} \varepsilon_{\mu,}{}^{\alpha}{}_{\alpha}, &= \partial^\alpha \partial_\alpha \varepsilon_\mu = \Box \varepsilon_\mu, \\ \varepsilon_{\alpha,}{}^{\alpha}{}_{\mu} &= \partial_\mu \partial^\alpha \varepsilon_\alpha = \partial_\mu \partial_\alpha \varepsilon^\alpha = \varepsilon^\alpha_{,\alpha\mu} \\ &= \varepsilon^\alpha_{,\mu\alpha}, \end{aligned}$$

we have

$$\begin{aligned} \Box h'_{\mu\nu} &= \theta'_{\mu,\nu} - \Box \varepsilon_{\mu,\nu} + \theta'_{\nu,\mu} - \Box \varepsilon_{\nu,\mu} + \Box \varepsilon_{\mu,\nu} + \Box \varepsilon_{\nu,\mu} \\ &= \theta'_{\mu,\nu} + \theta'_{\nu,\mu} . \end{aligned}$$

Hence $h'_{\mu\nu}$ is also a solution of Einstein's equation, i.e., the field equation satisfies gauge invariance.

b. Let

$$\gamma^\nu_\mu = h^\nu_\mu - \frac{1}{2}\delta^\nu_\mu h,$$

where $h = h^\alpha_\alpha$. Then

$$\begin{aligned} \gamma^\nu_{\mu,\nu} &= h^\nu_{\mu,\nu} - \frac{1}{2}\delta^\nu_\mu h_{,\nu} \\ &= h^\alpha_{\mu,\alpha} - \frac{1}{2} h^\alpha_{\alpha,\mu} = \theta_\mu . \end{aligned}$$

Use a gauge in which $\theta_\mu = 0$. Then

$$\Box h_{\mu\nu} = 0,$$

or

$$\Box h_\mu^\nu = 0.$$

Consider a solution representing a plane wave propagating in the $+x$ direction, i.e., h_μ^ν is a function of $u = x - t$ (velocity of light c is taken to be unity in the units employed). Then γ_μ^ν is also a function of u and $\theta_\mu = 0$ implies

$$\begin{aligned}
\gamma_{\mu,\nu}^\nu &= \frac{\partial \gamma_\mu^0}{\partial x^0} + \frac{\partial \gamma_\mu^1}{\partial x^1} \\
&= \frac{d\gamma_\mu^0}{du}\frac{\partial u}{\partial t} + \frac{d\gamma_\mu^1}{du}\frac{\partial u}{\partial x} \\
&= -\frac{d\gamma_\mu^0}{du} + \frac{d\gamma_\mu^1}{du} = 0.
\end{aligned}$$

Integrating and noticing that the constant of integration does not contribute to the wave motion and can be omitted, we have

$$\gamma_\mu^0 = \gamma_\mu^1, \qquad \mu = 0, 1, 2, 3.$$

We could also choose a coordinate transformation

$$x'^\mu = x^\mu + \varepsilon^\mu (x - t)$$

where ε^μ is a small quantity, and the transformation would still give rise to gauge invariance. Then as $\Box \varepsilon_\mu = 0$ we have

$$\theta'_\mu = \theta_\mu + \Box \varepsilon_\mu = \theta_\mu$$

We can choose such a transformation to make γ_1^0, γ_2^0, γ_3^0, $\gamma_2^2 + \gamma_3^3$ vanish. Now for $\mu \neq \nu$, $\gamma_\mu^\nu = h_\mu^\nu = h_\nu^\mu$ since as $g_{\mu\nu}$ is symmetric $h_{\mu\nu}$ is also symmetric. Hence all γ_μ^ν vanish except $\gamma_2^2 = -\gamma_3^3$, $\gamma_2^3 = \gamma_3^2$. As $h = h_Q^\alpha = 2\gamma_\alpha^\alpha = 0$, this implies that $h_{22} = -h_{33}$, $h_{23} = h_{32}$. Therefore, there are only two degrees of freedom for a plane wave solution. In other words, plane gravitational waves are transverse waves with two planes of polarization.

2022

a. Consider a spinning particle (perhaps a gyroscope) moving in a gravitational field. No non-gravitational forces are present. Write down and explain the equation which governs the behavior in time of the spin of the particle.

b. Consider a slowly rotating thin spherical shell of mass M, radius R and rotation frequency ω. The metric of the field due to this shell can be written as

$$d\tau^2 = -c^2H(r)dt^2 + \frac{1}{H(r)}\left[dr^2 + r^2d\theta^2 + r^2\sin^2\theta(d\varphi - \Omega dt)^2\right],$$

where $\Omega = 4GM\omega/3Rc^2$ for $r < R$, $\Omega \to 0$ for $r \to \infty$, and

$$H(r) = \begin{cases} 1 - 2GM/Rc^2, & r < R \\ 1 - 2GM/rc^2, & r > R. \end{cases}$$

This form of the metric is valid if $GM/Rc^2 \ll 1$. Consider a spinning particle at rest at the center of the sphere $r = 0$. Using the equation you have written in your answer to part (a), with what frequency will the spin of the particle precess? What is this precession frequency quantitatively, if ω is the rotational frequency of the Earth and M and R, the mass and radius of the Earth, are $M \approx 6.0 \times 10^{27}$gm. and $R \approx 6.4 \times 10^3$ km? A rough estimate is enough.

Note: $G = 6.67 \times 10^{-8}$ cm^3/gm sec^2. *You may wish to know*

$$\Gamma^{\mu}_{\alpha\beta} = \frac{1}{2}g^{\mu\rho}(\partial_\alpha g_{\rho\beta} + \partial_\beta g_{\rho\alpha} - \partial_\rho g_{\alpha\beta}).$$

(*Princeton*)

Sol:

a. The spinning motion of the particle is given by the equation

$$\frac{dS_\mu}{d\tau} = \Gamma^{\lambda}_{\mu\nu}S_\lambda\frac{dx^\nu}{d\tau},$$

where S_μ is the spin vector. The left-hand side is the time rate of change of the spin vector and the right-hand side gives the effect of the gravitational force, the whole equation describing the precession of the spinning body in free fall. If gravitational forces are absent, $\Gamma^{\lambda}_{\mu\nu} = 0$, giving $\frac{dS_\mu}{d\tau} = 0$. Hence a particle under no force will have a constant spin.

b. Let $x^0 = t$, $x^1 = r$, $x^2 = \theta$, $x^3 = \varphi$. As the particle is at rest at $r < R$, $H =$ constant and

$$\frac{dx^j}{d\tau} = 0 \text{ for } j = 1, 2, 3.$$

The equation of motion reduces to

$$\frac{dS_\mu}{d\tau} = \Gamma^\lambda_{\mu 0} S_\lambda \frac{dx^0}{d\tau},$$

or

$$\begin{aligned}\frac{dS_i}{dt} &= \Gamma^\lambda_{i0} S_\lambda \\ &= \frac{1}{2} g^{\lambda\rho} \left(\partial_{ig\rho 0} + \partial_{0g\rho i} - \partial_{\rho gi0}\right) S_\lambda \\ &= \frac{1}{2} S^\rho \left(\partial_{ig\rho 0} - \partial_{\rho gi0}\right), \ (i = 1, 2, 3)\end{aligned}$$

as $g_{\mu\nu}$ do not depend on x^0 explicitly. Note that $S_0 = 0$ as the particle is at rest. This can be written in the 3-dimensional form

$$\frac{d\mathbf{S}}{dt} = \frac{1}{2} \mathbf{S} \times (\nabla \times \boldsymbol{\xi}) \qquad (1)$$

with

$$\xi_i = \frac{g_{i0}}{\sqrt{g_{ii}}}.$$

In the given metric if we set $\varphi' = \varphi - \Omega t$, then the space and time parts of the line element separate out, showing that the spherical shell rotates with angular frequency Ω about the z-axis. Equation (1) shows that the spin vector **S** precesses with angular velocity

$$\Omega' = -\frac{1}{2} \nabla \times \xi.$$

As

$$g_{30} = -\frac{\Omega r^2 \sin^2\theta}{H}, \quad g_{33} = \frac{r^2 \sin^2\theta}{H},$$

we have

$$\begin{aligned}\xi_3 &= -\frac{\Omega r \sin\theta}{\sqrt{H}} \\ &\approx -\Omega r \sin\theta,\end{aligned}$$

Since

$$H^{-\frac{1}{2}} = \left(1 - \frac{2GM}{Rc^2}\right)^{-\frac{1}{2}} \approx 1 + \frac{GM}{Rc^2} \approx 1.$$

Hence we can write

$$\boldsymbol{\xi} = \mathbf{r} \times \boldsymbol{\Omega}.$$

Then

$$\boldsymbol{\Omega}' = -\frac{1}{2} \nabla \times (\mathbf{r} \times \boldsymbol{\Omega}) = \boldsymbol{\Omega}.$$

Hence the precession angular velocity is in the z direction and has magnitude

$$|\mathbf{\Omega}'| = \Omega = \frac{4GM\omega}{3Rc^2}.$$

For a rough estimate take the Earth as a spherical shell of

$$M = 6.0 \times 10^{27}\,\text{g}, \qquad R = 6.4 \times 10^{8}\,\text{cm},$$

then

$$\begin{aligned}\Omega' &= \frac{4 \times 6.67 \times 10^{-8} \times 6 \times 10^{27} \times 2\pi}{3 \times 6.4 \times 10^{8} \times (3 \times 10^{10})^2 \times 24 \times 3600} \\ &= 6.74 \times 10^{-14}\,\text{rad s}^{-1} \\ &= 1.39 \times 10^{-8}\,\text{second of arc/s.}\end{aligned}$$

Actually the coordinate transformation

$$t' = t,\ r' = r,\ \theta' = \theta,\ \varphi' = \varphi - \Omega t$$

would give rise to an Euclidean space-time in which **S** is constant. Then in the original frame **S** will precess with angular velocity Ω about the z-axis.

2023

The Robertson-Walker line element for absolutely empty space, $T_i^j = 0$ and $\wedge = 0$, is

$$ds^2 = dt^2 - a(t)^2\left(\frac{dx^2}{1+x^2} + x^2(d\theta^2 + \sin^2\theta d\varphi^2)\right)$$

with $a(t) \propto t$. Show that this describes flat space and find the coordinate transformation that brings it to the Minkowski form.

(*Princeton*)

Sol: The line element can be written in the form

$$(ds)^2 = A(dx^0)^2 + B(dx^1)^2 + C(dx^2)^2 + D(dx^3)^2,$$

where

$$x^0 = t, \qquad x^1 = x, \qquad x^2 = \theta, \qquad x^3 = \varphi,$$

with

$$A = 1,\ B = -\frac{t^2}{1+x^2},\ C = -t^2x^2,\ D = -t^2x^2\sin^2\theta,$$

$a(t)$ being taken as t for convenience. To prove that the space is flat we have to show that the curvature or Riemann–Christoffer tensor vanishes:

$$\begin{aligned}R_{\mu\nu\rho\sigma} &\equiv \frac{1}{2}\left(\frac{\partial^2 g_{\mu\sigma}}{\partial x^\nu \partial x^\rho} + \frac{\partial^2 g_{\nu\rho}}{\partial x^\mu \partial x^\sigma} - \frac{\partial^2 g_{\mu\rho}}{\partial x^\nu \partial x^\sigma} - \frac{\partial^2 g_{\nu\sigma}}{\partial x^\mu \partial x^\rho}\right) \\ &\quad + g_{\alpha\beta}\left(\Gamma^\alpha_{\mu\sigma}\Gamma^\beta_{\nu\rho} - \Gamma^\alpha_{\mu\rho}\Gamma^\beta_{\nu\sigma}\right) = 0,\end{aligned}$$

where $\Gamma^{\alpha}_{\mu\sigma} \equiv \frac{1}{2} g^{\alpha\lambda}\left(\frac{\partial g_{\lambda u}}{\partial x^{\sigma}} + \frac{\partial g_{\lambda\sigma}}{x^{\mu}} - \frac{\partial g_{\alpha\beta}}{\partial x^{\lambda}}\right)$, etc.

The curvature tensor has the symmetry properties

$$R_{\mu\nu\rho\sigma} = -R_{\nu\mu\rho\sigma}, \qquad R_{\mu\nu\rho\sigma} = -R_{\mu\nu\sigma\rho}, \qquad R_{\mu\nu\rho\sigma} = R_{\rho\sigma\mu\nu},$$

as well as satisfies the identities

$$R_{\mu\nu\rho\sigma} + R_{\mu\sigma\nu\rho} + R_{\mu\rho\sigma\nu} = 0.$$

Then it follows that all components $R_{\mu\nu\rho\sigma}$ in which $\mu = \nu$ or $\rho = \sigma$ are zero and that for four-dimensional space the curvature tensor has only 20 independent components

0101	0112	0202	0213	0312	1212	1313
0102	0113	0203	0223	0313	1213	1323
0103	0123	0212	0303	0323	1223	2323

with $R_{0123} + R_{0312} + R_{0231} = 0$.

The Robertson–Walker space-time has

$$g_{00} = A, \qquad g_{11} = B, \qquad g_{22} = C, \qquad g_{33} = D,$$
$$g_{\mu\nu} = 0 \quad \text{for } \mu \neq \nu.$$

Denote

$$\alpha = \frac{1}{2A}, \qquad \beta = \frac{1}{2B}, \qquad \gamma = \frac{1}{2C}, \qquad \delta = \frac{1}{2D},$$
$$A_{\mu} = \frac{\partial A}{\partial x^{\mu}}, \qquad B_{\mu} = \frac{\partial B}{\partial x^{\mu}}, \qquad A_{\mu\nu} = \frac{\partial^2 A}{\partial x^{\mu}\partial x^{\nu}}, \text{ etc.}$$

Then by direct computation we find

$$\begin{aligned}
R_{0123} &= 0 \\
R_{0102} &= \frac{1}{2}(-A_{12} + \alpha A_1 A_2 + \beta A_1 A_2 + \gamma A_2 C_1) = 0 \\
R_{0101} &= \frac{1}{2}\big[-A_{11} - B_{00} + \alpha(A_0 B_0 + A_1^2) \\
&\quad + \beta(A_1 B_1 + B_0^2) - \gamma A_2 B_2 - \delta A_3 B_3\big] \\
&= \frac{1}{2}(-B_{00} + \beta B_0^2) = \frac{1}{2}\left(\frac{2}{1+x^2} - \frac{2}{1+x^2}\right) = 0.
\end{aligned}$$

Since the assignment of the indices is arbitrary we can obtain the other components by interchanging indices. For example, R_{0202} can be obtained from R_{1010} by the interchange

$$(1,\ B,\ \beta) \leftrightarrow (2,\ C,\ \gamma)\ ,$$

which gives

$$\begin{aligned}
R_{2020} &= \frac{1}{2}\big[-A_{22} - C_{00} + \alpha(A_0 C_0 + A_2^2) \\
&\quad + \gamma(A_2 C_2 + C_0^2) - \beta A_1 C_1 - \delta A_3 C_3\big] \\
&= \frac{1}{2}(-C_{00} + \gamma C_0^2) = \frac{1}{2}(2x^2 - 2x^2) = 0.
\end{aligned}$$

In this way we find all $R_{\mu\nu\rho\sigma} = 0$, showing that the space is indeed flat.

Let $r = xt$ and the coefficient of $d\theta^2 + \sin^2\theta d\varphi^2$ becomes$-r^2$. The remaining part of the line element can be rewritten

$$\begin{aligned} dt^2 - \frac{(tdx)^2}{1+x^2} &= dt^2 - \frac{(tdr - rdt)^2}{t^2 + r^2} \\ &= \frac{(tdt + rdr)^2}{t^2 + r^2} - dr^2 \\ &= d\tau^2 - dr^2 , \end{aligned}$$

if we set

$$\tau = \sqrt{t^2 + r^2} .$$

Hence the transformation

$$r = xt, \ \tau = \sqrt{t^2 + r^2}$$

can reduce the Robertson–Walker line element to the Minkowski form.

2.3

RELATIVISTIC COSMOLOGY (2024–2028)

2024

Suppose that the universe is described by a $k = 1$ Robertson-Walker space-time with metric

$$ds^2 = -dt^2 + R^2(t)dx^2 + \sin^2 x\left(d\theta^2 + \sin^2\theta d\varphi^2\right)$$

with $R(t) = R_0 t^{2/3}$ at the present epoch.

An observer at $t = t_1$ observes a distant galaxy of proper size D perpendicular to the line of sight at $t = t_0$.

a. What is the observed red shift in terms of R_0, t_0, t_1?

b. What is the angular diameter of the galaxy, δ, in terms of the red shift?

c. Show that as the red shift increases δ reaches a minimum for fixed D and then starts to increase.

(Princeton)

Sol:

a. Suppose in the coordinate frame of the galaxy and observer successive crests of a light wave are emitted by the galaxy at times t_0, $t_0 + \Delta t_0$ and received by the observer at times t_1, $t_1 + \Delta t_1$. The world-line of each crest is a radial null geodesic along which θ and φ remain constant so that

$$0 = -dt^2 + R^2 dx^2,$$

or

$$dx = \frac{dt}{R(t)},$$

along each world-line. Integrating for each crest gives

$$x_0 = \int_0^{x_0} dx = \int_{t_0}^{t_1} \frac{dt}{R(t)} = \int_{t_0+\Delta t_0}^{t_1+\Delta t_1} \frac{dt}{R(t)}.$$

As Δt_0, Δt_1 are small quantities the above implies

$$\frac{\Delta t_0}{R(t_0)} = \frac{\Delta t_1}{R(t_1)}.$$

If λ_0, λ_1 are the emitted and received wavelengths then

$$\frac{\lambda_1}{\lambda_0} = \frac{\Delta t_1}{\Delta l_0} = \frac{R(t_1)}{R(t_0)}$$

and the red shift is

$$z = \frac{\lambda_1 - \lambda_0}{\lambda_0} = \frac{\lambda_1}{\lambda_0} - 1 = \left(\frac{t_1}{t_0}\right)^{\frac{2}{3}} - 1$$

since $R(t) = R_0 t^{\frac{2}{3}}$.

b. The angular diameter of the galaxy is by definition

$$\delta = \frac{D}{R(t_0)\, x_0} = \frac{D}{R_0 x_0} t_0^{-\frac{2}{3}}.$$

As

$$x_0 = \int_{t_0}^{t_1} \frac{dt}{R(t)} = \frac{1}{R_0} \int_{t_0}^{t_1} t^{-\frac{2}{3}} dt = \frac{3}{R_0}\left[\left(\frac{t_1}{t_0}\right)^{\frac{1}{3}} - 1\right] t_0^{\frac{1}{3}},$$

$$\left(\frac{t_1}{t_0}\right)^{\frac{1}{3}} = (z+1)^{\frac{1}{2}},$$

we have

$$\delta = \frac{D}{3\, t_0\left[(z+1)^{\frac{1}{2}} - 1\right]} = \frac{D(z+1)^{\frac{3}{2}}}{3\, t_1\left[(z+1)^{\frac{1}{2}} - 1\right]}.$$

c. Differectiating the above gives

$$\frac{d\delta}{dz} = A(z)\left[(z+1)^{\frac{1}{2}} - \frac{3}{2}\right],$$

where $A(z) > 0$. Hence $\frac{d\delta}{dz} \gtrless 0$ if $(z+1)^{\frac{1}{2}} \gtrless \frac{3}{2}$, or $z \gtrless \frac{5}{4}$. For $z < \frac{5}{4}$, $\frac{d\delta}{dz} < 0$ and the angular diameter decreases as the red shift increases. For $z > \frac{5}{4}$, $\frac{d\delta}{dz} > 0$ and the angular diameter increases as the red shift increases. Thus the angular diameter reaches a minimum at $z = \frac{5}{4}$.

2025

Consider a galaxy "A" in a flat, matter-dominated universe that emits a specific spectral line at a wavelength of 500 nm in its rest frame. Assume that the universe's scale factor at the time when Galaxy A emitted the light was a factor of 1/3 smaller than it is today. When this spectral line is observed from Earth using a powerful telescope, it's seen that the spectral line has been redshifted. However, there is another nearby galaxy, Galaxy "B," between Earth and Galaxy A, at a distance of 100 million light-years from Earth, which is receding from Earth at a speed of 0.6c, and its motion affects the observed redshift of Galaxy A's light.

a. Calculate the redshift expected due to the motion of Galaxy B relative to Earth.

b. Calculate the cosmological redshift expected for Galaxy A's light due to the expansion of the universe.

c. Find out the total observed redshift of Galaxy A's light as seen from Earth.

Sol:

a. **Redshift due to the motion of Galaxy B:**

The redshift caused by the motion of Galaxy "B" can be calculated using the relativistic Doppler effect formula for a receding object:

$$Z_{\text{motion}} = \sqrt{\frac{1 + v/c}{1 - v/c}} - 1$$

where v is the velocity of Galaxy "B" and "c" is the speed of light. Substitution of the values:

$$Z_{\text{motion}} = \sqrt{\frac{1 + 0.6c/c}{1 - 0.6c/c}} - 1 = 1$$

So, the redshift due to the motion of Galaxy B is $Z_{\text{motion}} = 1$.

b. **Cosmological Redshift**

The cosmological redshift can be calculated using the formula for redshift in an expanding universe.

$$Z_{\text{cosmological}} = \frac{a(t_0)}{a(t_e)} - 1$$

where $a(t_0)$ is the scale factor today and $a(t_e)$ is the scale factor at the time when Galaxy "A" emitted the light. Since the universe was 1/3 smaller at that time, $a(t_e) = a(t_0)/3$. Thus,

$$Z_{\text{cosmological}} = \frac{a(t_0)}{a(t_0)/3} - 1 = 3 - 1 = 2$$

So, the cosmological redshift due to the expansion of the universe $Z_{\text{cosmological}}$ =2.

c. **Total Redshift**

When both the motion of Galaxy "B" and the cosmological expansion are involved, the redshifts must be combined as shown in the following, because these two effects are independent of each other. By multiplying these two redshift factors together, the total observed redshift can be obtained

$$1 + Z_{\text{tot}} = (1 + Z_{\text{motion}})\left(1 + Z_{\text{cosmological}}\right)$$

Substituting the values,

$$1 + Z_{\text{tot}} = (1 + 1)(1 + 2) = 6$$
$$Z_{\text{tot}} = 6 - 1 = 5$$

So, the total observed redshift of Galaxy A's light is $Z_{\text{total}} = 5$.

The motion of a galaxy can cause a redshift or blueshift (depending on whether it's moving away or toward the observer), and the cosmological expansion uniformly stretches the wavelengths of light from all objects in the universe, causing a redshift.

2026

The Robertson-Walker metric

$$ds^2 = -dt^2 + a^2(t)\left(\frac{dr^2}{1 - \kappa r^2} + r^2 d\Omega^2\right)$$

(where $k = 0, +1, -1$, according to whether the three-dimensional space has zero, positive or negative curvature, respectively) gives rise to the first order Einstein field equation

$$\dot{a}^2 + k = \frac{8\pi G\rho}{3}a^2,$$
$$\rho a^3 = \text{constant}$$

for a matter-dominated universe of density ρ.

a. Calculate the distance $L_r(t)$ from the origin ($r = 0$) to a particle with coordinate r at time t, in terms of r and $a(t)$.

Alternatively, we can formulate the theory in purely classical Newtonian terms by ignoring curvature inside a spherical volume of sufficiently small radius, i.e., assume that the space is flat inside the sphere and that any isotropic distribution of matter outside has no effect on the curvature inside.

b. Write down Newton's equation for the acceleration of a particle towards the origin at a distance L away. (*Hint*: consider a uniform distribution of matter inside a sphere of radius L.)

c. To conserve matter, we must also have ρa^3 = constant. Combine this with your result in (b) to determine the equations satisfied by the expansion parameter $a(t)$ and compare your answer with the cosmological one.

(*Princeton*)

Sol:

a. The distance from the origin to the particle is

$$\begin{aligned} L_r(t) &= \int_0^r \sqrt{g_{rr}}\, dr \\ &= a(t) \int_0^r \frac{dr}{\sqrt{1 - kr^2}} \\ &= a(t)h(r), \end{aligned}$$

where

$$h(r) = \begin{cases} \arcsin r & (k = 1) \\ r & (k = 0) \\ \operatorname{arcsinh} r & (k = -1) \end{cases}.$$

b. As any isotropic distribution of matter outside a sphere has no effect on the field inside, the gravitational force per unit mass on a particle at radial distance L_r from the origin is

$$-\frac{4\pi}{3}\frac{L_r^3 G\rho}{L_r^2} = -\frac{4\pi}{3} G\rho L_r.$$

Then Newton's equation for radial motion is

$$\ddot{L}_r = -\frac{4}{3}\pi G\rho L_r.$$

Since from (a)

$$L_r(t) = a(t)h(r),$$

the equation becomes

$$\ddot{a} = -\frac{4\pi G\rho}{3}a.$$

c. As $\ddot{a} = \frac{1}{2}\frac{d\dot{a}^2}{da}$, and $\rho a^3 = b$, a constant, the last equation can be written as

$$d\dot{a}^2 = -\frac{8\pi Gb}{3a^2}da.$$

Integration gives

$$\dot{a}^2 + K = \frac{8\pi Gb}{3a} = \frac{8\pi G\rho a^2}{3}.$$

By proper scaling k can be made to have values ± 1 or 0, so that this equation becomes formally identical with the cosmological one. For convenience of interpretation write the equation as

$$\dot{a}^2 = \frac{\text{constant}}{a} - K.$$

When $K = -1$ or 0, any two stars in the universe can have infinite separation but then with a finite or zero relative velocity. When $K = +1$, any two stars can only have finite separation as $\dot{a}^2$ must be positive. Therefore, $K = -1$ or $+1$ will determine whether two stars can have a relative velocity greater or smaller than that required to escape from each other. In other words, if $K > 0$, the universe is closed; if $K < 0$, the universe is open. However, Newton's theory cannot properly give the value of K because it is a global property of the cosmological space-time, while Newton's theory is valid only for local regions.

2027

The metric of the expanding universe has the form $ds^2 = dt^2 - R^2(t)(dx^2 + dy^2 + dz^2)$, where the possible curvature of space has been neglected. The detailed form of the function $R(t)$ depends on the matter content of the universe.

a. A particle of mass m has energy E_0 and momentum p_0 at time t_0; assume $R(t_0) = R_0$. The particle thereafter propagates freely except for the effects of the metric written above. Calculate its energy and momentum as a function of time.

b. Suppose that the early universe contained a gas of non-interacting massless particles (perhaps photons) subject to gravitational effects only. Show that if at time t_0 they were in a thermal distribution at temperature T_0, they remained in a thermal distribution later, but with a temperature that depends on time in a fashion you should determine.

c. Show that, instead, a gas of non-interacting massive particles initially in a thermal distribution would not remain in a thermal distribution under the influence of the expansion of the universe.

d. Suppose that the early universe contained a non-interacting gas of massless photons and also a non-interacting gas of massive particles of mass m (massive neutrinos, to be definite). Suppose that at some early time the photons and neutrinos were both in a thermal distribution with a temperature $kT = mc^2$ (m being the neutrino mass) for both photons and neutrinos. It has been observed that in today's universe the photons are in a thermal distribution with a kT of about 3×10^{-4} eV. In terms of the neutrino mass m, what (roughly) would be the typical velocity and kinetic energy of a neutrino today? Assume $m \gg 3 \times 10^{-4}$ eV.

You may wish to know

$$\Gamma^{\mu}_{\alpha\beta} = \frac{1}{2} g^{\mu\sigma}\left(\partial_\alpha g_{\sigma\beta} + \partial_\beta g_{\sigma\alpha} - \partial_\sigma g_{\alpha\beta}\right).$$

(*Princeton*)

Sol:

a. In polar coordinates the metric is

$$ds^2 = dt^2 - R^2(t)\left(dr^2 + r^2 d\theta^2 + r^2 \sin^2\theta d\varphi^2\right).$$

Letting $x^0 = t$, $x^1 = r$, $x^2 = \theta$, $x^3 = \varphi$, we have

$$g_{00} = 1, \qquad g_{11} = -R^2, \qquad g_{22} = -R^2 r^2, \qquad g_{33} = -R^2 r^2 \sin^2\theta,$$
$$g_{\mu\nu} = 0 \qquad \text{for } \mu \neq \nu.$$

As $g_{\mu\nu}$ is diagonal, for which $g^{\mu\mu} = g_{\mu\mu}$, we have

$$\Gamma^{\mu}_{\alpha\beta} = 0 \qquad \text{for } \mu \neq \alpha \neq \beta,$$

$$\Gamma^{\mu}_{\mu\beta} = \Gamma^{\mu}_{\beta\mu} = \frac{1}{2} g^{\mu\mu} \frac{\partial g_{\mu\mu}}{\partial x^\beta} = \frac{1}{2 g_{\mu\mu}} \frac{\partial g_{\mu\mu}}{\partial x^\beta}$$
$$= \frac{\partial}{\partial x^\beta}\left(\log |g_{\mu\mu}|^{\frac{1}{2}}\right), \quad \text{(no summation)}$$

giving for the metric

$$\begin{aligned}
\Gamma^{\mu}_{0\mu} &= \Gamma^{\mu}_{\mu 0} = \frac{1}{R}\frac{\partial R}{\partial t}, \quad (\mu = 1,\, 2,\, 3)\\
\Gamma^{\mu}_{1\mu} &= \Gamma^{\mu}_{\mu 1} = \frac{1}{r}, \quad (\mu = 2,\, 3)\\
\Gamma^{\mu}_{2\mu} &= \Gamma^{\mu}_{\mu 2} = \Gamma^{\mu}_{3\mu} = \Gamma^{\mu}_{\mu 3} = \Gamma^{\mu}_{\mu\mu} = 0. \quad (\mu = 0,\, 1,\, 2,\, 3)
\end{aligned}$$

The motion of a particle is described by the geodesic equation

$$\frac{d^2x^{\mu}}{d\tau^2} + \Gamma^{\mu}_{\alpha\beta}\frac{dx^{\alpha}}{d\tau}\frac{dx^{\beta}}{d\tau} = 0,$$

where $d\tau = ds$, τ being the local proper time. For $\mu = 1$ the above becomes

$$\frac{d^2 r}{d\tau^2} + \frac{2}{R}\frac{dR}{dt}\frac{dt}{d\tau}\frac{dr}{d\tau} = 0.$$

Multiplying both sides by R^2 gives

$$R^2\frac{d}{d\tau}\left(\frac{dr}{d\tau}\right) + \frac{dR^2}{d\tau}\frac{dr}{d\tau} = 0,$$

or

$$\frac{d}{d\tau}\left(R^2\frac{dr}{d\tau}\right) = 0.$$

Hence

$$R^2\frac{dr}{d\tau} = \text{constant}.$$

The momentum 4-vector of the particle is by definition

$$\begin{aligned}
p^{\alpha} &= mu^{\alpha} = m\frac{dx^{\alpha}}{d\tau},\\
p_{\alpha} &= g_{\alpha\beta}p^{\beta} = g_{\alpha\beta}mu^{\beta}.
\end{aligned}$$

Then

$$p^{\alpha}p_{\alpha} = g_{\alpha\beta}p^{\alpha}p^{\beta} = m^2 g_{\alpha\beta}\frac{dx^{\alpha}}{d\tau}\frac{dx^{\beta}}{d\tau},$$

or

$$m^2\left(\frac{dt}{d\tau}\right)^2 - R^2 m^2\left(\frac{dr}{d\tau}\right)^2 = m^2,$$

making use of the metric equation. Writing the left-hand side as $E^2 - p^2$, we have the momentum

$$p = mR\frac{dr}{d\tau},$$

the energy

$$E = \sqrt{m^2 + p^2}.$$

Initially $p = p_0$, $E = E_0$, so that

$$R^2 \frac{dr}{d\tau} = \frac{R_p}{m} = \frac{R_0 p_0}{m},$$

or

$$p(t) = \frac{R_0}{R} p_0,$$

and thus

$$\begin{aligned} E(t) &= \sqrt{m^2 + \left(\frac{R_0}{R}\right)^2 p_0^2} \\ &= \sqrt{E_0^2 - p_0^2 + \left(\frac{R_0}{R}\right)^2 p_0^2} \\ &= \sqrt{E_0^2 - p_0^2\left[1 - \left(\frac{R_0}{R}\right)^2\right]}, \end{aligned}$$

where $R_0 = R(t_0)$.

b. If the gas of photons is in a thermal equilibrium at time t_0, then according to Planck's theory of black body radiation the number in volume $V(t_0)$ of photons with frequencies between ν and $\nu + d\nu$ are

$$dN(t_0) = \frac{8\pi\nu^2 V(t_0) d\nu}{c^3\left[\exp\left(\frac{h\nu}{kT_0} - 1\right)\right]},$$

where h and k are Planck's and Boltzmann's constants. At a later time t', photons with original frequency ν will have frequency ν' given by (**Problem 2025**)

$$\frac{\nu'}{\nu} = \frac{R(t_0)}{R(t')}.$$

Also, the volume of the gas has changed as the scaling factor $R(t)$ changes:

$$\frac{V(t')}{V(t_0)} = \frac{R^3(t')}{R^3(t_0)}.$$

Hence

$$dN(t') = dN(t_0) = \frac{\frac{8\pi}{c^3}\left(\frac{\nu' R(t')}{R(t_0)}\right)^2 \frac{R^3(t_0)}{R^3(t')} V(t') \frac{R(t')}{R(t_0)} d\nu'}{\exp\left[h\nu' R(t')/R(t_0) kT(t_0) - 1\right]}.$$

If we set

$$T(t') = \frac{R(t_0) T(t_0)}{R(t')}$$

the distribution will still retain the black body form:

$$dN(t') = \frac{8\pi V(t') \nu'^2 d\nu'}{c^3\left\{\exp\left[\frac{h\nu'}{kT(t')}\right] - 1\right\}}.$$

c. Consider non-interacting massive particles as an ideal gas. In thermal equilibrium the number of particles with momenta between p and $p + dp$ is

$$dN_p = \frac{gVp^2 dp}{2\pi^2 h^3}\left[\exp\left(\frac{E-\mu}{kT}\right) \pm 1\right]^{-1},$$

where $E = (m^2 + p^2)^{1/2}$ and μ the chemical potential, can be set to zero as the particles are non-interacting. Since the expansion of the gas is adiabatic, $TV^{\gamma-1} =$ constant, and hence $T \propto R^{-3(\gamma-1)}$ as $V \alpha R^3$. Since $p \alpha \frac{1}{R}$, $E = (m^2 + p^2)^{1/2}$, the initial thermal equilibrium distribution cannot be maintained as the universe expands.

d. Considering the photon gas, we have from (b)

$$\frac{R(t)}{R(t_0)} = \frac{T(t_0)}{T(t)} = \frac{mc^2}{3 \times 10^{-4}}.$$

For the neutrinos, we then have

$$\frac{p(t)}{p(t_0)} = \frac{R(t_0)}{R(t)} = \frac{3 \times 10^{-4}}{mc^2}.$$

At the initial time t_0, the neutrinos have kinetic energy $\approx mc^2$, i.e.,

$$\sqrt{p^2c^2 + m^2c^4} - mc^2 \approx mc^2.$$

Hence

$$p(t_0) \approx \sqrt{3}\, mc.$$

It follows that at the present time t,

$$p(t) \approx 3\sqrt{3} \times 10^{-4}\,\text{eV}/c,$$

$$v(t) = \frac{p(t)}{m} \approx \frac{3\sqrt{3} \times 10^{-4}}{mc^2} \times c$$

where mc^2 is in eV. As $v \ll c$, the kinetic energy is approximately

$$\frac{m}{2}v^2 = \frac{mc^2}{2}\left(\frac{v}{c}\right)^2 = \frac{1.35 \times 10^{-7}}{mc^2}\,\text{eV}.$$

2028

Assume that the geometry of the universe is described by the Robertson-Walker metric ($c = 1$)

$$ds^2 = -dt^2 + R^2(t)\left(\frac{dr^2}{1 - kr^2} + r^2 d\Omega^2\right).$$

A spaceship sets out with velocity v relative to cosmological observers. At a later time when the universe has expanded by a scale factor $(1 + z)$, find the velocity v' with respect to cosmological observers.

(*Princeton*)

Sol: Since k is a constant, by properly selecting the unit of r we can make $k = \pm 1$ or 0. Then using the transformations $r = \sin x$ and $r = \sinh x$ for $k = +1$ and $k = -1$ respectively, we have

$$ds^2 = \begin{cases} -dt^2 + R^2(t)[dx^2 + \sin^2 x d\Omega^2], & k = +1 \\ -dt^2 + R^2(t)[dx^2 + \sinh^2 x d\Omega^2], & k = -1 \end{cases}.$$

Suppose the spaceship is launched along a radial direction so that $\theta =$ constant, $\varphi =$ constant. We also introduce the local proper time τ by $-d\tau^2 = ds^2$. Then the metric becomes

$$d\tau^2 = dt^2 - R^2(t)dx^2.$$

As all $g_{\alpha\beta}$ do not depend on x explicitly, the geodesic equation

$$\frac{d}{d\tau}\left(g_{\mu\nu}\frac{dx^\nu}{d\tau}\right) - \frac{1}{2}g_{\alpha\beta,\mu}\frac{dx^\alpha}{d\tau}\frac{dx^\beta}{d\tau} = 0$$

gives for $\mu = 1$

$$\frac{d}{d\tau}\left(g_{11}\frac{dx}{d\tau}\right) = 0,$$

or

$$R^2(t)\frac{dx}{d\tau} = R^2(t)\frac{dx}{dt}\left(\frac{d\tau}{dt}\right)^{-1} = \frac{R^2(t)\frac{dx}{dt}}{\sqrt{1 - R^2(t)\left(\frac{dx}{dt}\right)^2}} = \text{constant}.$$

As the length element is $dl = R(t)dx$ and the velocity is thus $v = \frac{dl}{dt} = R(t)\frac{dx}{dt}$, the above can be written as

$$\frac{R(t)v}{\sqrt{1 - v^2}} = \text{constant},$$

or

$$\frac{R(t')v'}{\sqrt{1 - v'^2}} = \frac{R(t)v}{\sqrt{1 - v^2}}.$$

As $R(t') = (1 + z)R(t)$, we have

$$\frac{v'}{\sqrt{1 - v'^2}} = \frac{v}{\sqrt{1 - v^2}} \frac{1}{(1 + z)},$$

or

$$v'^2 = \frac{v^2}{(1 - v^2)(1 + z)^2 + v^2},$$

i.e.,

$$v' = \frac{v}{(1 + z)\sqrt{1 - v^2 + \left(\frac{v}{1+z}\right)^2}} \approx \frac{v}{1 + z}$$

for $v \ll 1$.

Part III
Miscellaneous Topics

3.1

HISTORY OF PHYSICS AND GENERAL QUESTIONS (3001–3025)

3001

State two major contributions to physics made by each of the following individuals:

- **a.** Rudolf Clausius
- **b.** Albert A. Michelson
- **c.** Enrico Fermi
- **d.** Murray Gell-Mann

Sol:

- **a.** (1) Gave a formulation of the second law of thermodynamics. (2) Introduced the concept of entropy.
- **b.** (1) Invented the Michelson interferometer for measuring small lengths such as light wavelengths. (2) Using this interferometer performed the Michelson–Morley experiment which disproved the existence of ether and paved the way for Einstein's theories of relativity.
- **c.** (1) Applying the Pauli exclusion principle to statistical mechanics, founded (independently with P. A. M. Dirac) the Fermi–Dirac statistics. (2) Studied the reactions of neutrons culminating in demonstrating the possibility of chain reactions, and led a team to construct the first nuclear reactor.

d. (1) Proposed the concept of fractionally charged quarks. (2) Suggested the eightfold way of classification of elementary particles.

3002

"Endless Void or Big Crunch"—Predict the fate of the universe.

a. Justify your answer in the context of dark energy.

b. Sketch a timeline based on your prediction.

Sol: Let's discuss about the two scenarios in detail and then we will come to a conclusion.

a. **Endless Void (Heat Death or Big Freeze)**

Dark energy is a mysterious force that appears to be causing the expansion of the universe to accelerate. Its primary effect is to counteract gravity, pushing galaxies and galaxy clusters apart. If dark energy continues to dominate and its repulsive effects grow stronger over time, it could eventually lead to an "Endless Void" scenario. In this scenario, the universe will continue to expand at an accelerating rate, causing galaxies to drift apart from each other. As the universe expands, galaxies will move away from one another, stars will burn out, and ultimately, the universe will become colder and darker. This scenario is often referred to as the "Heat Death" or "Big Freeze" of the universe. It implies that the universe will end in a state of maximum entropy, where all energy gradients are dissipated, and nothing interesting happens on cosmic scales.

Big Crunch (or Big Bounce)

In this scenario, dark energy's effects are counteracted by the gravitational attraction of matter and radiation in the universe. If the density of matter and radiation is sufficiently high and dark energy is relatively weak or absent, the universe's expansion could slow down and eventually reverse. This would lead to a contraction of the universe, ultimately resulting in a "Big Crunch." In the Big Crunch scenario, the universe would collapse in on itself, potentially leading to a singularity, similar to the Big Bang. This could result in a cyclic universe, where the universe goes through an endless cycle of expansion and contraction, known as the "Big Bounce."

The current observational evidence suggests that dark energy is indeed causing the universe to expand at an accelerating rate, which favors the **"Endless Void"** scenario.

b. Predicting the Precise Timeline of the Future Universe

Immediate Future (Next Billions of Years):

Cosmic Expansion Continues: Dark energy will dominate, causing the universe to expand at an accelerating rate. Galaxies will move further apart.

Galactic Evolution: Within galaxy clusters, gravitational interactions will lead to the formation of new stars and the eventual merging of galaxies.

Stellar Evolution: Stars will exhaust their nuclear fuel and go through various stages of evolution, potentially forming black holes, neutron stars, or white dwarfs.

Tens to Hundreds of Billions of Years

Heat Death: As galaxies drift apart, the universe becomes increasingly cold and dark.

Star Formation Ceases: All gas and dust needed for star formation will be depleted. Existing stars will eventually exhaust their fuel and go dark.

Trillions of Years

Stellar Remnants: Most stars would have died, leaving behind stellar remnants like white dwarfs, neutron stars, and black holes.

Rare Stellar Events: Extremely rare events like proton decay or black hole mergers might be the only notable occurrences in an otherwise dormant universe.

Tens of Trillions to Quadrillions of Years

Degenerate Era: Stellar remnants will slowly decay through processes like Hawking radiation (for black holes) and electron capture (for white dwarfs).

Isolation: Galaxies and star systems will become increasingly isolated from each other due to cosmic expansion.

Quadrillions of Years and Beyond

The End of Matter: Over an extremely long time, even the most stable objects, like black holes, may eventually decay due to quantum processes.

The Ultimate Fate: The universe will approach a state of maximum entropy and minimal energy gradients. It will become a cold, dark, and near-featureless expanse—an "Endless Void."

3003

Name four women who have made important contributions to physics research. Describe briefly the contributions of each.

Sol:

1. Marie Curie – discovered the elements radium and polonium.
2. Maria Goeppert Mayer – established a shell model for the structure of atomic nuclei.
3. Irene Joliot Curie – participated in the discovery of the neutron.
4. Chien-Shiung Wu – verified experimentally the nonconservation of parity in weak interactions by studying the β-decay of ^{60}Co.

3004

a. Identify the atom model shown below and point out its limitations and problems

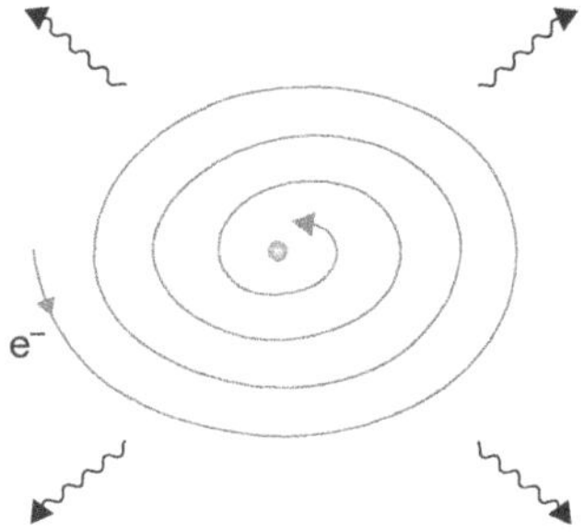

b. Did Planck contribute anything to resolve the problem with the previous model?

c. Using Planck's idea, how did Bohr propose the new atom model?

Sol:

a. The atom model shown is the one proposed by Ernest Rutherford in 1911. Here are the main problems with Rutherford's atomic model:

Lack of Explanation for Stability: Rutherford's model depicted the atom as a central nucleus with electrons orbiting around it, similar to the way planets orbit the sun in our solar system. However, classical electromagnetic theory predicts that accelerated charged particles, like electrons orbiting the nucleus, should continuously emit electromagnetic radiation (radiate energy) and spiral into the nucleus. This would result in the collapse of the atom, which contradicts the stability of matter.

Continuous Spectrum: Rutherford's model could not explain the discrete line spectra observed for hydrogen and other elements. The model predicted a continuous spectrum, whereas the observed spectra consisted of distinct lines.

Electron Path Predictions: Rutherford's model did not specify the exact paths or orbits that electrons would follow. Instead, it suggested that electrons move in elliptical orbits around the nucleus, which contradicted the quantized energy levels observed in atomic spectra.

Incompatibility with Quantum Mechanics: Rutherford's model was based on classical physics, which could not account for the wave-like properties of electrons, as described by quantum mechanics.

Angular Momentum Problem: Rutherford's model did not account for the quantization of angular momentum.

b. Max Planck's resolution in the context of Rutherford's model:

Max Planck's groundbreaking work in resolving the ultraviolet catastrophe in blackbody radiation was not directly related to Rutherford's atomic model but rather to the problem of understanding the spectral distribution of electromagnetic radiation emitted by hot objects (blackbodies). The ultraviolet catastrophe was a problem in explaining why classical Physics predicted that the energy radiated by a blackbody should increase without being bound at high frequencies, which contradicted experimental observations. Planck's solution to this problem was to introduce the concept of quantized energy levels for the oscillators responsible for emitting radiation within a blackbody. In this context, it didn't directly address the issues with Rutherford's atomic model. However, Planck's idea of quantized energy levels for oscillators became a foundational concept in the development of quantum mechanics, which later played a crucial role in resolving issues with atomic models, including Rutherford's.

c. Bohr's atomic model using Planck's ideas:

Niels Bohr's atomic model, proposed in 1913, was heavily influenced by the emerging principles of quantum mechanics, including Max Planck's idea of quantized energy levels. Bohr's model was specifically designed to address the problems with Rutherford's atomic model and to explain the behavior of electrons within atoms.

Bohr's key contributions based on Planck's ideas are

- Electrons were allowed to exist only in specific quantized energy levels or orbits around the nucleus. These energy levels were associated with fixed amounts of energy.
- Electrons could change from one energy level to another by either absorbing or emitting energy in discrete units or quanta. When an electron absorbs energy, it moves to a higher energy level, and when it emits energy, it moves to a lower energy level.
- Bohr used Planck's relationship ($E = h\nu$) to relate the energy of emitted or absorbed photons to the energy difference between the initial and final electron energy levels. This explained the discrete spectral lines observed in the hydrogen atom's emission spectrum.
- Bohr's model successfully explained the quantization of energy levels in the hydrogen atom and provided a framework for understanding the stability of atoms, as well as the observed spectral lines.

However, Bohr's model had limitations, particularly in explaining the behavior of electrons in more complex atoms. It was eventually replaced by the more comprehensive quantum mechanical model, which extended the principles of quantization to all atomic systems and provided a more accurate description of atomic structure.

3005

Name two physicists (one experimentalists and one theorist) who have made important contributions to our understanding of each of the following phenomena. State briefly (in one sentence) what each contributed and about when the work was done.

a. Scattering of photons by electrons
b. Nuclear fission
c. Parity violation in physics
d. Superconductivity or superfluidity
e. Strangeness or charm quantum number

Sol:

a. A. H. Compton in 1923 explained the scattering of a photon by a free electron resulting in a photon of longer wavelength. In 1927 A. A. Bless showed that the observed energy of the recoil electrons are in agreement with Compton's theory.

b. In 1938 Otto Hahn and Fritz Strassmann demonstrated the nuclear fission of uranium by finding that barium was produced in the bombardment of uranium by neutrons. In the same year Lise Meitner and Otto Robert Frisch theoretically explained the experimental results.

c. In 1956 T. D. Lee and C. N. Yang proposed the hypothesis of parity non-conservation in weak interactions. This was experimentally verified by Chien-Shiung Wu in 1957 by studying the β-decay of ^{60}Co.

d. In 1911 Onnes discovered the superconductivity of metals at low temperatures. In 1957 Bardeen, Schrieffer and Cooper proposed a microscopic theory of superconductivity.

e. In 1954 Yasuo Nagano, Kazuhiko Nishijima and M. Gellmann independently suggested that certain elementary particles have an additional quantum number called strangeness, besides spin, charge and isospin.

3006

Which of the following physicists does not belong in this list of Nobel laureates? Explain your answer and highlight the work for which the Nobel laureates received their prize.

Richard Feynman, Werner Heisenberg, Ernest Rutherford, Galileo Galilei, Max Planck, Enrico Fermi, Pierre Curie, James Clerk Maxwell, Niels Bohr, Marie Curie, and Erwin Schrödinger.

Sol: There are two physicists in the given list who did not receive any Nobel Prize. They are

Galileo Galilei: He is primarily known for his contributions to astronomy that predate the Nobel Prize, and he is not a Nobel laureate.

James Clerk Maxwell: He is well-known for his work in classical electromagnetism, particularly the formulation of Maxwell's equations, which do not fall within the scope of Nobel-recognized physics as it was developed in the 20th century.

The other physicists mentioned in the list received Nobel Prizes for their significant contributions to various fields of physics, such as quantum mechanics, nuclear Physics, and the study of radiation phenomena.

a. **Richard Feynman**: Nobel Prize in physics for his contributions to quantum electrodynamics (QED).

b. **Werner Heisenberg**: Nobel Prize in Physics for the creation of quantum mechanics.

c. **Ernest Rutherford**: Nobel Prize in Chemistry for his investigations into the disintegration of the elements and the chemistry of radioactive substances.

d. **Max Planck**: Nobel Prize in Physics for his discovery of energy quanta and the development of quantum theory.

e. **Enrico Fermi**: Nobel Prize in Physics for his work on induced radioactivity.

f. **Marie Curie and Pierre Curie**: Nobel Prize in Physics for their joint research on the radiation phenomena.

g. **Niels Bohr**: Nobel Prize in Physics for his model of the atom and the structure of the atom's nucleus.

h. **Erwin Schrödinger**: Nobel Prize in Physics for his development of the Schrödinger equation in quantum mechanics.

3007

Below are listed three sets of names. For each set indicate the physical problem they commonly resolved. List their names according to the chronological order of their contributions. For one of the sets, discuss carefully the contribution of each person.

a. Maxwell, Einstein, Newton, Huygens, Young, Fresnel.

b. Schrödinger, Bohr, Planck, Dirac, Einstein, Heisenberg.

c. Jensen and Mayer, Chadwick, Fermi, Rutherford.

Sol:

a. Huygens → Newton → Young → Fresnel → Maxwell → Einstein: Wave and corpuscular nature of light. Specifically,

Huygens: formulation of Huygens' principle in the wave theory of light,

Newton: formulation of a corpuscular theory of light,

Young: experimental study of the interference of light,.

Fresnel: explanation of the diffraction of light,

Maxwell: formulation of the electromagnetic theory of light,

Einstein: hypothesis of light quantum.

b. Planck → Einstein → Bohr → Heisenberg → Schrödinger → Dirac: Development of quantum mechanics. Specifically,

Planck: hypothesis of energy quantum,

Einstein: hypothesis of light quantum,

Bohr: use of quantum idea in atomic model,

Heisenberg: formulation of matrix mechanics,

Schrödinger: formulation of wave mechanics,

Dirac: formulation of relativistic wave equation of electron.

c. Rutherford → Fermi → Chadwick → Jensen and Mayer: Theory of nuclear structure. Specifically,

Rutherford: proposing an atomic model with a massive, positively charged nucleus surrounded by electrons,

Chadwick: experimental discovery of the neutron, which together with protons constitutes the atomic nucleus,

Fermi: formulation of the theory of β-decay of atomic nuclei,

Jensen and Mayer: formulation of the shell model of atomic nucleus.

3008

With what advances in physics are the following persons associated? 1. H. Hertz 2. T. Young 3. J. Henry 4. J. W. Gibbs 5. J. Chadwick 6. W. Lamb 7. M. Gell-Mann 8. H. A. Kramers 9. H. Becquerel

Sol:

1. H. Hertz: Experimental verification of the existence of electromagnetic waves.
2. T. Young: Experimental study of the interference of light.
3. J. Henry: Discovery of self-induction and mutual induction.
4. J. W. Gibbs: Development of analytical thermodynamics and the representation of a system by graphic means.
5. J. Chadwick: Discovery of the neutron.

6. W. Lamb: Discovery of the Lamb shift for hydrogen atom: the $2^2S_{\frac{1}{2}}$ level being 1058 MHz higher than the $2^2P_{\frac{1}{2}}$ level.
7. M. Gell-Mann: Formulation of the quark model of hadrons.
8. H. A. Kramers: Formulation of the WKB method for approximately solving the Schrödinger equation.
9. H. Becquerel: Discovery of radioactivity.

3009

Give a physical explanation of the following phenomena. Before explaining the phenomena describe as precisely as you can the nature of them and the circumstances under which they are normally observed. (Relative position of the Sun and observer, direction of observation, state of atmosphere, etc.)

1. Reddening of the Sun at sunset (or sunrise).
2. The blue sky.
3. The rainbow arcs (primary and secondary).
4. The halo (at 22°) around the Sun.

Sol:

1, 2. Clear sky assumes the light blue color as a result of the scattering of sunlight by the molecules of the atmosphere which has a nonhomogeneous density distribution arising from irregular molecular motions. Rayleigh's law states that the scattered light intensity is inversely proportional to the fourth power of wavelength. Hence the light-blue and blue color light which has shorter wavelengths is scattered more strongly than the yellow and red color light with longer wavelengths. As a result clear sky, which is seen by scattered light, looks light blue.

At sunrise and sunset, the sunlight is almost parallel to the Earth's surface and thus has to traverse a thicker atmosphere to reach the observer. Most of the light with shorter wavelengths like the blue light is scattered out of its path, only longer wavelength light like the red light remains in its path to reach the observer.

Besides the greater layer of air to traverse, the mist and dust in the air near the surface of the earth further enhances the scattering of shorter wavelengths. Hence the Sun, when looked at directly, appears red at sunrise and sunset.

3. After rain there remain suspended in the air numerous tiny water droplets, which behave as small prisms and refract the incident sunlight. The resulting dispersion causes the formation of rainbow. As the refraction angle is nearly the same as the incident angle, the Sun, the centre of curvature of the rainbow and the observer must be in a straight line for the rainbow to be seen. The secondary rainbow is formed by secondary refraction by the water droplets.

4. The halo is caused mainly by refraction of sunlight by tiny ice crystals in the atmosphere. Most common is the ring of 22° diameter, corresponding to refraction of an ice crystal with faces meeting at 60°. Reflection of light from the surfaces of the crystal also plays a part.

3010

Give a physical explanation for the following phenomena. Be as quantitative as possible.

1. Salts of transition metals and rare earths are often highly colored and paramagnetic while salts of alkali or alkali-earth metals are usually colorless and diamagnetic.
2. The mantle of a gasoline lantern glows white even though the temperature of the mantle does not correspond to a "white" visible spectrum. Furthermore, the color of the unheated mantle is white rather than black, which would be required for a good radiator of light.
3. A solution of cane sugar rotates plane polarized light to the right even in the absence of a magnetic field.
4. A mixture of helium and neon gases will lase when excited by an electrical discharge.

Sol:

1. Salts of alkali or alkali-earth metals are mostly ionic crystals in which an alkali or alkali-earth atom loses its valence electrons and become a positive ion while the remaining electrons form closed shells. The total spin of the acid radical ion is supplied by the spins of the constituent atoms and the angular momentum of their relative motions (contributions from nuclear spins are very small since as $m_{\text{nucleus}} \approx 10^3 m_{\text{electron}}$ the nuclear magnetic moments are extremely small). But as the shells of each atom in an acid radical ion are fully filled, the total spins of both the metallic ion and the acid radical ion are zero. It follows that this kind of salt is diamagnetic. The fact that both the metallic and acid radical ions are of closed shells also means that the gap between the first excited state and the ground state is large so that it is not possible

for visible light to excite a molecule to an excited state and the salt is usually colorless.

For atoms of transition metals and rare earths, the d or f electrons of the second outer shell do not form a closed shell and so the binding is not very strong. These inner electrons may sometimes be excited so that such atoms are frequently multivalent. A more important point is that when the valence electrons are lost the outer shell of a metallic ion is usually not closed. This means that the metallic ions generally have nonvanishing spins and consequently the salts are generally paramagnetic. The fact that the outer shell of a metallic ion is incomplete means that the inner electrons can be excited easily. There may be several energy levels whose spacings from the ground state are within the visible light range, giving rise to absorption peaks in the visible region. Hence these salts are often colored.

2. The mantle of a gasoline lantern is usually coated with a salt of the rare earth group. When the lantern is lighted, the energy released by chemical reaction excites the valence electrons of the metallic ions. Many electrons are excited within the visible light range and the different colors emitted on their return to lower levels mix to become white light, a situation not unlike that of black body radiation.

3. Sugar is an optically active substance which rotates the plane of polarization of plane polarised light passing through it. In particular its optical activity is right-handed (when looking toward the oncoming light the rotation is clockwise). The activity arises from the fact that the molecule of cane sugar contains an asymmetric carbon in the dextro group which rotates polarized light to the right. Even though the orientation of sugar molecules in a solution is irregular, the absence of left-rotating molecules makes the rotation right-handed.

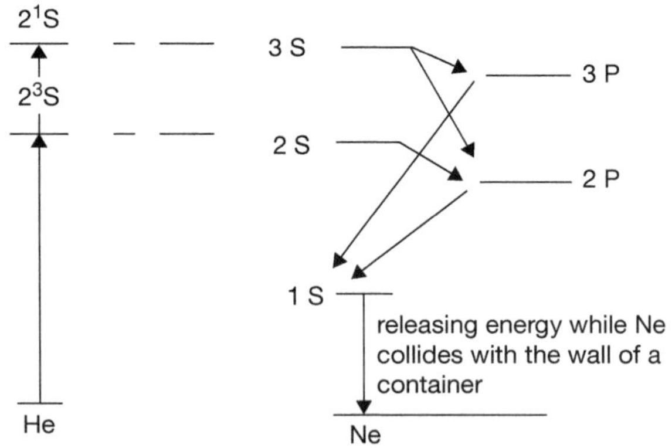

Fig. 3.1

4. Fast electrons in the plasma formed by electric discharge in a mixture of He and Ne gases can easily collide with the He atoms and excite them to the metastable energy levels 2^3S and 2^1S, which match well with the laser energy levels of Ne atom, the later being slightly lower as shown in Fig. 3.1. The energy of excitation of He atoms can very effectively transfer to the laser energy levels of Ne to produce population inversion and subsequently lase. Thus the role of the helium is to promote effectively the pumping process, thereby ensuring the realization of population inversion.

3011

Give the best guess you can for each of the following quantities:

a. Voltage to produce a 5 cm spark in air.

b. Earth's magnetic field.

c. Top magnetic field of a standard laboratory iron-core electromagnet.

d. Cosmic ray flux at sea level.

e. Pressure in a metal vacuum system with oil diffusion pump.

f. Safe current-carrying capacity of insulated copper wire with conductor 1 mm in diameter.

g. Boiling point of liquid nitrogen (1 atmosphere).

Sol:

a. 150 kV.

b. 5.0×10^{-3} Wb m^{-2} or 0.5 G.

c. 1714 G (at room temperature).

d. 0.5 cm^{-2}s^{-1}.

e. 10^{-8}–10^{-6} Torr (vapor pressure at 25°C).

f. 12 A (with 600-volt rubber or ethylene insulating coating).

g. 77.36 K.

3012

For any 3 of the areas of physics listed below give an example of a significant discovery or development which has occurred within the last 5 years. In one or two sentences describe the significance, who did the work, and where the work was done.

a. High Energy Physics
b. Astrophysics or Space Physics
c. Plasma Physics
d. Atomic Physics
e. Nuclear Physics
f. Solid State Physics

Sol:

a. Using the CERN proton-antiproton collider a group of over one hundred experimentalists led by C. Rubbia discovered the charged intermediate vector bosons W^+ and W^- during November–December 1982, and the neutral intermediate vector boson Z^0 during April–May 1983. These discoveries are of great significance as they experimentally verified the unification of weak and electromagnetic interactions.

b. On January 26, 1983, USA, Britain and Holland jointly launched an infrared astronomical satellite (IRAS), which marked the beginning of the era of infrared astronomy in space. For the first time infrared observations of celestial bodies could be made without obstruction of the Earth's atmosphere.

c. The Tokamak Fusion Test Reactor (TFTR) of Princeton University, USA was completed in 1982, and the Joint European Toras (JET) of Kirham laboratory in UK was completed a year later. These facilities will be used for the study of thermonuclear fusion for energy production.

d. An important advance in the research of single atoms was made in 1986 when Swiss scientists developed a new type of microscope which could observe single atoms. In the meantime, US scientists had been able to "catch" single atoms using a laser trap and US and German scientists claimed that they had observed quantum transitions in a single atom which was first suggested by Bohr.

e. Following the discovery of the element of atomic number 107 in 1981 and the element of atomic number 109 in 1982, element 108 was found by fourteen physicists led by G. Mungenberg at the GSI Laboratory in Germany in March, 1984. Studies of these elements may show new nuclear structures or effects beyond those of the old periodic table.

f. In 1982, H. Stormer, A. C. Gossard and Dan Tsui of the Bell Laboratory, USA, who were studying the high-mobility junction $Al_xGa_{1-x}/GaAs$, discovered quite unexpectedly a new fractional quantum Hall effect. It has been speculated that this effect, which had not been predicted by theorists, might have something to do with the Coulomb interaction between electrons. On the practical side it might lead to the development of a series of new devices.

3013

Each of following phrases is common in the recent physics literature. With a very few brief and precise sentences identify and characterize each and indicate scientific and industrial applications: Rydberg atoms, synchrotron radiation, fiber optics, neutral beams, colliding beams, jets.

Sol:

1. *Rydberg atoms*

   These are atoms in highly excited states with the principal quantum number n as large as 10 to 10^2. In the interstellar medium hydrogen atoms with $n > 250$ have been found. One can also prepare hydrogen and other atoms with $n \sim 100$ in the laboratory. Since the energy levels of such highly excited states lie close to the ionization level, these atoms can be easily ionized. Meanwhile, since the atomic radius is proportional to n^2, the linear sizes of these atoms are almost of the order of a micron. They may remain unperturbed when other atoms pass through them. Furthermore they have very long natural lives. Their other properties are still a subject of further research. A knowledge of the Rydberg atoms is important for research in radio astronomy and on new types of laser.

2. *Synchrotron radiation*

   This refers to electromagnetic radiations emitted by relativistic electrons moving in a magnetic field which are strongly directional, being emitted in directions close to the electron trajectory. Usually produced by a synchrotron or storage ring, they serve as a source of strong electromagnetic radiation for research in atomic physics, materials science, astrophysics, etc.

3. *Fiber optics*

   An optical fiber, often made of quartz glass, has the form of a long cylinder with a transparent core and shell with the latter having a smaller index of refraction. It may also be a dielectric cylinder with the refraction index decreasing radially outward from the axis. Because of total reflection at the boundary surface propagation of light is limited to the core. In fact, light propagates with the velocity of light in free space along the fiber which provides a flexible guiding tube with little leakage or absorption. It is widely used in communications and as a flexible source of light.

4. *Neutral beams*

The Glashow-Weinberg-Salam model of weak electromagnetic unification predicted the existence of neutral-beam weak interaction, which had not been observed experimentally at that time, in addition to successfully explaining the basic laws of the known weak and electromagnetic interactions. Neutral beams are currents of particles that have the same charge at the initial and final states. The existence of neutral-beam weak interaction was detected experimentally in 1973, giving support to the GWS model. In 1978, nine independent experimental groups made measurements on the intensity and properties of neutral-beam weak interaction. The results confirmed the correctness of the model. In particular, the Weinberg angle could be accurately determined using neutral beams.

5. *Colliding beams*

Positively and negatively charged particles orbiting in opposite directions in a storage ring are allowed to collide headon after they reach maximum energy. These are called colliding beams. The particles are usually electrons and positrons, or protons and antiprotons. For these the laboratory frame is identical with the center-of-mass frame of the colliding particles.

Suppose a particle of rest mass m can be accelerated to a total energy E in a synchrotron. If it collides with a stationary particles of the same rest mass, the collision energy, which is the energy of the colliding particles in the center of mass frame, is $\sqrt{2(E + mc^2)mc^2} \approx \sqrt{2Emc^2}$. On the other hand, in the colliding beam arrangement, in which each particle has a total energy E, the collision energy is $2E$, which is much higher. Here lies the advantage of using colliding beams for the study of hight-energy interactions.

6. *Jets*

Collision of high-energy particles produces a large number of secondary particles, which tend to be emitted in certain well-defined directions. They are said to form jets. For example, high-energy e^+e^- annihilations, collisions of high-energy hadrons and deep inelastic scatterings of leptons by hadrons can produce jets. Jets are generally considered to be produced in collisions involving a deeper level, and their study may yield new insight into the structure of hadrons and their interactions.

3014

Discuss the source of the Sun's energy. What controls (or limits) the rate of energy release? What observations of the Sun made on the Earth could give direct information concerning the reactions involved deep in the Sun's interior?

Sol: The Sun's energy is supplied by the thermonuclear fusion that takes place continuously inside the Sun. There are two kinds of thermonuclear reaction, proton-proton and carbon-nitrogen, each having the final effect of combining four hydrogen nuclei (protons) into a helium nucleus. In the process each hydrogen nucleus transforms 7% of its mass into energy. Thermonuclear reactions can only be initiated at temperatures at least as high as about 10^7 K and must therefore take place in the core region of the Sun where the temperature is sufficiently high. The energy released in the reaction appears initially mainly in the form of X-rays and γ-rays. It reaches the surface of the Sun after numerous absorbtion and scattering processes, where it is radiated as light and heat. The rate of radiation is governed by mechanical equilibrium in the Sun's gravitational field and thermal equilibrium, so that there is a balance between the energy produced by the thermonuclear reaction and the energy radiated from the Sun's surface. As hydrogen nuclei combine to form helium nuclei, a large number of neutrinos are emitted. Since the interaction of neutrinos with matter is very weak, they can penetrate easily through the outer layers of the Sun to reach the earth. A study of the neutrino flux, theoretically estimated at 10^{11} cm^{-2}s^{-1} near the Earth's surface, can yield detailed information on the process of energy production in the core region of the Sun.

3015

Write briefly on what you know about pulsars. What is the range of periods and in what wavelength regions are they known to emit? What is the likely origin of the pulsed radiation and what is the supposed energy source?

Sol: Pulsars are celestial bodies that emit radio pulses at highly regular intervals of time. The pulses may be in the γ-ray, X-ray, or even optical frequency range, and have periods usually less than 4.3 s, even as low as 0.033 s for the Crab Nebula pulsar. Recently radio pulsars have been found with millisecond pulse periods. X-ray pulsars have relatively large periods, from 0.7 s to 8.35 s. It is now generally believed that the pulsar is a kind of magnetic neutron star which rotates rapidly and possesses magnetic field so strong that the helical orbit of an electron is stretched into almost a straight line. At the magnetic poles, electrons are ejected in streams along the magnetic field lines, and the synchrotron radiation emitted by the electrons also form fine beams along such directions. Usually the magnetic axis of a neutron star does not coincide with the rotational axis. As the neutron star rotates, a fine beam of synchrotron radiation sweeps across space and appears as a pulse each time it encounters our telescope. Thus the period of the pulses mirrors the period of rotation of the neutron star. The radiant energy of a pulsar possibly originates from the rotational energy of the neutron star: the rotation gradually slows down, releasing its kinetic energy to maintain the observed pulsed radiation. How the rotational energy is converted into the energy of the fast electrons is still unknown.

3016

A counting experiment has an average counting rate r of 10^3 counts/s, part of which is due to random background. The average background counting rate r_B is 900 counts/s.

a. What is the average counting rate r_T of "true" event's?

b. How long must be the counting time interval Δt in order to achieve a signal-to-noise ratio of 1 (the signal is the number of true events in Δt)?

c. If the counting efficiency for true events is increased a factor of 4, the background counting rate remaining unchanged, what Δt achieves the same signal-to-noise ratio as in part (b)?

Sol:

a. The average counting rate of "true" events is

$$r_T = r - r_B = 1000 - 900 = 100 \text{ events/s.}$$

b. To achieve a signal-to-noise ratio of 1, we require $r_T \Delta t = 900$ counts.

Hence the counting time interval required is

$$\Delta t = \frac{900}{100} = 9 \text{ s.}$$

c. Γ_T is now 400 counts/s. Hence

$$\Delta t = \frac{900}{400} = 2.25 \text{ s.}$$

3017

A photomultiplier tube is used to measure the intensity of a light source. Assume the light source is constant, the current gain of the photomultiplier is 10^6, and the average anode current measured is 10^{-8} amperes. This current measurement is made by averaging the currents for 10 ms, that is, the measurement time constant is 10^{-2} s.

a. How many photoelectrons are ejected per second from the photocathode to produce the 10^{-8} amp anode current?

b. Assuming random ejection of electrons, what is the rms variation in charge collected during each 10-millisecond measuring interval?

c. What is the signal-to-noise ratio of the measurement?

Sol:

a. The anode current is given by $I = \epsilon re$, where ϵ is the photomultiplier gain factor, r the electron ejection rate and e the electronic charge. Thus

$$r = \frac{I}{\epsilon e} = \frac{10^{-8}}{10^6 \times 1.6 \times 10^{-19}} = 6.25 \times 10^4 \text{ electrons/s.}$$

b. Within the measurement time $\tau = 10^{-2}$ s, the number N of the photoelectrons emitted and its fluctuation ΔN are

$$N = r\tau = 625, \quad \Delta N = \sqrt{N} = 25.$$

The number of electrons collected by the anode and its fluctuation are

$$\begin{aligned} N^* &= \epsilon N = 6.25 \times 10^8, \\ \Delta N^* &= \sqrt{N^*} = 2.5 \times 10^4. \end{aligned}$$

Hence the rms variation in the charge collected is

$$\Delta Q = e\Delta N^* = 4 \times 10^{-15} C.$$

c. The signal-to-noise ratio of the measurement is

$$\frac{N^*}{\Delta N^*} = 2.5 \times 10^4.$$

3018

State the area of Physics in which you plan a thesis.

(*For reference only*)

3019

Nominate a candidate, citing recent accomplishments (with a description of their significance), for next year's Nobel prize in physics.

(*For reference only*)

3020

a. What is your are of specialization?

b. Describe briefly one of the more important and exciting contributions to your chosen area made within the last ten or fifteen years. Identify the person(s) and lab(s) involved. Give the name of the journal(s) where the work was reported.

(For reference only)

3021

a. What is your field of specialization? Name the two journals in this area most likely to have promptly published original articles, also the journal (or book series) with the best review or summary articles in your field. In what journal would your Russian counterpart publish?

b. Describe briefly an important experimental research project now underway in the physics department of your university but outside your area (e.g., low temperature, biophysics, chemical physics, atomic or molecular spectroscopy, high energy, nuclear, plasma, solid state physics, etc.).

(For reference only)

3022

Write, for not more than five minutes, on each of ten of the following subjects:

1. The quark structure of nucleons and mesons
2. Nuclear magnetic resonance induction signals
3. Bose-Einstein condensation
4. Very long baseline interferometry
5. Excimer lasers
6. Ortho and para hydrogen molecules
7. Proportional counters
8. Isotropic spin
9. Strangeness and "associated production"

10. Ortho and para helium spectra
11. The third law of thermodynamics
12. Measurement of the magnetic moment of the neutron
13. Pulsars
14. Source of energy released in nuclear fission
15. Radiofrequency detection of galactic hydrogen

Sol:

1. *Quark structure of nucleons and mesons*

There are believed to be six types of quark, three light (u, d, s) and three heavy (c, b, t), each having its own antiparticle represented by ($\bar{u}$, $\bar{d}$, $\bar{s}$, $\bar{c}$, $\bar{b}$, $\bar{t}$) respectively. The quarks have baryon number $\frac{1}{3}$, fractional charges, either $-\frac{1}{3}$ or $\frac{2}{3}$ of the electronic charge, spin $\frac{1}{2}$, and other quantum numbers such as hypercharge, isospin, strangeness and charm. The nucleons are believed to be made up of three light quarks each:

$$p = (u\ u\ d), \quad n = (d\ d\ u),$$

and the π-mesons, bound states of a quark and an antiquark each:

$$\pi^+ = (\bar{d}\ u), \quad \pi^- = (d\ \bar{u}),$$
$$\pi^0 = \frac{1}{\sqrt{2}}(\bar{u}\ u - \bar{d}\ d).$$

2. *NMR induction signals*

In general, atomic nuclei have an angular momentum $\mathbf{I}\hbar$ and hence a magnetic moment $\boldsymbol{\mu} = g\mu_n\mathbf{I}$, where μ_n is the nuclear magneton and g is a constant. In an external field $\mathbf{H}$ the magnetic moment will precess with the Larmor angular frequency $\boldsymbol{\omega} = -\ g\mu_n\mathbf{H}$. As the orientation of the magnetic moment is quantized the energy level is split into $2I + 1$ sublevels with $I_z = 2I, 2I - 1, 2I - 2, \ldots, 2I + 1, -2I$. Since the energy of the dipole is $-\boldsymbol{\mu} \cdot \mathrm{H} = -\mu_z H$, the spacing between the sublevels is $g\mu_n H$. When an external ac signal of angular frequency ω is applied, a transition between adjacent sublevels is induced when $\hbar\omega = g\mu_n H$, causing resonance absorption. The resonance frequency is usually in the radio frequency range. When the rf oscillating magnetic field is not on between pulses, *NMR* signals are emitted.

Experimentally, a pulse Fourier-transform nuclear magnetic resonance apparatus, which employs short periodic pulses to induce a sample to emit NMR signals, is used. The number of pulses can be controlled to enhance the signal-to-noise ratio of the accumulated signals.

3. *Bose-Einstein condensation*

For an ideal Bose–Einstein gas, transition occurs at a certain low temperature above which the atoms distribute themselves among the various energy levels in the usual way, but below which a certain fraction of atoms falls into the ground state, this fraction increasing from zero at the transition temperature to unity as temperature approaches absolute zero. This phase transition is known as Bose–Einstein condensation.

It has been suggested that the properties of liquid helium II might be related to this effect. However, because of interatomic forces, which have been ignored, the theory is not able to make good quantitative predictions.

4. *Long-baseline interferometry*

In radio astronomy, signals from a source are received by a number of radio telescopes, and the detected signals from different telescopes are superimposed to derive the structure of the source. The angular resolving power of the system depends on the length of the baseline and the working wavelength. If a baseline of intercontinental length is employed, the angular resolving power for centimeter waves can attain 10^{-3} to 10^{-4} second of arc.

5. *Excimer lasers*

Short-lived molecules such as diatomic molecules formed by an inert gas and the halogen family (e.g., XeF) are called excimer molecules. Lasers emitted from such molecules are mostly in the vacuum ultraviolet wave band and are of short pulses.

6. *Ortho and para hydrogen molecules*

A hydrogen molecule consists of two protons, each of spin $\frac{1}{2}$ and thus satisfying Fermi–Dirac statistics, and hence requires a total wave function which is antisymmetric with respect to the interchange of the protons. If spin energies can be neglected, spatial symmetry and spin symmetry can be treated separately. States of the molecule that have parallel nuclear spins are symmetric in the spin function and are called ortho states. States that have antiparallel nuclear spins and are thus antisymmetric in the spin function are called para states. A very strong selection rule forbids the interconversion between the two types of states unless the molecules are first dissociated.

7. *Proportional counters*

It is a type of gas-filled radiation counter in which the magnitude of the pulse generated per count is proportional to the energy of the particle or photon being counted. It is thus able to distinguish α-particles and protons from β-rays and γ-rays. To count fast neutrons, a proportional counter contains a hydrogenous gas such as methane or has walls containing materials having hydrogen in it. Fast neutrons react with the hydrogen ejecting recoil protons which produce large pulses.

8. *Isotopic spin*

The quantum number isotopic spin or isospin I was first proposed by Heisenberg in 1932. In analogy with the spin in ordinary space, isospin is a spin in isospin space with components in the z-direction having the quantized values $I, I-1, I-2, \ldots, -I+1, -I$, a total of $2I+1$ values each representing a charge state. If the small mass difference between proton and neutron is neglected, they can be considered as the same particle, the nucleon, in the isospin states $I_z = \frac{1}{2}$ and $I_z = -\frac{1}{2}$ respectively. The idea also applies to other hadrons, e.g. $\Sigma(I=1)$, $\Xi(I=\frac{1}{2})$. In a strong interaction, the total isospin of a system is conserved, reflecting the charge independence of the strong force. For a particle I_z can be related to the charge (in units of electronic charge) Q, baryon number B and strangeness S by

$$Q = I_z + \frac{1}{2}(B + S).$$

Total I_z of a system is conserved in strong and electromagnetic interactions.

9. *Strangeness and associated production*

Certain baryons and mesons are produced in strong interaction, of time scale 10^{-24} s, but decay through weak interaction, of time scale 10^{-10}–10^{-8} s. To explain such a strange behavior, an additive quantum number called strangeness was introduced by Gell-Mann and others in 1954, which is conserved in strong interaction but can change by 0 or ± 1 in weak interaction. The strangeness quantum number is either an integer or zero, and can be either positive or negative. Particles with the strange behavior are assigned nonzero strangeness, and those without, zero strangeness. The former are known as strange particles. For examples $S = 0$ for π, $S = 1$ for K, $S = -2$ for Ξ, $S = -3$ for Ω. In the production of strange particles from non-strange particles by strong interaction, as the total strangeness is conserved, the strange particles must be produced multiply so that the total strangeness may cancel out, thus leading to the so-called associated production.

10. *Ortho and parahelium spectra*

The total spin of a helium atom is ether zero or one. Correspondingly in the optical spectrum of helium there are two groups of energy levels, singlet and triplet states, respectively. A selection rule forbids transition between the two groups, so that they behave as two different types of atom when seen from the spectrum. The two groups are named ortho and para helium atoms respectively.

11. *Third law of thermodynamics*

The third law of thermodynamics, also called Nernst heat theorem, states that if a chemical change occurs between pure crystalline solids at absolute zero, there is no change in entropy. This was extended to saying that the value of the entropy for each condensed phase is zero at absolute zero. From this law it may be inferred that the absolute zero cannot be reached for any system through any finite process.

12. *Measurement of the magnetic moment of the neutron*

The neutron has spin $\frac{1}{2}$ and thus a magnetic moment parallel or antiparallel to the spin. To measure it a beam of slow neutrons from a reactor is polarized and passed through a radio frequency coil situated in a fixed magnetic field. The frequency necessary for depolarization is compared with the proton depolarization frequency in the same set up. In this way the magnetic moment of the neutron is found to be −1.9132 nuclear Bohr magnetons, the negative sign indicating that the magnetic moment and spin are antiparallel.

13. *Pulsars*

See **Problem 3015**

14. *Source of energy released in nuclear fission*

In nuclear fission, the sum of the rest masses of the fragments produced is less than the rest mass of the original nucleus. This difference in mass is released as the kinetic energy of the fragment and as radiation according to Einstein's relation $E = mc^2$. Thus the energy released in nuclear fission originates from mass-energy conversion.

15. *Radio frequency detection of galatical hydrogen*

Neutral hydrogen can be detected by the 1420 MHz (21-cm) hyperfine transitions within the ground 1^2S state by a frequency modulation method. Since hydrogen is the most abundant element in the universe, observations of its characteristic spectra and the relative intensities of these lines could yield significant information concerning the structure and temperature of the universe, as well as the distribution of the interstellar gas. The resonance Lyman α and β lines have also been used for detection of excited hydrogen atoms and the 18-cm lines for the OH radical.

3023

Write a brief (1–2 pages in 10 minutes) but thorough description of ten of the following:

1. Mössbauer effect
2. Quarks
3. Isotopic spin
4. Rayleigh scattering
5. Strangeness
6. Parity conservation
7. Optical activity, i.e., rotation of plane polarized light
8. ^{14}C dating
9. Cooling by adiabatic demagnetization
10. Čerenkov radiation
11. Magnetic confinement of plasma
12. Statistical interpretation of entropy
13. Coherence of light
14. Electromagnetic mass of electron
15. Fermi–Dirac statistics
16. Gravitational collapse
17. Microwave black body background
18. Viscosity of a gas
19. Neutron star
20. Bose–Einstein condensation
21. Lamb shift
22. Ortho and para states of hydrogen

Sol:

1. *Mössbauer effect*

 This effect, discovered by Mössbauer in 1958, is the phenomenon of recoilless resonance fluorescence γ-rays from nuclei bound in solids. The extreme sharpness of the recoil γ-transitions and the relative ease and accuracy in observing small energy differences make the effect an important tool in nuclear and solid state physics.

 Resonance fluorescence involves the excitation of a quantized system from its ground state 0 to an excited state 1 by absorption of a photon emitted from an identical system decaying from state 1 to state 0. To conserve energy and momentum in the emission and absorption processes, the source and absorber systems must each acquire a recoil energy R and resonance fluorescence can be achieved only if $2R$ is not larger than the widths of the levels involved. However, if the source and absorption nuclei are embedded in a crystal, the lattice as a whole can absorb the recoil momenta so as to produce a recoil-free situation. Nevertheless, the influence of the atomic environment can still cause hyperfine splitting. In order to achieve resonance fluorescence, the source must have a small relative velocity, thus adding a Doppler frequency shift. For example, Fe_3O_4 and Fe_2O_3 cannot usually absorb the γ-photons emitted by ^{57}Fe; absorption occurs only when a small relative velocity is applied between the source and the absorber. The resulting fluorescence gives the Mössbauer spectrum.

2. Quarks

Quarks are particles proposed by Gell-Mann and Zweig independently in 1964 as constituents of strongly interacting "elementary" particles. There are generally believed to be six kinds of quarks, three light quarks u, d, s and three heavy quarks c, b, t, and their antiquarks. Each quark has fractional electric charge $-\frac{1}{3}$ or $\frac{2}{3}$, spin $\frac{1}{2}$, baryon number $\frac{1}{3}$ and such quantum numbers as isospin, hypercharge, strangeness, charm and color. A baryon consists of three quarks, e.g., $p = (u\ u\ d)$, $n = (d\ d\ u)$, while a meson is a bound state of a quark and an antiquark, e.g., $\pi^+ = (\bar{d}\ u)$, $\pi^- = (d\bar{u})$, $\pi^0 = \frac{1}{\sqrt{2}}(\bar{u}\ u - \bar{d}\ d)$.

Although a quark cannot exist in a free state, existence of the quarks has been confirmed experimentally.

3. *Isotopic Spin*

See **Problem 3022** (8).

4. *Rayleigh scattering*

This refers to the scattering of light by small particles of dimensions small compared with the wavelength of light. It is characterized by the scattered intensity being inversely proportional to the fourth power of the wavelength, which means that blue light is more strongly scattered than red light from a medium containing very fine particles, and accounts for the bluish appearance of smoke and of clear sky when the observation is not along the direction of illumination.

5. *Strangeness*

See **Problem 3022** (9).

6. *Parity conservation*

A wave function that changes sign under space inversion

$$x \to -x, \quad y \to -y, \quad z \to -z, \quad t \to t$$

has odd parity, −1; a wave function that does not change sign under space inversion has even parity, +1. Parity can also be assigned to hadrons. A system of hadrons has a total parity which is the sum of the parities of the individual particles. If parity is conserved for a reaction, the total parity does not change in the course of the reaction. Physically this means that the process is identical with that seen in a mirror. It was first suggested, later confirmed experimentally, by C. N. Yang and T. D. Lee in 1956 that parity is not conserved in weak interaction. Of the four fundamental interactions, parity is not conserved only for the weak interaction.

7. *Optical activity*

Certain crystals and solutions of certain substances rotate the plane of polarization of plane-polarized light passing through them in proportion to their thickness. When looking toward the oncoming light, if the rotation is clockwise the optical activity is called right-handed or dextro, if anticlockwise, left-handed or laevo. Crystals or substances having such property are said to be optically active.

8. ^{14}C *dating*

Radioactive isotope ^{14}C is produced by the neutrons in cosmic rays in the upper atmosphere through the reaction ^{14}N (n,p) ^{14}C and has a half-life of 5600 years. It decays back into ^{14}N, and an equilibrium is established in the atmosphere so that the air contains a stable fraction of radioactive ^{14}C per gram of carbon. The atmospheric carbon in the form of CO_2, including the fixed fraction of ^{14}C, is incorporated into all living objects through a biological carbon cycle. After death the ^{14}C activity decays exponentially. Then by determining the ^{14}C activity the archaeological age of the remains of living things can be deduced.

9. *Cooling by adiabatic demagnetization*

For a paramagnetic salt, its entropy at 1 K is still fairly large due almost entirely to nonalignment of magnetic moments, while the entropy of lattice vibration is very small. If the electron spins are aligned by application of a magnetic field, the entropy of the salt drops to a low value and the heat of magnetization can be extracted isothermally. If the salt is then thermally isolated and demagnetized adiabatically, its temperature will fall. Temperatures of the order of 0.01 K can be readily reached by this method.

10. *Čerenkov radiation*

A charged particle moving uniformly does not generally radiate energy. However if it moves in a medium with a speed u exceeding the phase velocity of light v in the medium, electromagnetic radiation is emitted in a continuous spectrum including the visible range. The radiation, similar to the shock wave generated when an airplane breaks the sound barrier, was first discovered by Čerenkov in 1937. It is highly directional, being confined mainly to the surface of a forward cone of half angle $\cos^{-1}\frac{v}{u}$ with the axis coinciding with the trajectory, which property may be used to determine the velocity of the particle.

11. *Magnetic confinement of plasma*

For a thermonuclear reaction to take place the nuclei involved must have sufficiently high energies, corresponding to millions of degrees of temperature. At such temperatures, atoms are ionized by collisions and electrons are not bound to any particular nucleus; the ions and electrons form a plasma. To confine the plasma for fusion reaction, mechanical methods are obviously out and magnetic confinement must be employed, in which a suitably designed magnetic field is applied to confine the charged particles to a small space to allow fusion to take place.

12. *Statistical interpretation of entropy*

The Boltzmann entropy hypothesis relates the entropy S of a system in a particular state with the statistical probability W of finding it in that state: $S := k_B$ In W, where k_B is Boltzmann's constant. Thus a system will have maximum entropy in its most probable state. As a system tends to go from a state of order to a state of disorder, entropy is a measure of disorder.

13. *Coherence of light*

Two wave sources are said to be coherent if the phase difference between a pair of points, one on each source, remains constant. Waves from coherent sources are called coherent waves. A light source usually consists of many emitters, all sending out finite wave trains enduring no longer than about 10^{-8} s. As an observation usually takes much longer, separate light sources or different parts of the same source are not coherent, and to get stable interference patterns one usually requires waves from a common luminous origin (lasers being an exception).

The superposition of two light waves of the same frequency of intensities I_1, I_2 from a common source many be expressed as

$$I = I_1 + I_2 + 2\sqrt{I_1 I_2}\,\mathrm{Re}\,r_{12}.$$

The degree of partial coherence r_{12} depends on the characteristics of the source and the relative path lengths. We have

$$\begin{array}{ll} \text{complete coherence} & \\ \text{Partial coherence} & \text{if } |r_{12}| \begin{cases} = 1 \\ < 1 \\ = 0 \end{cases} \\ \text{complete incoherence} & \end{array}$$

14. *Electromagnetic mass of electron*

That part of the total inertia of a charged body which arises from its electric charge is called its electromagnetic mass. The motion of an electron generates an electromagnetic field, whose total energy minus the self-energy of a static electron is the energy due to the electromagnetic mass.

15. *Fermi-Dirac statistics*

This is the statistics satisfied by microscopic identical particles that have half-integer spins and obey Pauli's exclusion principle which states that no two particles can occupy the same state. Such particles are characterized by the asymmetry of their wave functions and their distribution function is $\bar{n}_i = \left\{\exp\left[(\epsilon_i - \mu)/kT\right] + 1\right\}^{-1}$, where μ is the chemical potential or the Fermi energy.

16. *Gravitational collapse*

At the later stage of evolution of a star, the thermonuclear reaction that takes place in its core almost ceases, and as the radiation pressure produced by the thermonuclear reaction, now greatly reduced, can no longer balance the gravitational attraction by the core on the outer shell, the latter rapidly contracts toward the center, its radius possibly even falling below the Schwarzschild radius of the star. As a result of this process, called gravitational collapse, the excess gravitational potential energy is completely released, causing the outer atmosphere to expand rapidly with greatly increased luminous intensity. Finally the star explodes and becomes a nova or supernova.

17. *Microwave black body background*

Pehzias and R. W. Wilson discovered in 1964 the isotropic microwave background of the universe, which corresponds to black body radiation of temperature about 3 K. Their observations showed that the universe is everywhere filled with microwave photons in an isotropic, uniform distribution. This observation supports strongly the theory that the universe started in a big-bang explosion, which also produced black body radiation.

18. *Viscosity of a gas*

On the interface between gas and solid, gas and liquid, or different layers of a gas which have relative motion, there is an interaction which tends to reduce the relative motion, slowing down the fast-moving layer and speeding up the slow-moving layer. This phenomenon shows that the flow of gas is subject to viscosity, which is similar to friction for solids.

19. *Neutron star*

A neutron star is a compact star constituted mainly by neutrons, with a mass between 0.1 and 2 solar masses and typically a radius of 10 km. A neutron star that possesses a strong magnetic field and rotates rapidly is called a pulsar. A neutron star is probably the result of a supernova explosion. When a star suffers gravitational collapse, its size decreases and when it falls below a critical value, the internal pressure of the star will increase, as well as the probability for inverse β-decay of protons and electrons producing neutrons. At this time the degeneracy pressure of the neutrons and the gravitational force will reach a mechanical equilibrium, forming a neutron star.

20. *Bose-Einstein condensation*

See **Problem 3022** (3).

21. *Lamb shift*

Dirac's relativistic theory of the hydrogen atom shows that states with the same principal quantum number $n$and total angular momentum quantum number j (**j**= **1**+ s) have equal energies, giving rise to degeneracy. However for $n = 2$, Lamb and Rutherford found experimentally that the level of $2^2 P_{\frac{1}{2}}$ is as predicted but the $2^2 S_{\frac{1}{2}}$ level is slightly higher, by 1058 MHz. This effect is known as Lamb shift. It can be explained by the interaction of the electron with its virtual photon field, which removes the degeneracy.

22. *Ortho and para states of hydrogen*

See **Problem 3022** (6)

3024

Multiple Choice Problems

1. The equation $x^5+x^4+x^3+1=0$:

a. has no roots,

b. has 5 roots,

c. has 6 roots.

Sol: A polynomial equation of the nth degree in one variable has n roots. Hence the correct answer is (b).

---

2. The binary representation of the decimal 137 is:

a. 10001001,

b. 1001001,

c. 101001.

Sol: As $137 = 2^7 + 2^3 + 2^0 = (10001001)_2$, the correct answer is (a).

---

3. 0 divided by 0 is:

a. 0,

b. infinity,

c. undefined.

Sol:

c. $\frac{0}{0}$ represents the ratio of two infinitesimal numbers; its value depends on the situation in which it occurs. For example, when $x \to \infty$, $\frac{1}{x}$ and e^{-x} will both approach zero but $\lim_{x\to\infty} \frac{\frac{1}{x}}{e-x} = \lim_{x\to\infty} \frac{e^x}{x} = \lim_{x\to\infty} \left(\frac{1}{x} + 1 + \frac{x}{2!} + \frac{x^2}{3!} + \dots\right)$ $= \infty$. On the other hand, when $x \to 0$, x and x^2 will both approach zero but $\lim_{x\to 0} \frac{x^2}{x} = \lim_{x\to 0} x = 0$. Hence the correct answer is (c).

---

4. The number of possible ways to assign 6 people to 6 seats is:

a. 602,

b. 720,

c. 1024.

Sol: Since people are not identical and order is essential, this is a permutation of 6 different objects taken all at a time: $6! = 720$. Thus the correct answer is (b).

---

5. The number of potential ways to pick 3 things out of 7 is:

a. 35,

b. 60,

c. 50.

Sol: Since order is not essential in this case, the problem is one of combination: $C_3^7 = \frac{7!}{3!} = 35$. Thus the correct answer is (a).

---

6. A continuous function always has:

a. a well-defined first derivative,

b. a well-defined second derivative,

c. neither of the above.

Sol: A curve can contain a point at which the tengent depends on the side from which it is approached and is still continuous. Hence a continuous function need not be differentiable. Thus the correct answer is (c).

---

7. If the probability of occurrence for event 1 is P and that for event 2 is Q, then the probability that they both occur is:

a. $P + Q$,

b. PQ,

c. neither.

Sol: Provided events 1 and 2 are independent, the correct answer is (b).

---

8. The Poisson equation $\nabla^2 F(x, y, z) = g(x, y, z)$ with a given $g(x, y, z)$ and a physical bounding condition can have:

a. more than one solution,

b. one and only one solution,

c. occasionally no solution.

Sol: If we consider solutions that differ only by a constant as the same physically, then according to the uniqueness theorem there is only one solution under a given set of boundary conditions. Hence the correct answer is (b).

---

9. The age of the universe is approximately:

a. 10^{20} years,

b. 10^{10} years,

c. 10^8 years.

Sol: It is generally accepted that the age of the universe is about 1.8×10^{10} years. Hence the correct answer is (b).

---

10. Ocean waves break near the shore because:

a. it is usually windy near the shore,

b. the surface wave velocity of water depends on the water depth,

c. neither of the above.

Sol: The correct answer is (b).

---

11. Commonly used digital computers work with binary numbers represented by:

a. 1000 bits,

b. 32 bits,

c. 4 bits.

Sol: One byte of a binary digital computer is equal to 8 bits, and character inputs to the computer are in units of byte. The answer a) and c) are not integral multiples of eight. Hence the correct answer is (b).

---

12. A sequence of random numbers generated by a digital computer:

a. is truly random,

b. will repeat itself eventually,

c. is useful only for playing games.

Sol: Random numbers are artificially generated by a digital computer according to a certain program. After a large number of repeated use a sequence of random numbers will eventually repeat itself. Hence the correct answer is (b).

---

13. Stars twinkle, whereas planets do not, because:

a. planets merely reflect light, whereas stars emit light,

b. stars are more pointlike to us,

c. stars pulsate.

Sol: Stars twinkle because of the thermal fluctuation of the air density of the atmosphere which starlight has to penetrate. The twinkling is obvious to the human eye only for light sources of small apparent size. Hence the correct answer is (b).

---

14. Astronomers now believe the age of the universe is:

a. 10^{12} years,

b. 10^{10} years,

c. 10^{6} years,

d. 6000 years.

Sol: The correct answer is (b).

---

15. Density of matter between the star is:

a. 1 particle/cm^3 **b.** 10^5 particles/cm^3 **c.** 10^{15} particles/cm^3

Sol: In the interstellar space, the mean density of molecules and atoms is about 1 particle/cm^3. The correct answer is (a).

---

16. Total volume of water in the oceans is:

a. 10^6 km^3, **b.** 10^7 km^3, **c.** 10^8 km^3,

b. 10^9 km^3, **e.** 10^{10} km^3.

Sol: The correct answer is (d).

3025

Multiple Choice Questions

1. The mass a hot (120°C) air balloon of 10 m^3 volume will lift at sea level when the temperature is 20°C is closest to:

a. 300 kg, **b.** 30 kg, **c.** 3 kg, **d.** 300 g, **e.** 30 g.

Sol: The buoyancy of the balloon is equal to the difference in weight of 10^7 cm^3 of air at 293 K and at 393 K, at one atmospheric pressure. As

$$\begin{aligned} m_0 - m_1 &= \frac{pVM}{R}(T_0^{-1} - T_1^{-1}) \\ &= \frac{76 \times 13.6 \times 980 \times 10^7 \times 29}{8.31 \times 10^7}(293^{-1} - 393^{-1}) = 3070\,\mathrm{g} \approx 3\,\mathrm{kg}, \end{aligned}$$

the correct answer is (c).

---

2. The charge on the "up" quark is (2/3)e. If free "up" quark were found, its ionization in matter would be λ times that of a proton of the same velocity. $\lambda =$

a. $\left(\frac{3}{2}\right)^2$, **b.** $\frac{2}{3}$, **c.** $\frac{1}{3}$, **d.** 1, **e.** $\left(\frac{2}{3}\right)^2$.

Sol: The proton and the up quark are both heavy charged particles ($m \gg m_e$). Hence, the rate of loss of energy due to ionization in traversing matter is

$$-\frac{dT}{dx} = \frac{4\pi z^2 e^4}{m_e v^2} NZ\left[\mathrm{In}\frac{2m_e v^2}{I} - \mathrm{In}(1-\beta^2) - \beta^2\right] \propto z^2,$$

ze being the charge of the particle, for a given velocity v. Thus

$$\lambda = \frac{z_q^2}{z_p^2} = \left(\frac{2}{3}\right)^2.$$

Hence the correct answer is (e).

---

3. The binding energy of the ground state of positronium is a factor f times that of a hydrogen atom. $f =$

a. 1, **b.** $\frac{1}{2}$, **c.** 2, **d.** $\frac{1}{4}$, **e.** 4.

Sol: An electron moving in the field of a proton or positron has potential energy $-\frac{e^2}{r}$, for which the radial Schrödinger equation gives eigenvalue

$$E_n = -\frac{\mu e^4}{2\hbar^2 n^2}, \quad n = 1, 2, 3, \ldots,$$

where μ is the reduced mass of the orbiting electron. For the positronium $\mu = \frac{1}{2}m_e$, for hydron atom $\mu \approx m_e$. Hence

$$f = \frac{E_P}{E_h} = \frac{\frac{1}{2}m_e}{m_e} = \frac{1}{2}.$$

The correct answer is (b).

---

4. The scale height of the atmosphere (change in altitude to cause a pressure drop by a factor of e) is

a. 4 km, **b.** 8 km, **c.** 12 km, **d.** 16 km, **e.** 20 km.

Sol: If we consider air near the Earth's surface as an isothermal layer, the pressure change with altitude is given by $p = p_0 e^{-mgz/k_B T}$. The scale height z of the atmosphere is then given by $\frac{mgz}{k_B T} = 1$. Thus

$$z = \frac{k_B T}{mg} = \frac{1.38 \times 10^{-23} \times 300}{\frac{29 \times 10^{-3}}{6.02 \times 10^{23}} \times 9.8} = 8.6 \times 10^3 \, \text{m} = 8.6 \, \text{km}.$$

Hence the correct answer is (b).

---

5. A system contains 4 non-identical spin 1/2 particles. Disregard any other degrees of freedom. In classifying the wave function in terms of total spin, what statistical weight should be assigned to states with a total spin = 1?

a. 0, **b.** 3, **c.** 6, **d.** 9, **e.** 12.

Sol: Coupling two non-identical spin-$\frac{1}{2}$ particles results in 2 states of total spins S equal to

$$\frac{1}{2} + \frac{1}{2} = 1, \quad \frac{1}{2} - \frac{1}{2} = 0.$$

Adding another spin-$\frac{1}{2}$ particle to the above states gives 3 states of total spin S equal to

$$1 + \frac{1}{2} = \frac{3}{2}, \quad 1 - \frac{1}{2} = \frac{1}{2}; \quad 0 + \frac{1}{2} = \frac{1}{2}.$$

Adding another spin-$\frac{1}{2}$ particle to the above states gives 6 states of total spin S equal to

$$\frac{3}{2} + \frac{1}{2} = 2, \quad \frac{3}{2} - \frac{1}{2} = 1; \quad \frac{1}{2} + \frac{1}{2} = 1,$$
$$\frac{1}{2} - \frac{1}{2} = 0; \quad \frac{1}{2} + \frac{1}{2} = 1, \quad \frac{1}{2} - \frac{1}{2} = 0.$$

Thus the coupling of four particles has total-spin eigenfunction

$$2|0\rangle + 3|1\rangle + |2\rangle,$$

and the statistical weight of states with total spin $S = 1$ (in terms of probability) is $3 \times 3 = 9$. Hence the correct answer is (d).

---

6. The mean free path of a 5 eV neutron in vacuum is closest to:

a. 1 km, **b.** 10 km, **c.** 100 km, **d.** 1000 km, **e.** 10,000 km.

Sol: The mean free path in vacuum is the distance the neutron travels in its lifetime, from generation to decay. Lifetime of the neutron is about 10^3 s. As its energy 5 eV is much less than its rest energy 940 MeV, nonrelativistic approximation may be used and its velocity is

$$v = \sqrt{\frac{2E}{m}} = c\sqrt{\frac{2E}{mc^2}} = \sqrt{\frac{2 \times 5 \times 10^{-6}}{940}} \times 3 \times 10^8 = 10^4 \,\text{m/s}.$$

Thus $S = vt = 10^4$ km. Hence the correct answer is (e).

---

7. The Bohr radius of an atom of nuclear charge Z is of order

a. $\frac{\hbar}{Z\alpha mc}$, **b.** $\frac{Z\alpha\hbar}{mc}$, **c.** $\frac{mc}{Z\alpha\hbar}$.

Note α is the fine structure constant $e^2/\hbar c$.

Sol: The Bohr radius for a hydrogen-like atom of nuclear charge Z and one electron outside complete shells is $\frac{\hbar^2}{me^2 Z} = \frac{\hbar}{Z\alpha mc}$. Thus the correct answer is (a).

---

8. The binding energy of the deuteron is of order

a. 10^6 eV, **b.** 10^8 eV, **c.** 10^{10} eV.

Sol: The uncertainty principle gives $2pr \sim \hbar$, where p is the momentum of each nucleon and r is their separation, or

$$p \sim \frac{\hbar}{2r}.$$

The binding energy is then

$$\frac{p^2}{2m} \sim \frac{\hbar^2 c^2}{8r^2 mc^2} = \frac{(6.6 \times 10^{-16} \times 3 \times 10^{10})^2}{8 \times (1.4 \times 10^{-13})^2 \times 940 \times 10^6}$$
$$= 2.7 \times 10^6 \,\text{eV}.$$

Hence the correct answer is (a).

---

9. A piece of organic material contains N_e electrons per gram. N_e is about

a. $3{\times}10^{22}$, **b.** $3{\times}10^{23}$, **c.** $3{\times}10^{24}$, **d.** $3{\times}10^{25}$, **e.** $3{\times}10^{26}$.

Sol: One mole of a substance contains 6.02×10^{23} molecules or atoms. The molecule of organic matter consists of carbon and hydrogen atoms for which the number of electrons is equal to the atomic number. As the molecular weight of organic matter is generally greater than 13, the number of electrons per gram is

$$N_e \leq \frac{6.02 \times 10^{23}}{13} = 4.6 \times 10^{22}.$$

Hence the correct answer is (a).

---

10. The wavelength of a neutron with energy 1 eV is closest to:

a. 1 cm, **b.** 10^{-2} cm, **c.** 10^{-4} cm, **d.** 10^{-6} cm, **e.** 10^{-8} cm.

Sol: The wavelength λ of a particle of momentum p is

$$\lambda = \frac{h}{p} = \frac{h}{\sqrt{2mE}} = \frac{hc}{\sqrt{2mc^2 E}} = \frac{4.14 \times 10^{-15} \times 3 \times 10^8}{\sqrt{2 \times 940 \times 10^6}} = 2.9 \times 10^{-11}\,\text{m}.$$

Hence the correct answer is (e).

3.2

MEASUREMENTS, ESTIMATIONS AND ERRORS (3026–3048)

3026

a. An observer weighs a body repeatedly with the following results:

10.1, 10.6, 10.2, 10.5, 10.3, 10.5, 10.4, 10.6 grams.

The standard deviation of a single measurement (from the average of a very large number of measurements) is __________.

The standard deviation of a mean of the measurements (from the average of a very large number of measurements) is __________.

If we call this last quantity $\overline{\sigma}$, the probability that the true value of the mass is within $2\overline{\sigma}$ of the mean value obtained is about __________.

(Make whatever assumptions you consider reasonable)

b. The mean result of measurements of the mass of the body by observer A is 10.7 ± 0.1 grams; by observer B is 10.1 ± 0.2 grams.

The weighted average of these measurements is __________.

The standard deviation of the average based upon the quoted errors is __________.

The standard deviation of the average based upon the disagreement between the two determinations is __________.

What does the difference between the two standard deviations indicate about the measurements? (Make whatever assumptions you consider reasonable)

Sol:

a. The mean of the eight measurements is

$$\begin{aligned}\overline{m} &= \frac{1}{8}(10.1 + 10.6 + 10.2 + 10.5 + 10.3 + 10.5 + 10.4 + 10.6) \\ &= 10.4\,\text{g}.\end{aligned}$$

As

$$\begin{aligned}\sigma^2 &= \frac{1}{8}\sum_{i=1}^{8}(m_i - \overline{m})^2 \\ &= \frac{1}{8}(0.3^2 + 0.2^2 + 0.2^2 + 0.1^2 + 0.1^2 + 0.1^2 + 0.2^2) \\ &= 0.03\,\text{g}^2,\end{aligned}$$

the standard deviation of a single measurement is

$$\sigma = 0.17\,\text{g},$$

and the standard deviation of the mean of the measurements is

$$\overline{\sigma} = \frac{\sigma}{\sqrt{N}} = \frac{0.03}{\sqrt{8}} = 0.061\,\text{g}.$$

If we assume that the measured values, which are continuous, obey the normal distribution, then the mean has distribution function

$$p(\overline{m}) = \frac{1}{(2\pi)^{1/2}\overline{\sigma}}e^{-(\overline{m}-m_0)^2/2\overline{\sigma}^2},$$

where m_0 is the true value of the mass. The probability that $\overline{m}$ is within $2\overline{\sigma}$ of m_0 is

$$\int_{m_0-2\overline{\sigma}}^{m_0+2\overline{\sigma}} p(\overline{m})d\overline{m},$$

which is the area under the normal curve for $m_0 = 0, \overline{\sigma} = 1$, between $x = -2$ and $x = +2$. This is given by standard table as $2 \times 0.4773 \simeq 95\%$.

b. The weight of a measurement is inversely proportional to the square of its error. Thus the weighted mean of the two measurements is

$$\left(\frac{10.7}{0.1^2} + \frac{10.1}{0.2^2}\right)\Bigg/\left(\frac{1}{0.1^2} + \frac{1}{0.2^2}\right) = 10.58,$$

and the standard deviation of the weighted average based upon the quoted errors is

$$\bar{\sigma} = \left(\frac{0.1}{0.1^2} + \frac{0.2}{0.2^2}\right) \bigg/ \left(\frac{1}{0.1^2} + \frac{1}{0.2^2}\right) = 0.12.$$

The standard deviation of the average based upon the disagreement between the two determinations is, as

$$\bar{\sigma}'^2 = \frac{1}{2}[(10.7 - 10.58)^2 + (10.1 - 10.58)^2] = 0.122,$$

$$\bar{\sigma}' = 0.35.$$

The disagreement between $\bar{\sigma}$ and $\bar{\sigma}'$ indicates that the measurements by the two observers probably involve certain systematic errors.

3027

To within a factor of ten, estimate, calculate roughly, or remember the following (say or show how you got your answer):

a. The de Broglie wavelength of the Earth

b. The total binding energy of an Fe nucleus

c. The magnetic dipole moment of the Earth

d. The vapor pressure of water at 20° C

e. The average speed of a conduction electron in a bulk sample of sodium metal

Sol:

a. The Earth revolves around the Sun once a year and its orbit has a mean radius 150×10^6 km. Hence its average orbital speed is

$$v = \frac{2\pi \times 150 \times 10^9}{365 \times 24 \times 3600} = 3 \times 10^4 \,\mathrm{m\,s^{-1}}.$$

As the Earth has mass 6×10^{24} kg, its average momentum is

$$p = mv = 1.8 \times 10^{29} \,\mathrm{kg\,m\,s^{-1}},$$

and so its de Broglie wavelength is

$$\lambda = \frac{h}{p} = \frac{6.63 \times 10^{-34}}{1.8 \times 10^{29}} = 3.7 \times 10^{-63}\,\mathrm{m},$$

if the Sun can be considered stationary.

b. The mean binding energy per nucleon for middle-range elements is about 8.8 MeV. Hence the binding energy of ^{56}Fe is

$$56 \times 8.8 = 493 \ MeV.$$

c. Consider the Earth as a magnetic dipole of moment M. The horizontal component of the magnetic field at distance r from the geomagnetic axis is

$$B = \frac{M}{r^3}\cos\lambda,$$

where λ is the geomagnetic latitude. At $\lambda = 0°$, $B \approx 0.3$ gauss. As $r \approx 6400$ km, we have

$$\begin{aligned} M &\simeq 0.3 \times (6.4 \times 10^8)^3 \\ &= 8 \times 10^{25}\,\text{gauss cm}^3. \end{aligned}$$

d. For rough estimation, consider water vapor as an ideal gas. The Clapeyron equation

$$\frac{dp}{dT} = \frac{L}{T(v_\beta - v_\alpha)} \approx \frac{L}{Tv},$$

where p is the vapor pressure, T the temperature, L the molar latent heat in energy units, and v, v_l are the specific volumes of the water in the vapor and liquid phases ($v \gg v_l$) respectively, and the ideal gas equation

$$pv = RT$$

give

$$\frac{dp}{dT} = \frac{Lp}{RT^2},$$

or

$$p = p_0{}^{e^{-L/RT}},$$

i.e.

$$\text{In}\left(\frac{p}{p_0}\right) = \frac{L}{R}\left(\frac{1}{T_0} - \frac{1}{T}\right),$$

assuming L to be independent of T.

At the triple point of water, $p_0 = 4.581$ mmHg, $T_0 = 273.16$ K, $L = 590$ cal/g $=$ $590 \times 4.186 \times 18$ J/mol, $R = 8.31$ J/mol·K.

Thus

$$\frac{L}{R}\left(\frac{1}{273} - \frac{1}{293}\right) = 1.34$$

and

$$p = p_0 e^{1.34} = 17.45\ \text{mmHg}.$$

e. The conduction electrons in a metal at absolute temperature T fill up states to within ~ $k_B T$ of the Fermi energy $E_F(0)$ at absolute temperature zero given by

$$E_F(0) = \frac{\hbar^2}{2m}(3\pi^2 n)^{\frac{2}{3}},$$

where n is the number of conduction electrons per unit volume of the metal. For sodium metal, $n = 2.5 \times 10^{28}\ \text{m}^{-3}$, and

$$\begin{aligned} E_F(0) &= \frac{\hbar^2 c^2}{2mc^2}(3\pi^2 n)^{\frac{2}{3}} \\ &= \frac{(6.58 \times 10^{-16} \times 3 \times 10^8)^2}{2 \times 0.51 \times 10^6} \times (3\pi^2 \times 2.5 \times 10^{28})^{\frac{2}{3}} \\ &= 3.13\ \text{eV}. \end{aligned}$$

At room temperature, $T \approx 300$ K and we have

$$k_B T \approx \frac{1}{40} \ll 3.13,$$

which means that we can treat the problem as if the temperature were 0 K.

At thermal equilibrium, the number of electrons with energies between E and $E + dE$ is

$$dn \propto f(E) E^{\frac{1}{2}} dE,$$

where $f(E)$ is the Fermi-Dirac distribution function. At $T = 0$ K, $f(E) = 1$ for $E \leq E_F(0)$ and $f(E) = 0$ for $E > E_F(0)$. As $E = \frac{p^2}{2m}$, the average momentum is

$$\bar{p} = \frac{\int_0^{E_F(0)} pE^{\frac{1}{2}} dE}{\int_0^{E_F(0)} E^{\frac{1}{2}} dE} = \frac{\int_0^{p_F} p^3 dp}{\int_0^{p_F} p^2 dp} = \frac{3}{4} p_F,$$

where $p_F = \sqrt{2mE_F(0)}$. The average speed is then

$$\begin{aligned} \bar{v} &= \frac{\bar{p}}{m} = \frac{3\sqrt{2mE_F(0)}}{4m} = \frac{3c}{2}\sqrt{\frac{E_F(0)}{2mc^2}} \\ &= \frac{3}{2}\sqrt{\frac{3.13}{2 \times 0.51 \times 10^6}} \times 3 \times 10^8 = 7.9 \times 10^5\ \text{m s}^{-1}. \end{aligned}$$

Note that since

$$\bar{v} = \frac{3}{2}\sqrt{\frac{E_F(0)}{2m}} = \frac{3\hbar}{4m}(3\pi^2 n)^{\frac{1}{3}}$$

the average speed is determined solely by the electron concentration n.

3028

An astrophysicist found a binary star system consisting of a visible star and an invisible massive nonluminous companion. The star, which is 20 times more massive than the Sun, was found to be orbiting the invisible companion with an orbital period of about 2.83 years. The average distance between the visible star and the center of mass is 9.67×10^{11} m. Short X-ray bursts for an average duration of 2.8 ms from the invisible companion were also recorded.

a. Estimate the mass of the binary system and that of the invisible companion.

b. Identify the invisible companion based on your estimate.

c. Estimate the radius of the invisible companion.

Sol:

a. Estimating the Total Mass

Using Kepler's third law of planetary motion, the total mass of the system can be determined. Kepler's third law relates the period of revolution (T_{orb}), the semimajor axis (a), and the total mass (M_{tot})

$$T_{orb}^2 = \frac{4\pi^2}{GM_{tot}}a^3 \rightarrow M_{tot} = \frac{4\pi^2}{G\,T_{orb}^2}a^3$$

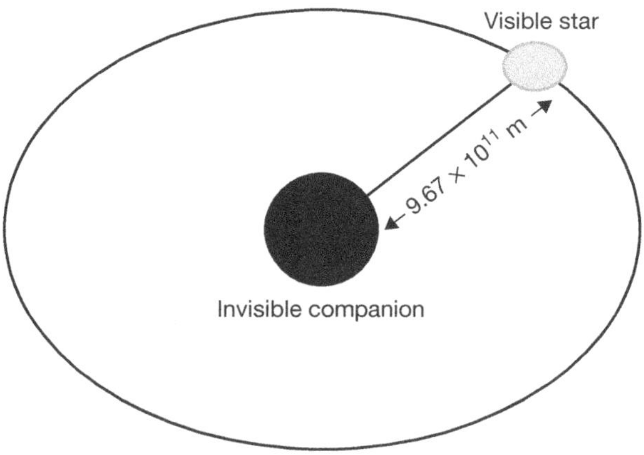

where,

G, gravitational constant $= 6.67 \times 10^{-11}$ m^3/s^2 kg

a, semimajor axis, that is, the distance between the star and the center of mass $= 9.67 \times 10^{11}$ m

$T_{orb} = 2.83$ years $= 2.83 \times 365 \times 24 \times 60 \times 60$ s $= 89{,}246{,}880$ s

Substituting these values in the formula gives,

$$M_{tot} = \frac{4\pi^2}{(6.67 \times 10^{-11})(89246880)^2}(9.67 \times 10^{11})^3 = 6.7125 \times 10^{31}\,\text{kg}$$

We know that one solar mass $1M = 1.988 \times 10^{30}$ kg, so M_{tot} in solar mass is given by

$$= \frac{6.7125 \times 10^{31}}{1.988 \times 10^{30}} = 33.76M \approx 34M$$

Estimating the Mass of the Invisible Companion

Using the total mass, one can now estimate the mass of the invisible companion (M_{IC}) using the following relation: $M_{IC} = M_{tot} - M_S$, where M_S is the mass of the visible star.

$$M_{IC} = 34M - 20M = 14M$$

b. Identifying the Invisible Companion

Both neutron star and black hole are massive, nonluminous, and emit X-rays. But the mass of the neutron stars cannot exceed $3M$. So the only possibility is a black hole and based on its mass it's a stellar black hole and not the super massive ones.

c. Estimating the Radius of the Invisible Companion

The width of the X-ray-emitting region (invisible companion) can be determined using the formula,

Distance = speed of light × time $= 3 \times 10^8 \times 2.8 \times 10^{-3} = 8{,}400$ km

Estimated radius = 4,200 km

But we also know that the radius of a black hole can be calculated using the formula, $R = 3M$, where R is the radius of the black hole in km and M is the mass of the black hole in solar mass.

That is, $R = 3 \times 14 = 42$ km

The actual radius is 100 times smaller than the estimated radius of 4,200 km. This can be accounted for if we include the relativity factor of 1/100 in the calculation that is due to the "beaming effect."

3029

In the Thomson-Lorentz model of an atom, an electron is harmonically bound to a force center, i.e., it is a (3-dimensional) harmonic oscillator of natural frequency ω_0. (Short answers, as indicated by the blanks which follow, are sufficient, but you may expand on them if you wish)

a. Classical statistical mechanics (CSM) states that the mean energy (kinetic plus potential) of each oscillator is ________ in thermal equilibrium.

b. Quantum statistical mechanics (QSM) states that the mean energy is ________.

c. For an assembly of oscillators, the specific heat is ______ (CSM) or ______ (QSM). CSM is valid when ________ $\gg$ ____________.

d. A static electric field **E** is applied, producing a mean dipole moment $\mathbf{P} = \alpha\mathbf{E}$. Classical mechanics (CM) predicts that $\alpha =$ _______. Quantum mechanics (QM) predicts that $\alpha =$ _______ in the ground state and $\alpha =$ _______ at thermal equilibrium.

Sol:

a. The energy of an oscillator can be expressed as $\varepsilon = \dfrac{p^2}{2m} + \dfrac{1}{2}\omega^2 q^2$, where there are six squared terms in all. The law of equipartition of energy of classical statistical mechanics gives for each degree of freedom an average energy $\dfrac{1}{2}kT$, k being Boltzmann's constant, at thermal equilibrium. Hence the mean energy of an oscillator is $\bar{\varepsilon}_{\text{CSM}} = 3kT$.

b. The oscillator can be considered a boson and the problem treated, in principle, by using Bose-Einstein statistics. However, this is rather complicated and we shall assume, without loss of generality, that the motion of the oscillator is localized. Then the oscillators are distinguishable, and we may apply the quantum Boltzmann statistics. The partition function for each dimension of the oscillator is

$$Z = \sum_{n=0}^{\infty} e^{-\beta(n+\frac{1}{2})\hbar\omega_0} = e^{-\frac{1}{2}\beta\hbar\omega_0} \Big/ \left(1 - e^{-\beta\hbar\omega_0}\right),$$

where $\beta = \dfrac{1}{kT}$. Hence the average energy is

$$\bar{\varepsilon}_{\text{QSM}} = -3\frac{\partial}{\partial \beta}\ln Z = \frac{3\hbar\omega_0}{2} + \frac{3\hbar\omega_0}{e^{\beta\hbar\omega_0} - 1},$$

where $\frac{3\hbar\omega_0}{2}$ is the zero point energy of the oscillator.

c. The specific heat per oscillator is

$$C_{\text{CSM}} = \frac{\partial \bar{\varepsilon}_{\text{CSM}}}{\partial T} = 3k$$

or

$$C_{\text{QSM}} = \frac{\partial \bar{\varepsilon}_{\text{QSM}}}{\partial T} = \frac{3k(\beta\hbar\omega_0)^2 e^{\beta\hbar\omega_0}}{\left(e^{\beta\hbar\omega_0} - 1\right)^2}.$$

Note that when $\beta\hbar\omega_0 \ll 1$, i.e., $kT \gg \hbar\omega_0$, $C_{\text{QSM}} \to 3k = C_{\text{CSM}}$. In other words, classical statistical mechanics is valid at high temperatures.

d. For the CM case, two methods will be used to find α.

(1) The restoring force on the electron is $-m\omega_0^2\mathbf{r}$, where $\mathbf{r}$ is the displacement from equilibrium position. Then

$$-m\omega_0^2\mathbf{r} - e\mathbf{E} = 0,$$

or

$$\mathbf{r} = -\frac{e\mathbf{E}}{m\omega_0^2}.$$

By definition

$$\mathbf{P} = -e\mathbf{r} = \alpha\mathbf{E}.$$

Hence

$$\alpha = \frac{e^2}{m\omega_0^2}.$$

The negative sign indicates that **P** is opposite in direction to **E**.

(2) The second method is based on classical statistical mechanics. Take the z-axis along the direction of **E**. As $E = -\frac{\partial V}{\partial z}$, $V = -Ez$ and the total energy of the oscillator is

$$\begin{aligned}\varepsilon &= \frac{p^2}{2m} + \frac{m\omega_0^2}{2}z^2 + eEz \\ &= \frac{p^2}{2m} + \frac{m\omega_0^2}{2}\left(z + \frac{eE}{m\omega_0^2}\right)^2 - \frac{e^2E^2}{2m\omega_0^2}.\end{aligned}$$

By definition $P = -ez$, and Boltzmann statistics gives

$$\overline{P} = \frac{\int (-ez)e^{-\varepsilon/kT} d^3\mathbf{P}\, d^3\mathbf{r}}{\int e^{-\frac{\varepsilon}{kT}} d^3\mathbf{P}\, d^3\mathbf{r}}$$

$$= \frac{\int_{-\infty}^{\infty} (-ez) e^{-m\omega_0^2\left(z+\frac{eE}{m\omega_0^2}\right)^2/2kT} dz}{\int_{-\infty}^{\infty} e^{-m\omega_0^2\left(z+\frac{eE}{m\omega_0^2}\right)^2/2kT} dz}$$

$$= \frac{\int_{-\infty}^{\infty} \left(-ez'+\frac{e^2E}{m\,\omega_0^2}\right) e^{\frac{-m\omega_0^2}{2kT}z'^2} dz'}{\int_{-\infty}^{\infty} e^{-\frac{m\omega_0^2}{2kT}z'^2} dz'} = \frac{e^2 E}{m\,\omega_0^2},$$

where we have set $z' = z + \dfrac{eE}{m\,\omega_0^2}$. Hence

$$\alpha = \frac{\overline{P}}{E} = e^2/m\,\omega_0^2.$$

For QM calculations, consider the electron in its ground state. As the ground state is nondegenerate, we use the nondegenerate perturbation method to find the wave function to the first order:

$$|\psi\rangle = |0\rangle + \sum_k{}' \frac{\langle k|H'|0\rangle}{\varepsilon_0^{(0)} - \varepsilon_k^{(0)}}|k\rangle,$$

where

$$\widehat{H}' = eEz,$$
$$\widehat{H}_0 = -\frac{\hbar^2}{2m}\nabla^2 + \frac{1}{2}m\,\omega_0^2 z^2, \quad \varepsilon_n^0 = \left(n + \frac{3}{2}\right)\hbar\,\omega_0,$$
$$|n\rangle \equiv |n_x\rangle|n_y\rangle|n_z\rangle, \quad |0\rangle \equiv |0\rangle|0\rangle|0\rangle.$$

As

$$\widehat{H}'|0\rangle = eE|0\rangle|0\rangle\sqrt{\frac{\hbar}{m\,\omega_0}}\left(\sqrt{\frac{1}{2}}|1\rangle\right)$$
$$= eE\sqrt{\frac{\hbar}{2m\,\omega_0}}|0\rangle|0\rangle|1\rangle,$$

we have

$$|\psi\rangle = |0\rangle + \frac{eE\sqrt{\frac{\hbar}{2m\omega_0}}}{\varepsilon_0^{(0)} - \varepsilon_1^{(0)}}|0\rangle|0\rangle|1\rangle$$
$$= |0\rangle - \frac{eE}{\hbar\,\omega_0}\sqrt{\frac{\hbar}{2m\,\omega_0}}|0\rangle|0\rangle|1\rangle,$$

and

$$P = \langle \psi | -ez | \psi \rangle = 2\frac{e^2 E}{\hbar \omega_0}\sqrt{\frac{\hbar}{2m\omega_0}} \cdot \sqrt{\frac{\hbar}{2m\omega_0}} = \frac{e^2}{m\omega_0^2}E.$$

Hence $\alpha = e^2/m\,\omega_0^2$, identical with that from classical mechanics.

Actually this problem is amenable to precise calculation. Even though the precise wave function is not identical with the approximate one given by the perturbation method, α has the same value in both cases. Precise calculation gives the energy levels

$$\varepsilon_n = \left(n + \frac{3}{2}\right)\hbar\omega_0 - \frac{e^2 E^2}{2m\omega_0}$$

and eigenfunctions

$$\psi_n'(\mathbf{r}) = \psi_n(\mathbf{r}'),$$

where $\psi_n(\mathbf{r})$ are the eigenfunctions in the absence of the applied electric field and

$$\mathbf{r}' = \mathbf{r} + \frac{e\mathbf{E}}{m\,\omega_0^2}.$$

The average displacement of the electron in any state is

$$\begin{aligned}\langle \psi | \mathbf{r} | \psi \rangle &= \left\langle \psi \left| \mathbf{r}' - \frac{eE}{m\,\omega_0^2} \right| \psi \right\rangle \\ &= -\frac{eE}{m\,\omega_0^2} + \langle \psi | \mathbf{r}' | \psi \rangle = -\frac{eE}{m\,\omega_0^2}.\end{aligned}$$

Boltzmann statistics then gives

$$\bar{z} = \frac{\sum_{n=0}^{\infty} -\frac{eE}{m\,\omega_0^2} e^{-E_n/kT}}{\sum_{n=0}^{\infty} e^{-E_n/kT}} = -\frac{eE}{m\,\omega_0^2},$$

whence

$$P = -e\bar{z} = \frac{e^2}{m\,\omega_0^2}E,$$

$$\alpha = e^2/m\,\omega_0^2.$$

Note that the four results obtained are all the same.

3030

1. A beam of electrons is fired in free space over a distance of 10^4 km. If the transversal width of the initial packet is 1 μm, what will be the minimum spread upon arrival if the kinetic energy of the electrons is 13.6 eV?

2. For surface tension waves in shallow water, the relation between frequency and wavelength is

$$v = \left(\frac{2\pi T}{\rho\lambda^3}\right)^{1/2},$$

where T is the surface tension and ρ the density. What is the group velocity of these waves?

3. The resistive force on a very-high-velocity object of area A passing through a gas of density ρ at a velocity v is

$$\text{Force} \sim A^r \rho^s v^t.$$

Estimate r, s, t.

4. Which of the following reactions are allowed? If forbidden, state the reason.

a. $\pi^- + \mathrm{P} \to \mathrm{K}^- + \Sigma^+$

b. $\mathrm{d} + \mathrm{d} \to \mathrm{He}^4 + \pi^0$

c. $\mathrm{K}^- + \mathrm{P} \to \Xi^- + \mathrm{K}^+$

What is the ratio of reaction cross sections

$$\frac{\sigma(\mathrm{P} + \mathrm{P} \to \pi^+ + \mathrm{d})}{\sigma(\mathrm{n} + \mathrm{p} \to \pi^0 + \mathrm{d})}$$

at the same center-of-mass energy?

5. If the average number of counts in a second from a radioactive source is 4, what is the probability of recording 8 counts in one second?

Sol:

1. The initial wave packet of an electron in the beam has uncertainties Δp_x, Δx in a transverse direction related by the uncertainty relation

$$\Delta p_x \cdot \Delta x \geq \frac{\hbar}{2}.$$

If it has momentum p, the angle of divergence is

$$\theta = \frac{\Delta p_x}{p} \gtrsim \frac{1}{2\Delta x}\frac{\hbar}{p} = \frac{1}{2\Delta x}\frac{\hbar c}{pc}.$$

As

$$p = \sqrt{2mT} = \frac{1}{c}\sqrt{2mc^2 T} = \frac{1}{c}\sqrt{2 \times 0.51 \times 10^6 \times 13.6} = 3.72 \times 10^3\,\text{eV/c},$$

$$\Delta x \approx 10^{-6}\,\text{m},$$

we have

$$\theta \gtrsim \frac{6.6 \times 10^{-16} \times 3 \times 10^8}{2 \times 10^{-6} \times 3.72 \times 10^3} = 2.66 \times 10^{-5}\,\text{radian}$$

and the minimum transverse spread is

$$L\theta_{\text{min}} = 266\text{m}.$$

2. The group velocity is by definition

$$v_g = \frac{d\omega}{dk} = \frac{dv}{d(\frac{1}{\lambda})} = \frac{3}{2}\sqrt{\frac{2\pi T}{\rho\lambda}}.$$

3. The dimensions of the quantities are

$$[F] = \text{MLT}^{-2}, \qquad [A] = \text{L}^2, \qquad [\rho] = \text{ML}^{-3}, \qquad [v] = \text{LT}^{-1}.$$

The equation must be valid for the dimensions also:

$$[F] = [A]^r[\rho]^s[v]^t,$$

or

$$\text{MLT}^{-2} = \text{L}^{2r}\,\text{M}^s\,\text{L}^{-3s}\,\text{L}^t\text{T}^{-t} = \text{L}^{2r+t-3s}\text{M}^s\,\text{T}^{-t}.$$

Equating the powers of each dimension gives

$$2r + t - 3s = 1, \qquad s = 1, \qquad t = 2,$$

i.e.,

$$r = 1,\ s = 1,\ t = 2.$$

4. To find out if the following reactions are forbidden by the strong force, we first consider the isospin I and its third component I_3 which must both be conserved.

a.

$$\begin{array}{lccccccc} & \pi^- & + & p & \rightarrow & K^- & + & \Sigma^+ \\ I_3: & -1 & & \frac{1}{2} & & -\frac{1}{2} & & 1 \end{array}$$

As $\Delta I_3 = \frac{1}{2} - \left(-\frac{1}{2}\right) = 1 \neq 0$, the reaction is forbidden.

b.

$$\begin{array}{lcccc} & d & + \; d & \rightarrow \; He^4 & + \; \pi^0 \\ I: & 0 & 0 & 0 & 1 \end{array}$$

As $\Delta I = 1 - 0 = 1$, the reaction is forbidden.

c.

$$\begin{array}{lcccc} & K^- & + \; P & \rightarrow \; \Xi^- & + \; K^+ \\ I_3: & -\frac{1}{2} & \frac{1}{2} & \frac{1}{2} & \frac{1}{2} \end{array}$$

As $\Delta I_3 = 1 - 0 = 1$, the reaction is forbidden.

Since all the reactions are already forbidden it is not necessary to consider other quantum numbers.

To estimate

$$\frac{\sigma_a}{\sigma_b} \equiv \frac{\sigma(p + p \rightarrow \pi^+ + d)}{\sigma(n + p \rightarrow \pi^0 + d)},$$

we note that, as $m_n \approx m_p$, $m_{\pi^+} \approx m_{\pi^0}$, the two reactions have identical center-of-mass dynamics so that the ratio is independent of energy and momentum. Since no strange particles are involved we consider the isospin states $|II_3\rangle$ of the initial and final systems.

With

$$\begin{array}{lcccc} a: & p & + \; p & \rightarrow \; \pi^+ & + \; d \\ I: & \frac{1}{2} & \frac{1}{2} & 1 & 0 \\ I_3: & \frac{1}{2} & \frac{1}{2} & 1 & 0 \\ b: & n & + \; p & \rightarrow \; \pi^0 & + \; d \\ I: & \frac{1}{2} & \frac{1}{2} & 1 & 0 \\ I_3: & -\frac{1}{2} & \frac{1}{2} & 1 & 0 \end{array}$$

the initial states are

for a, $(p + p): |11\rangle$,

for b, $(n + p)$: both $|1\,0\rangle$ and $|0\,0\rangle$ are possible, so the

state is $\frac{1}{\sqrt{2}}(|10\rangle + |00\rangle)$;

the final states are

for a, $(\pi^+ + d) : |1\,1\rangle$,

for b, $(\pi^0 + d) : |1\,0\rangle$.

Since $I = 1$ for the final states, only the $I = 1$ combination in the initial states can contribute. Thus the scattering amplitudes are in the ratio

$$A_a : A_b = 1 : \frac{1}{\sqrt{2}},$$

and the cross sections are in the ratio

$$\sigma_a : \sigma_b = A_a^2 : A_b^2 = 1 : \frac{1}{2}.$$

Hence

$$\frac{\sigma_a}{\sigma_b} = 2.$$

5. The number of counts follows the Poisson distribution:

$$P(n, \lambda) = \frac{\lambda^n e^{-\lambda}}{n!}.$$

The mean value is

$$\bar{n} = \sum_n^{\infty} nP = \sum_{n=0}^{\infty} \frac{n\lambda^n e^{-\lambda}}{n!} = \sum_{n=1}^{\infty} \frac{n\lambda^n e^{-\lambda}}{n!}$$
$$= \lambda e^{-\lambda} \sum_{n=1}^{\infty} \frac{\lambda^{n-1}}{(n-1)!} = \lambda e^{-\lambda} \sum_{k=0}^{\infty} \frac{\lambda^k}{k!} = \lambda e^{-\lambda} e^{\lambda} = \lambda = 4.$$

Therefore,

$$P(8, 4) = \frac{4^8 e^{-4}}{8!} = 0.030.$$

3031

A physicist sets up an experiment to determine the refractive index of a material using a Michelson interferometer and a laser source of wavelength (620 ± 5) nm. When he introduces a glass plate of thickness (0.140 ± 0.001) mm, a fringe shift of 250 ± 5 is observed.

a. Calculate the refractive index of the glass plate.

b. Estimate the uncertainty propagation in the refractive index calculation.

c. Determine which quantity contributes the most to the uncertainty propagation.

d. Discuss whether the measurement is consistent with the accepted value within the uncertainty.

Sol:

a. To determine the refractive index of the glass plate using the Michelson interferometer, let's use the formula:

$$t = \frac{n\lambda}{2(\mu - 1)}$$

where, μ is the refractive index of the material (glass plate),

t is the thickness of the glass plate,

λ is the wavelength of light used, and

n is the fringe shift

Rearranging the formula to determine the refractive index,

$$(\mu - 1) = \frac{n\lambda}{2t}$$

$$\mu = \left(\frac{n\lambda}{2t}\right) + 1$$

Substituting the given data,

$$\mu = \left(\frac{250 \times 620 \times 10^{-9}}{2 \times 0.140 \times 10^{-3}}\right) + 1 = 1.554$$

b. Uncertainty Calculation

The formula for the uncertainty propagation of a function of multiple variables in multiplication can be determined by considering the relative error as shown in the following.

$$\frac{\delta\mu}{\mu} = \left(\frac{\delta n}{n} + \frac{\delta\lambda}{\lambda} + \frac{\delta t}{t}\right) \rightarrow \delta\mu = \mu\left(\frac{\delta n}{n} + \frac{\delta\lambda}{\lambda} + \frac{\delta t}{t}\right)$$

$$= 1.55\left(\frac{5}{250} + \frac{5}{620} + \frac{0.001}{0.140}\right)$$

$$= 1.55\ (0.02 + 0.008 + 0.007) = 0.054$$

Therefore, the calculated refractive index based on the experimental data is

$\mu = 1.554 \pm 0.054$

c. Quantity That Contributes the Most in Error Propagation

Let's determine the percentage uncertainty for each measurement. The quantity with the highest percentage error will contribute the most toward error propagation.

$$\text{Percentage error in fringe shift} = \frac{\delta n}{n} \times 100 = \frac{5}{250} \times 100 = 2\%$$

$$\text{Percentage error in wavelength} = \frac{\delta\lambda}{\lambda} \times 100 = \frac{5}{620} \times 100 = 0.8\%$$

$$\text{Percentage error in thickness} = \frac{\delta t}{t} \times 100 = \frac{0.001}{0.140} \times 100 = 0.7\%$$

Based on the calculations, it is evident that the uncertainty in the measurement of fringe shift contributes the most toward error propagation.

d. Error Analysis

$$\%\ \text{Error} = \frac{\mu_{\text{calculated}} - \mu_{\text{standard}}}{\mu_{\text{standard}}} \times 100\% = \frac{1.554 - 1.5}{1.5} \times 100\% = 3.6\%$$

The small percentage error of approximately 3.6% indicates that the measurement is relatively precise. The refractive index represents how much light is slowed down when passing through the glass plate compared to its speed in vacuum (air). In this case, the glass plate has a refractive index slightly greater than 1, indicating that light slows down as it passes through the glass. This result is consistent with our expectations for typical glass materials. Overall, the experiment provides a reasonable measurement of the refractive index of the glass plate with acceptable precision.

3032

Explain how a breeder reactor can produce more fuel than it consumes. What are the physical reasons for the statement that breeder reactors are less safe than ordinary reactors.

Sol: A breeder produces power due to fission by fast neutrons and in the process regenerates more fissionable material than it consumes. For example, ^{232}Th and ^{238}U are not fissionable by thermal neutrons and cannot be used as reactor fuel, but they can absorb fast neutrons produced in thermal fission of their fissionable isotopes and become the fissionable ^{233}U and ^{239}Pu after two β^- decays:

$$^{232}\text{Th} + \text{n} \rightarrow {}^{233}\text{Th} \xrightarrow{\beta-} {}^{233}\text{Pa} \xrightarrow{\beta-} {}^{233}\text{U},$$

$$^{238}\text{U} + \text{n} \rightarrow {}^{239}\text{U} \xrightarrow{\beta-} {}^{239}\text{Np} \xrightarrow{\beta-} {}^{239}\text{Pu}.$$

As each fission produces two to three fast neutrons, more fissionable isotopes can be produced than those consumed. No moderator is used in the core of a breeder as fast neutrons are needed to breed fissionable fuel. It is difficult to control the density of neutrons in the reactor and should the amount of fissionable fuel exceed the critical mass, nuclear explosion may take place. Furthermore, liquid sodium usually has to be used as the coolant in a breeder reactor and it is a less safe material to handle than coolants such as water that are used in a thermal reactor.

3033

Decide whether the following statements can be correct. Explain your answer.

a. A focused beam of 100 eV electrons is used to probe molecular structure with a resolution of 0.1 Å (10^{-11} m).

b. Two states 10^{-3} eV apart, each with lifetime 10^{-10} sec, can be clearly resolved. The two states do not decay into each other.

c. A new narrow-band pulsed laser system operates at 600 nm and produces pulses of 2×10^{-10} sec duration with a bandwidth of 10^{-3} nm. ($\hbar = 6.58 \times 10^{-16}$ eV-sec, $m_e = 9.11 \times 10^{-31}$ kg, 1 eV $= 1.6 \times 10^{-19}$ joules)

Sol:

a. As the de Broglie wavelength of the electrons is

$$\lambda = \frac{2\pi\hbar}{\sqrt{2mE}} = \frac{2\pi \times 6.58 \times 10^{-16} \times 1.6 \times 10^{-19}}{\sqrt{2 \times 911 \times 10^{-31} \times 100 \times 1.6 \times 10^{-19}}}$$

$$= 1.23 \times 10^{-10}\,\text{m} = 1.23\ \text{Å} > 0.1\ \text{Å},$$

this is not possible.

b. Since the lifetime of the energy levels is 10^{-10}s each will have a width given by the uncertainty principle $\Delta E \cdot \Delta t \approx \frac{\hbar}{2}$ of about 3.3×10^{-6} eV. As this width is much smaller than the separation 10^{-3} eV, there is no overlapping of the two energy levels and they can be resolved.

c. The duration and width of a pulse of the laser must satisfy the uncertainty principle $\Delta t \cdot \Delta E \gtrsim \frac{\hbar}{2}$. As

$$\Delta E = h\Delta v \approx \frac{hc\Delta\lambda}{\lambda^2} = \frac{2\pi \times 3 \times 10^8}{600 \times 10^{-9}} \times \left(\frac{10^{-3}}{600}\right)\hbar$$

$$= \frac{\pi\hbar}{6} \times 10^{10},$$

$$\Delta t \approx 2 \times 10^{-10}\,\text{s},$$

$$\Delta E \cdot \Delta t = \frac{\pi}{3}\hbar > \frac{\hbar}{2}.$$

Hence the required pulse can be produced.

3034

A small perfectly reflecting particle is levitated by the force of a laser beam. The particle has a mass of 10^{-3} gram. The laser light is reflected over a wide spread of angles θ relative to the incoming beam; take the average value of $\cos\theta$ to be 0. What is the power (in watts) of the laser?

Sol: The weight of the particle mg is balanced by the upward force f exerted by the laser beam. Suppose the beam has power W and is vertically upward. Then it carries an upward vertical momentum $\frac{W}{c}$ per unit time. On reflection through angle θ, its downward vertical momentum is $-\frac{W}{c}\cos\theta$ per unit time. Hence the upward force is

$$f=\frac{W}{c}-\left(-\frac{W}{c}\cos\theta\right)=\frac{W}{c}(1+\cos\theta)\ ,$$

and so

$$\frac{W}{c}(1+\overline{\cos\theta})=\frac{W}{c}=mg,$$

giving

$$W=mgc=10^{-6}\times 9.8\times 3\times 10^{8}\approx 3\times 10^{3}\,\text{W}.$$

Thus the power of the laser beam is 3 kilowatts.

3035

Consider an electron in an atmospheric gas atom to be a bound harmonic oscillator with natural angular frequency ω_0. A monochromatic plane electromagnetic wave of frequency ω and wavelength much greater than atomic dimensions is incident on the atom. Ignore damping effects.

a. Find the dependence on ω of the total cross section for scattering by the atom when $\omega \ll \omega_0$.

b. The atoms are now ionized so that all the electrons are free. Find the dependence on ω of the total cross section for scattering by the free electrons.

NOTE: In both (a) and (b) you are not required to derive a complete formula. It is sufficient to determine only the dependence upon frequency.

Sol:

a. Under the action of the electric field $\mathbf{E}_0 e^{-i\omega t}$ of the incident plane electromagnetic waves, the oscillator has the equation of motion

$$\ddot{\mathbf{r}} - \tau \dddot{\mathbf{r}} + \omega_0^2 \mathbf{r} = -\frac{e}{m}\mathbf{E}_0 e^{-i\omega t},$$

where $\tau = \dfrac{e^2}{6\pi\epsilon_0 mc^3}$ is a characteristic time arising from the reaction of the secondary radiation emitted by the oscillator. In the steady state

$$\mathbf{r} = \mathbf{r}_0 e^{-i\omega t}.$$

Substitution gives

$$\mathbf{r} = -\frac{e}{m}\left(\frac{1}{\omega_0^2 - \omega^2 - i\omega^3\tau}\right)\mathbf{E}_0 e^{-i\omega t}.$$

The oscillator radiates secondary radiation of the same frequency whose energy per unit time per unit solid angle is

$$\frac{dP}{d\Omega} = \frac{e^2 \sin^2\theta}{16\pi^2\epsilon_0 c^3}{}^2\ddot{\mathbf{r}}^2.$$

Macroscopically the incident waves appear to be scattered by the oscillator into various directions. The differential scattering cross section is defined as

$$\frac{d\sigma}{d\Omega} = \frac{\text{average radiated power/unit solid angle}}{\text{incident energy/unit area/unit time}}$$

The denominator is the magnitude of the Poyting vector $\mathbf{N} = \mathbf{E} \times \mathbf{H}$ of the incident radiation. For plane electromagnetic waves, $\sqrt{\epsilon_0}|\mathbf{E}| = \sqrt{\mu_0}|\mathbf{H}|$, $\mathbf{E} \perp \mathbf{H}$, so that

$$N = \epsilon_0 c|\mathbf{E}|^2 = \frac{\epsilon_0}{2}c\mathbf{E}_0^2,$$

independent of frequency. Hence, considering the frequency dependence alone, we have

$$\frac{d\sigma}{d\Omega} \propto \left|\frac{\omega^2}{\omega_0^2 - \omega^2 - i\tau\omega^3}\right| = \frac{\omega^4}{(\omega_0^2 - \omega^2)^2 + \tau^2\omega^6}.$$

The radiation reaction term in the above is the square of

$$\begin{aligned}\tau\omega^3 &= \frac{2}{3}\left(\frac{e^2}{4\pi\epsilon_0 mc^2}\right)\frac{\omega^3}{c} = \frac{2\gamma_0\omega^3}{3c}\\ &= \frac{2}{3} \times \frac{2.8 \times 10^{-15}}{3 \times 10^8}\omega^3\\ &\approx 6 \times 10^{-24}\omega^3 \ll \omega^2\end{aligned}$$

if $\omega \ll 10^{23}$ s, and can usually be neglected except when $\omega \approx \omega_0$. Therefore, if $\omega \ll \omega_0$,

$$\frac{d\sigma}{d\Omega} \propto \left(\frac{\omega}{\omega_0}\right)^4$$

from which follows the total cross section

$$\sigma \propto \left(\frac{\omega}{\omega_0}\right)^4 .$$

b. The electrons are now free so that the equation of motion becomes

$$\ddot{\mathbf{r}} - \tau\dddot{\mathbf{r}} = -\frac{e\mathbf{E}}{m} e^{-i\omega t},$$

which has the steady state solution

$$\mathbf{r} = \frac{e}{m} \frac{\mathbf{E}_0 e^{-i\omega t}}{\omega^2 + i\omega^3\tau} \approx \frac{e}{m\omega^2} \mathbf{E}_0 e^{-i\omega t} .$$

As

$$\frac{dP}{d\Omega} \propto |\ddot{\mathbf{r}}|^2$$

the differential cross section is

$$\frac{d\sigma}{d\Omega} \propto \left(\frac{\omega^2}{\omega^2}\right)^2 = \text{constant.}$$

Hence the total cross section for scattering by free electrons is independent of frequency.

3036

The atmosphere shields the Earth from much of the electromagnetic radiation incident from external sources. The following processes are primarily responsible for this shielding in various regions of the electromagnetic spectrum. Give an approximate wavelength or frequency or the corresponding photon energy for each.

1. Photoelectric absorption by nitrogen K-shell electron
2. Reflection from plasma in the ionosphere
3. Photodisintegration of ozone
4. Compton scattering
5. Electron–positron pair production
6. Absorption by H_2O rotational excitations

Sol:

1. A photon that can eject an electron from the K-shell of a nitrogen atom must have an energy greater than the binding energy of a nitrogen K-shell electron, 402 eV. The corresponding angular frequency is $\omega = \frac{E}{\hbar} = \frac{402}{6.6 \times 10^{-16}} = 6 \times 10^{17} \mathrm{s}^{-1}$.

2. An electromagnetic wave incident on a plasma will be reflected if its angular frequency ω is less than the plasma (angular) frequency

$$\omega_\mathrm{p} = \sqrt{\frac{Ne^2}{\epsilon_0 m}},$$

where N is the number of free electrons per unit volume of the plasma. The maximum N for ionospheric layers is about 10^{12} m^{-3}, giving

$$\begin{aligned}\omega_\mathrm{p} &= \sqrt{\frac{4\pi N c^2 e^2}{4\pi\epsilon_0 m c^2}} = c\sqrt{4\pi N r_0} \\ &= 3 \times 10^8 \sqrt{4\pi \times 10^{12} \times 2.8 \times 10^{-15}} = 5.6 \times 10^7 \mathrm{s}^{-1},\end{aligned}$$

$r_0 = \frac{e^2}{4\pi\epsilon_0 m c^2}$ being the classical radius of the electron, 2.8×10^{-15} m^{-3}. Hence we require that $\omega < 6 \times 10^7$ s^{-1}.

3. Ozone, O_3, is a trace constituent of the upper atmosphere, with a maximum concentration at about 25 km above Earth's surface. It can photodisintegrate by absorbing a photon:

$$\mathrm{O_3} + h\nu \to \mathrm{O} + \mathrm{O_2}.$$

This happens most strongly for radiation of wavelengths 2000–3000 Å, and less strongly for radiation of 4500–8000 Å.

4. Compton scattering is important for radiation shielding when the wavelength of the incident radiation is comparable with the Compton wavelength of the electron

$$\lambda_\mathrm{C} = \frac{h}{m_\mathrm{e} c} \approx 0.024\,\text{Å}.$$

The cross section for Compton scattering decreases and the process of pair production of electrons gains importance with increasing frequency.

5. Electron-positron pair production can take place only when the energy ε of a photon is greater than the rest energy of the pair:

$$\varepsilon \geq 2m_\mathrm{e}c^2 = 1.02 \text{ MeV}.$$

The corresponding photon (angular) frequency is $\omega = \frac{1.02 \times 10^6}{6.6 \times 10^{-16}} = 1.5 \times 10^{21} \mathrm{s}^{-1}$.

6. The molecular rotational spectrum consists of lines

$$E_n = 2nB, \ (n = 0, 1, 2, \ldots)$$

with $B = h^2/8\pi^2 I$, I being the moment of inertia of the molecule. As the principal moment of inertia of H_2O is ~ 10^{-47} J·s^2, B ~ 10^{-21} J. The energy of a photon causing a transition $n = 0$ to $n = 1$ is $\varepsilon = \Delta E_{1,0} = 2B$ ~ 10^{-21} J, the corresponding angular frequency being

$$\omega = \frac{10^{-21}}{10^{-34}} = 10^{13} \mathrm{s}^{-1}.$$

3037

In preparation for a landing on the bright side of the Moon, you are asked to estimate the surface temperature. You may assume that the lunar surface material is a good insulator. Make as quantitative an estimate as you can. For your information, the solar constant is 1353 watts/m^2. What additional information would you need to improve your estimate?

Sol: Let the absorptivity of the Moon's surface be α. The absorption of heat and radiation from the surface are in equilibrium for a stable moon:

$$\varepsilon \sigma T^4 = \alpha I_0,$$

where I_0 is the solar constant (solar energy flux on lunar surface), $\varepsilon \sigma T^4$ is the radiation emitted per unit area per second by the lunar surface, ε being a correction factor introduced to account for the Moon not being a black body, σ the Stefan-Boltzmann constant, 5.67×10^{-8} W/m^2, and T the surface temperature. For a rough estimate of T, we take $\varepsilon \approx \alpha$. Then

$$T = \left(\frac{I_0}{\sigma}\right)^{\frac{1}{4}} = \left(\frac{1353}{5.67 \times 10^{-8}}\right)^{\frac{1}{4}} = 393\mathrm{K} = 120^\circ \mathrm{C}.$$

To improve the estimate of T, we have to have some knowledge of the ratio α/ε.

3038

Describe briefly how to measure 8 of the following quantities. Also make an order of magnitude estimate of the expected value.

1. Vacuum in a cyclotron
2. Magnetic field in a spectrometer magnet used for elementary particle experiments
3. Temperature of a sample of liquid hydrogen
4. Lifetime of the muon
5. The Hubble constant
6. Planck's constant
7. Distance to nearby stars
8. Distance to galaxies
9. Avogadro's number
10. Size of a nucleus
11. Size of an atom
12. Lamb shift
13. Temperature of the cosmic black body radiation
14. Nuclear g-factors
15. Mass of a meson or baryon resonance

Sol:

1. *Vacuum in a cyclotron*

   Measurement of the ultra-high vacuum in a cyclotron can be made with a Bayard–Alpert gauge shown schematically in Fig. 3.3. Electrons emitted by the cathode and accelerated by a high voltage ionize the residual gas molecules by collision. The resulting positive ions are collected by the collector electrode. The cathode current I_e, collector current I_i and the pressure are related by

$$\frac{I_i}{I_e} = kp,$$

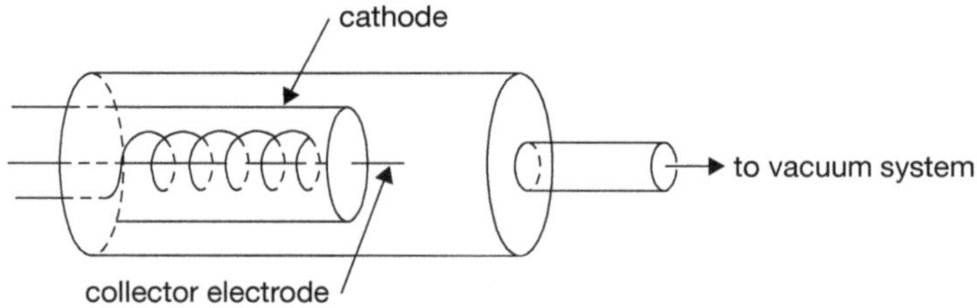

Fig. 3.3

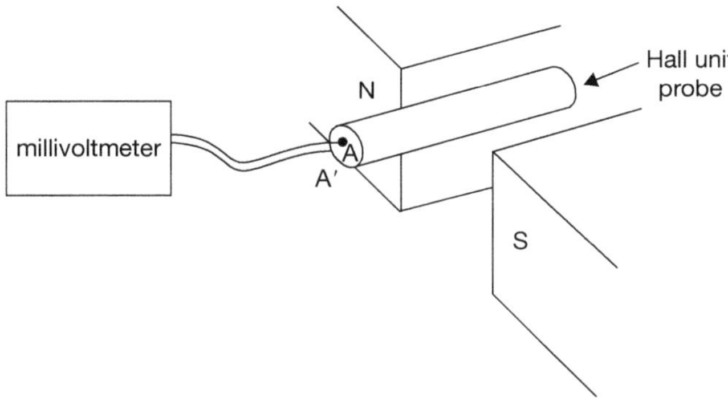

Fig. 3.4

where k is a proportional constant. By measuring the currents the pressure is obtained. It is about 10^{-8}–10^{-9} atmosphere in a cyclotron.

2. *Magnetic field in a spectrometer magnet*

The magnetic field can be measured as follows. A probe consisting of a Hall unit is placed in the magnetic field and a known current is allowed to flow through it, producing a Hall potential V_{AA}', which can be read off a millivoltmeter, as shown in Fig. 3.4. With the known Hall coefficient K the formula

$$V_{AA'} = K\frac{IB}{d}$$

gives the magnetic induction B of the field. For a spectrometer for elementary particle experiments, $B \sim 10$ kGs.

3. *Temperature of liquid hydrogen*

The temperature of liquid hydrogen is about 20 K. It can be determined by means of a hydrogen vapor pressure thermometer.

4. *Lifetime of muon*

A muon decays according to

$$\mu^{\pm} \to e^{\pm} + \nu_e + \nu_{\mu}.$$

As μ^+, μ^- are antiparticles of each other, they have the same mean lifetime, $\tau = 2.2 \times 10^{-6}$ s, which is defined as

$$\tau = \frac{1}{\Gamma},$$

where Γ is the decay rate, the probability per unit time that any given muon will disintegrate. As

$$\frac{dN}{dt} = -\Gamma N,$$

or

$$N(t) = N(0)e^{-\Gamma t},$$

Γ and hence τ can be obtained by observing the change of cosmic ray muon flux with altitude or by determining the time distribution of muon decays using a counter arrangement.

5. *The Hubble constant*

Hubble's cosmological law relates the distance R and receding velocity v of a distant galaxy: $R = \frac{v}{H}$, where H is Hubble's constant. Using v inferred from the red shift of spectral lines and R from other astronomical observations, H can be determined. $H = 100$ km/s/Mpc.

6. *Planck's constant*

The Compton wavelength $\lambda_C = \frac{h}{mc}$ of the electron can be determined from the Compton shift $\Delta\lambda = \lambda_C\ (1 - \cos\theta)$ resulting from Compton scattering. Knowing the electron mass, Planck's constant h can be deduced: $h \sim 6.626 \times 10^{-34}$ J·s.

7. *Distance to nearby star*

By using the trigonometric parallax method, the distance to a nearby star within about 30 parsecs of the solar system can be determined. The semimajor axis of the Earth's orbit is used as the base line. The angle between the lines drawn from the star to each end of the base line is the angle of parallax from which the distance of the star can be deduced by a geometric calculation.

8. *Distance to galaxies*

The apparent luminosity (observed optical flux) of a star or galaxy depends on its distance. Distances to galaxies within a few megaparsecs can be measured by using cepheid variable stars, for which there is a relation between the period of light variation and the apparent luminosity, the relation being calibrated using cepheids in our own galaxy. For farther galaxies, the distance is estimated by its red shift using Hubble's relation.

9. *Avogadro's number*

The amount of chemical reaction occurring at an electrode during electrolysis can be expressed in moles as $n = Q/zF$, where Q is the total charge supplied, z the charge of an ion and F the faraday. Thus F can be determined accurately by electrochemical means. Then Avogadro's number is obtained as

$$N_A = \frac{F}{e} = 6.023 \times 10^{23} \text{mol}^{-1}$$

using the electronic charge which can be accurately determined by Millikan's oil drop method.

10. *Size of a nucleus*

In Rutherford's experiment on scattering of α-particles by a heavy nucleus it was found that, for moderate α-particle energies, the experimental results were consistent with Rutherford's assumption of a simple inverse-square (Coulomb) force of interaction. As the energy was increased, deviation from this assumption appeared, particularly at large angles. The deviation is associated with the fact that the α-particle trajectory reaches closer to the nucleus at higher energies at which the inverse-square law fails. This distance at which the interaction changes character may be considered as the nuclear radius. It varies from a few times 10^{-15}m for light nuclei to about 10^{-14}m for heavy nuclei. A rough estimate is given by the closest approach of an α-particle incident head-on on a nucleus of charge Ze:

$$T = \frac{1}{2}mv^2 = \frac{2Ze^2}{4\pi\epsilon_0 r},$$

or

$$r = \frac{2Ze^2}{4\pi\epsilon_0 T} = \frac{2\times 100\times(1.6\times10^{-19})^2\times 9\times 10^9}{6\times10^6\times1.6\times10^{-19}} \simeq 5\times10^{-14}\,\text{m},$$

using $T = 6$ MeV, $Z = 100$.

11. *Size of an atom*

Take an element whose atoms are arranged in a cubic lattice. If it has atomic weight A and the density is measured to be ρ, the atomic radius r is given by

$$\frac{4}{3}\pi r^3 N_A = A/\rho,$$

N_A being Avogadro's number, or

$$r = \left(\frac{3A}{4\pi\rho N_A}\right)^{\frac{1}{3}} \sim 10^{-10}\,\text{m}.$$

12. *Lamb shift*

In 1947 Lamb and Rutherford measured the energy interval between the states $E_{2,0,1/2}$ and $E_{2,1,1/2}$ of the hydrogen atom, $\Delta E = 4.37516 \times 10^{-6}$ eV, by using a radio-frequency spectrometer. The correct theoretical expression for the energy difference was first given by Lamb and the effect is known as the Lamb shift.

13. *Temperature of cosmic black body radiation*

The spectral distribution of the cosmic background radiation is measured and the results compared with the black body spectrum to find the appropriate temperature. Or the Wien displacement law is used to calculate the temperature from the wavelength λ_{max} corresponding to the maximum intensity. $T \approx 3$ K.

14. *Nuclear g-factors*

By measuring the Larmor precession frequency of a nucleus in a magnetic field H.

$$\omega = \frac{\mu H}{I\hbar} = \frac{g\mu_N H}{\hbar},$$

where μ_N is the nuclear magneton, H is the applied magnetic field and $\mu = g\mu_N I$ is the nuclear magnetic moment, the nuclear g-factor can be deduced. It can be positive or negative and its magnitude falls in the range 0.1–10.

15. *Mass of a meson or baryon resonance*

A resonance state has a lifetime of only about 10^{-23} to 10^{-22} s and its mass is deduced by measuring the energies E_1, E_2, E_3,... and momenta $\mathbf{P}_1$, $\mathbf{P}_2$, $\mathbf{P}_3$, ... of the resonant particles and using the relativistic equation

$$m^2c^4 = (E_1 + E_2 + E_3 + \ldots)^2 - (\mathbf{p}_1 + \mathbf{p}_2 + \mathbf{p}_3 + \ldots)^2c^2.$$

For example, in

$$\begin{aligned} p + \pi^- &\to n + \rho^0 \\ &\to \pi^+ + \pi^- \end{aligned}$$

the mass of ρ^0 is given by

$$mc^2 \sqrt{(E_1 + E_2)^2 - (\mathbf{P}_1 + \mathbf{P}_2)^2c^2}.$$

Experimentally the interaction is observed as

$$p + \pi^- \to n + \pi^+ + \pi^-.$$

If the above calculation gives a peak in the distribution of m, then the resonance state is confirmed and its mass determined. In this way it is found that $m_{\rho^0} \approx 776\,\mathrm{MeV/c^2}$.

3039

A. Describe briefly how you could make as accurate as possible a measurement of 4 (four) of the following quantities. In each case, give a description of the apparatus and a careful explanation of the physical principles by which it works.

1. The relative natural abundance of the two stable isotopes of copper.
2. The pressure in the vacuum chamber of a cyclotron (approximately 10^{-8} atm.)
3. The magnetic moment of the deuterium nucleus.
4. The temperature of the surface of the Sun.
5. The lifetime of the short-lived ($\sim 10^{-22}$ sec) nuclear state.

B. Estimate very roughly (i.e., to within an order of magnitude) for the following. Use any method, including intuition, but credit may be given only if some method is shown.

1. Energy released in a nuclear fission explosion in which 1 kg of U^{235} undergoes fission.
2. Radius of the mumesic atom which is a hydrogen atom with the electron replaced by a muon.
3. Radius of a 10 TeV proton accelerator (1 TeV = 10^{12} eV).

Sol:

A. 1. The relative natural abundance of an isotope is its number expressed as a percentage in a natural mixture of the element. First a sample of the naturally occurring element copper, which is a mixture of ${}^{63}_{29}\text{Cu}$ and ${}^{65}_{29}\text{Cu}$, is prepared. Then with a mass spectrometer the mass m for each type of isotope is measured from the curvature of its trajectory, R, in a magnetic field B using the relation

$$R = \frac{mv}{qB}.$$

Suppose the masses of the isotopes measured are $63u$ and $65u$, where u is a constant, respectively. Measure also the mass M per mole of the sample by a chemical method. Suppose the mean mass of an atom is found to be $M/N_A = 63.54u$. Letting x denote the relative abundance of ${}^{63}_{29}\text{Cu}$, and $1 - x$ that of ${}^{65}_{29}\text{Cu}$, we have

$$63x + 65(1 - x) = 63.54,$$

and hence

$$x = 68.94\,\%\,,\ 1 - x = 31.06\,\%\,.$$

2. See **Problem 3038** (1).

3. The total angular momentum of a nucleus is the vector sum of the spin momenta of the constituent nucleons and is known as the nuclear spin momentum.

As a nucleus carries charge its spin produces a nuclear magnetic moment $\mu = g\mu_n T$, where μ_n is the nuclear magneton and I is the nuclear spin. In an applied magnetic field H_0 the nucleus will have $2I + 1$ spin energy levels. A radio-frequency alternating electromagnetic field is applied at right angles to the external magnetic field to induce transitions between the neighboring spin energy levels. Let two neighboring energy levels of the deuterium nucleus be E_I and E_{I+1}. Transition between them will produce a resonance line of angular frequency

$$\omega = \frac{E_I - E_{I+1}}{\hbar} = \frac{g\mu_n H_0}{\hbar}$$

Since $E_I = -g\mu_n I H_0$. This gives the magnetic moment of the deuterium nucleus as

$$\mu(^2_1\mathrm{H}) = 0.85742\mu_n,$$

where the nuclear magneton μ_n has the value 5.051×10^{-27} J/T. Experimentally, the measurements are carried out using a nuclear magnetic resonance apparatus.

4. The solar-radiation spectrum obtained outside the Earth's atmosphere us very nearly that of a black body. Then from the wavelength found for the maximum intensity, $\lambda_m \approx 4900$ Å, Wien's displacement law $\lambda_m T = 2.898 \times 10^7$ Å·K gives $T = 5.9 \times 10^3$ K.

5. A nuclear state of lifetime $\sim 10^{-22}$ s is a resonance state whose existence is ascertained only by the appearance of a narrow peak in the frequency distribution of the effective mass calculated from the energies and momenta of the secondary particles presumably produced in its decay (see **Problem 3038** (15)). Its lifetime is estimated using the uncertainty principle

$$\Delta t \cdot \Delta E \gtrsim \frac{\hbar}{2},$$

where $\Delta E = \Delta mc^2$, Δm being the width of the peak in the effective-mass distribution.

B. 1. For rough estimation assume

$$^{235}_{92}\mathrm{U} + ^1_0\mathrm{n} \to ^{236}_{92}\mathrm{U} \to \mathrm{X} + \mathrm{Y} + 2^1_0\mathrm{n},$$

where X and Y are nuclei with $A = 117$ each. The mean binding energy of each nucleon in a nucleus of mass number A is 7.7 MeV for $A \approx 236$ and 8.6 MeV for $A \approx 117$. So the total binding energy for ^{236}U is $236 \times 7.7 = 1817$ MeV and that for X and Y is $2 \times 117 \times 8.6 = 2012$ MeV, and thus about 195 MeV of energy is released per fission. As 1 kg of ^{235}U contains

$$\frac{10^3 \times 6.02 \times 10^{23}}{235} = 2.56 \times 10^{24}$$

^{235}U atoms, their fission gives rise to

$$2.56 \times 10^{24} \times 195 = 5 \times 10^{26}\,\text{MeV}$$

of energy.

2. The first radius a of the Bohr atom is inversely proportional to the mass m of the orbiting particle:

$$\frac{a_\mu}{a_e} = \frac{m_e}{m_\mu}.$$

As $a_e \approx 0.5$ Å, $m_\mu = 210\, m_e$,

$$a_\mu = 0.5 \times \frac{1}{210} \approx 0.0023\,\text{Å}.$$

3. The radius R of the orbit of a proton of mass m, charge e, velocity v, Lorentz factor γ in a magnetic field of flux density B is given by

$$\frac{m\gamma v^2}{R} = evB.$$

Hence

$$R = \frac{pc}{eBc},$$

where p is the momentum of the proton. Taking $B \approx 5$ T, as $pc \approx 10$ TeV $= 10^{13}$ eV $= 1.6 \times 10^{-6}$ J, we have

$$R = \frac{1.6 \times 10^{-6}}{1.6 \times 10^{-19} \times 5 \times 3 \times 10^8} = 6.7 \times 10^3\,\text{m} = 6.7\,\text{km}.$$

3040

A photomultiplier tube is to be used to detect light pulses each of which consists of a small but fixed number of photons. The average photoelectric efficiency is 10%. That is, a photon has a 10% probability of causing the emission of a detectable photoelectron. Assume the photomultiplier gain is 10^6 and that the output current as a function of time can be approximated as shown in Fig. 3.6.

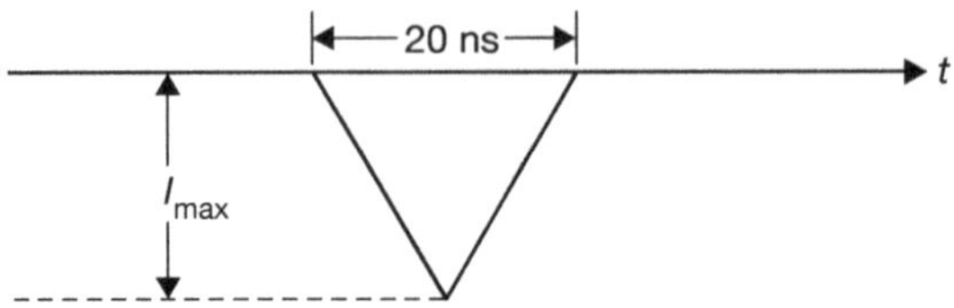

Fig. 3.6

a. I_{max}, when averaged over many pulses, is 80 μA. What is N, the average number of photons in each pulse?

b. What, in terms of N, is the probability that a light pulse will go undetected (i.e., that no photoelectrons are produced)?

c. What effects are responsible for the fluctuation in the value of I_{max}?

Sol:

a. The total quantity of charge carried by one pulse of current is

$$Q = \int I\,dt,$$

which is the area of the triangle in Fig. 3.6. Thus

$$Q = \frac{1}{2} \times 20 \times 10^{-9} \times 80 \times 10^{-6} = 8 \times 10^{-13}\,\text{C}$$

and the number of electrons carried by one pulse is

$$n = \frac{Q}{e} = \frac{8 \times 10^{-13}}{1.6 \times 10^{-19}} = 5 \times 10^{6}.$$

Then the number of photoelectrons emitted per light pulse is

$$n' = n/10^{6} = 5,$$

and hence the number of photons in one light pulse is

$$N = n'/0.1 = 50.$$

b. The probability that all photons of a light pulse will go undetected is

$$\left(\frac{90}{100}\right)^{N} = (0.9)^{50} = 5.15 \times 10^{-3} = 0.52\,\%.$$

c. In each pulse there is a finite probability for a certain number of photons not being detected. Thus the number of photons detected will vary from pulse to pulse. Consequently the maximum current I_{max} will fluctuate about a mean value. The greater the number of photons in a pulse, the smaller will be the fluctuation.

3041

Two parallel conducting plates of area A are separated by distance L_0 (from the inner surface of one plate to the of the other). They are connected by a conducting spring of relaxed length L_0 and spring constant k. The whole system is neutral. The spring is relaxed. (L_0 is small compared with the transverse dimensions of the plates) The system is now subjected to a uniform electric field E_0 normal to the plates. The equilibrium separation between the plates is L. Find L as a function of E_0, L_0, k, and A.

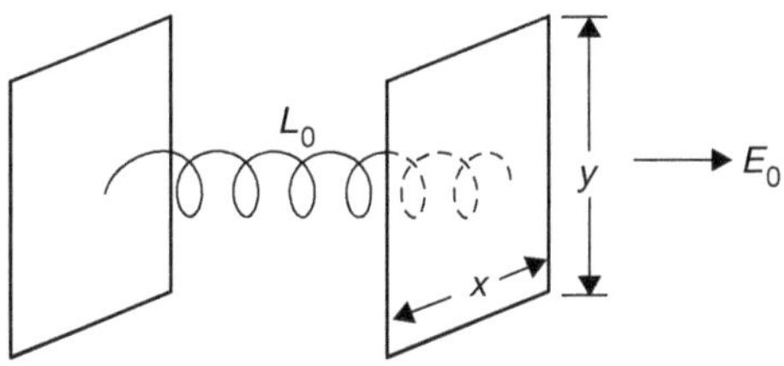

Fig. 3.7

Sol: As the gap between the plates is small compared with the transverse dimensions, we can consider shielding as complete and take the internal electric field as $E_{\text{in}} = 0$. Apply Guass' flux theorem

$$\oint_S \mathbf{E} \cdot d\mathbf{S} = \frac{Q}{\epsilon_0},$$

taking for S a closed surface closely fitting one plate. Then as $E_{\text{in}} = 0$ and any edge effect can be neglected we have

$$E_0 A = \frac{Q}{\epsilon_0}, \qquad \text{or} \qquad Q = \epsilon_0 E_0 A,$$

where Q is the magnitude of the charges carried by the plates. The force exerted by the field on each plate is

$$F = \int E_0 \sigma \, dS = E_0 \sigma A = E_0 \left(\frac{Q}{A} \right) A = \epsilon_0 E_0^2 A.$$

Let z be the extension of the spring caused by this force. Hooke's law gives

$$F = kz,$$

or

$$z = \frac{\epsilon_0 E_0^2 A}{k}.$$

The equilibrium separation is therefore

$$L = L_0 + z = L_0 + \frac{\epsilon_0 E_0^2 A}{k}.$$

3042

The Shiva laser at Livermore delivers 10 kJ of 1 μ wavelength radiation in 10^{-9} seconds onto a focal spot of 10^{-3} cm^2 area.

a. Find the power density, the peak electric field, the energy density in the electromagnetic field, and the radiation pressure, in practical MKS units.

b. If all the energy were absorbed and converted into thermal energy uniformly distributed over a sphere whose size just matches that of the focal spot, containing 5×10^{18} hydrogen atoms (fully ionized), estimate the pressure of this material and show that the radiation pressure of the laser light, although huge, is much too low to hold the cloud together, and therefore permits free expansion.

Sol:

a. The mean power density is

$$\langle S\rangle = \frac{\text{energy}}{At} = \frac{10 \times 10^3}{10^{-3} \times 10^{-4} \times 10^{-9}} = 10^{20}\ \text{J/s} \cdot \text{m}^2.$$

The electromagnetic radiation has mean energy density $\epsilon_0\langle E^2\rangle$, so

$$\langle S\rangle = \epsilon_0 c\langle E^2\rangle = \frac{\langle E^2\rangle}{\mu_0 c}.$$

Hence

$$\sqrt{\langle E^2\rangle} = \sqrt{\langle S\rangle \mu_0 c} = \sqrt{10^{20} \times 4\pi \times 10^{-7} \times 3 \times 10^8} = 1.94 \times 10^{11}\ \text{V/m},$$

and the peak value of the electric field is

$$E_0 = \sqrt{2\langle E^2\rangle} = 2.75 \times 10^{11}\text{V/m}.$$

The mean energy density is

$$\langle U\rangle = \varepsilon_0\langle E^2\rangle = \frac{\langle S\rangle}{c} = 3.33 \times 10^{11}\text{J/m}^3,$$

and the radiation pressure is

$$P = \frac{\langle U \rangle}{c} \cdot c = \langle U \rangle = 3.3 \times 10^{11} \text{ Pa}.$$

b. As all the absorbed energy is converted into the kinetic energy of the thermal motion of the hydrogen atoms, we have

$$W = \frac{3}{2} NkT,$$

where N is the number of hydrogen atoms involved and k is Boltzmann's constant. Considering the hydrogen atoms as an ideal gas, we have

$$pv = NkT,$$

and thus

$$p = \frac{NkT}{v} = \frac{2W}{3v} = \frac{2W}{3}\left(\frac{4}{3}\pi R^3\right)^{-1}.$$

As the radius R of the sphere is related to the area A of the focal spot by $\pi R^2 = A$, the above becomes

$$p = \frac{W}{2\pi}\left(\frac{\pi}{A}\right)^{\frac{3}{2}}.$$

With $W = 10^4$ J, $A = 10^{-7}$ m^2, $p = 2.80 \times 10^{14}$ *Pa*. Since this pressure is much higher than the radiation pressure generated by the laser light, the latter is unable to hold the cloud of atoms together and this permits free expansion.

3043

Light of unknown polarization is examined with a polaroid and found to exhibit maximum transmitted intensity $I_{max} = I_0$ for vertical polarization plane and minimum intensity $I_{min} = 2I_0/3$ for horizontal polarization plane.

a. Can one, from this information, determine the intensity transmitted through the polaroid at any angle θ relative to the vertical? If so, state $I(\theta)$ in terms of I_0 and θ.

b. The light is passed through a quarter-wave plate with axes vertical and horizontal, and then through the polaroid. After the polaroid the maximum intensity is at an angle of 30° to the vertical. State as much as you can about the polarization of the incident light (unpolarized fraction, locations of major and minor axes of elliptic polarization, relative intensities along major and minor axes).

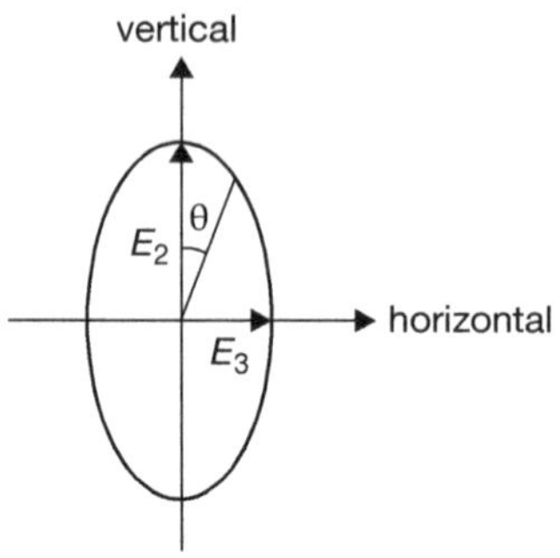

Fig. 3.8

Sol:

a. Suppose the incident light is a mixture of unpolarized and elliptically polarized light. An elliptically polarized light can be considered as consisting of two linearly polarized components of mutually perpendicular planes of polarization with difference amplitudes and a phase difference of 90°. Let the intensities of the unpolarized component, the polarized component with plane of polarization parallel to the vertical direction, and that with plane of polarization parallel to the horizontal direction be I_1, I_2, I_3 respectively. Then

$$I_1 + I_2 = I_0,$$
$$I_1 + I_3 = \frac{2I_0}{3}.$$

If the polaroid is now rotated about its axis through an angle θ, as shown in Fig. 3.8, the intensity transmitted is

$$\begin{aligned} I(\theta) &= I_1 + I_2\cos^2\theta + I_3\sin^2\theta \\ &= (I_1 + I_2)\cos^2\theta + (I_1 + I_3)\sin^2\theta \\ &= I_0\cos^2\theta + \frac{2}{3}I_0\sin^2\theta \\ &= I_0\left(1 - \frac{1}{3}\sin^2\theta\right). \end{aligned}$$

b. After passing through a quarter-wave plate, the two linearly polarized components will undergo a change of phase of –90°, so that the elliptically polarized light will become linearly polarized, with the plane of polarization making an angle θ with the vertical given by

$$\tan\theta = \sqrt{\frac{E_3^2}{E_2^2}} = \sqrt{\frac{I_3}{I_2}}.$$

With $\theta = 30°$, we have $I_2 = 3I_3$. Combining the above gives

$$I_3 = \frac{1}{6}I_0,\ I_1 = \frac{1}{2}I_0,\ I_2 = \frac{1}{2}I_0\,.$$

The unpolarized fraction of the light is $\frac{I_1}{I_0} = \frac{1}{2}$ As for the elliptically polarized light, the major axis is along the vertical direction, the minor axis is along the horizontal direction, and the ratio of the intensities along the major and minor axes is

$$I_2/I_3 = 3.$$

3044

An experimental result y is computed from the formula $y = \frac{(a-3b)\sqrt{x}}{g}$. The values for the right-hand side of the equation are measured to be $a \pm \Delta a$, $b \pm \Delta b$, $x \pm \Delta x$, $g \pm \Delta g$, where Δa, Δb, Δx, and Δg are estimates for the root-mean-square values of the random and uncorrelated errors in a, b, x, and g respectively.

Find an expression for $\Delta y/y$ in terms of a, b, x, g, Δa, Δb, Δx and Δg, where Δy is the r.m.s. value of the random error in y.

Sol: Taking logarithm of both sides of the equation, we have

$$\ln y = \ln(a-3b) + \frac{1}{2}\ln x - \ln g.$$

Differentiating we find

$$\frac{\delta y}{y} = \frac{\delta a - 3\delta b}{a-3b} - \frac{1}{2}\frac{\delta x}{x} - \frac{\delta g}{g}.$$

Considering root-mean-square deviations, the above becomes

$$\frac{\Delta y}{y} = \left[\left(\frac{\Delta a}{a-3b}\right)^2 + \left(\frac{3\Delta b}{a-3b}\right)^2 + \left(\frac{\Delta x}{2x}\right)^2 + \left(\frac{\Delta g}{g}\right)^2\right]^{\frac{1}{2}}.$$

3045

A long, narrow (right circular) cylinder stands on its end on a flat horizontal surface in a gravitational field as shown in Fig. 3.9. The cylinder has density ρ, height H and radius R. This configuration is not absolutely stable if $H/2 > R$. For given ρ and H use the uncertainty principle to give an order of magnitude estimate of how large R must be for the cylinder to remain upright for an appreciable time.

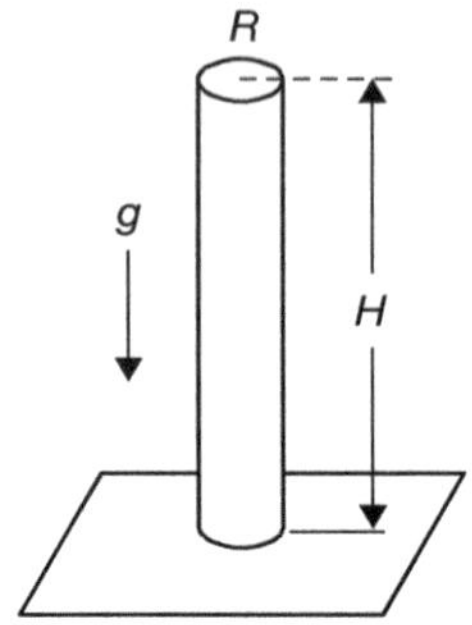

Fig. 3.9

Sol: When the cylinder lies flat on the horizontal surface, it is in a stable state of gravitational potential energy

$$E_0 = mgR = \pi\rho g R^3 H.$$

When it stands on an end surface, it is in a state of gravitational potential energy

$$E_1 = mg\frac{H}{2} = \frac{\pi}{2}\rho g R^2 H^2.$$

The difference in energy between the two states is

$$\Delta E = E_1 - E_0 = \pi\rho g R^2 H\left(\frac{H}{2} - R\right).$$

If $\Delta E > 0$, or $\frac{H}{2} > R$, the standing state E_1 is not absolutely stable as the cylinder can transit from this state to the lying state. The uncertainty principle

$$\Delta E \Delta t \sim \hbar,$$

where Δt is the time scale of transition from standing state to lying state, gives

$$\pi\rho g R^2 H\left(\frac{H}{2} - R\right)\Delta t \sim \hbar.$$

For Δt large, $\Delta E \sim 0$, giving $R \sim \frac{H}{2}$. A better estimate is

$$R \sim \frac{H}{2} - \frac{\hbar}{\pi\rho g R^2 H \Delta t} \sim \frac{H}{2} - \frac{4\hbar}{\pi\rho g H^3 \Delta t}.$$

3046

Suppose that the ionosphere is in thermal equilibrium at temperature *T*. It consists of electrons (mass *m*) and singly charged ions (mass *M*). If there were no large-scale electric field, the electrons and ions would have very different scale heights, kT/mg and kT/Mg respectively. This would produce a large electric field, thereby modifying the density vs. height relation.

Taking into account the electrostatic field, find

a. the density-height relation,

b. the magnitude (and direction) of the self-consistent electric field **E**,

c. the numerical value of E if the ions are protons. (You may ignore the curvature of the Earth's surface. it may be helpful to know that the electric field turns out to be uniform.)

Sol:

a. Use a coordinate system with origin at the Earth's surface and the z-axis vertically upward. The self-consistent electric field **E** is along the z direction:

$$E = -\frac{\partial V}{\partial z},$$

or

$$V = -Ez$$

assuming E to be uniform. Then an electron and an ion at height z will respectively have energies

$$\varepsilon_e = mgz + eEz,$$
$$\varepsilon_i = Mgz - eEz,$$

and number densities

$$n_e = n_{e0}e^{-\varepsilon_e/kT} = n_{e0}e^{-(mg+eE)z/kT},$$
$$n_i = n_{i0}e^{-\varepsilon/kT} = n_{i0}e^{-(Mg-eE)z/kT}$$

as functions of height.

b. Since

$$\Delta \cdot \mathbf{E} = \frac{\rho}{\epsilon_0} = 0$$

for a uniform field, the net charge density $\rho = 0$ everywhere. This means that

$$-en_e + en_i = 0,$$

or

$$n_{e0} = n_{i0},\ mg + eE = Mg - eE.$$

The last equation gives

$$E = \frac{(M - m)g}{2e} \approx \frac{Mg}{2e}$$

as $M \gg m$.

c. If the ions are protons, we have $M = 1.672 \times 10^{-27}$ kg, $e = 1.602 \times 10^{-19}$ C, and

$$E = \frac{1.672 \times 10^{-27} \times 9.807}{2 \times 1.602 \times 10^{-19}} = 5.12 \times 10^{-8}\text{Vm}^{-2}.$$

3047

Consider a more or less ordinary type of photomultiplier that you might find in the laboratory. It has for example five stages, and the manufacturer recommends a maximum supply voltage of 700 volts. Suppose the sensitivity versus wavelength is as shown in Fig. 3.10.

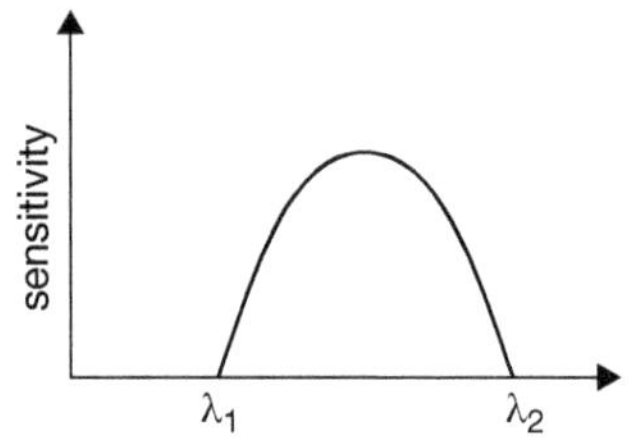

Fig. 3.10

a. Why is the phototube not sensitive for wavelengths less than λ_1?

b. Why is the phototube not sensitive for wavelengths greater than λ_2?

c. Sketch the electrical circuit you would use with this tube to observe light pulses of 20 nanosecond duration with am oscilloscope. Your sketch should include the electrical connections to the cathode, dynodes, anode, and oscilloscope.

Sol:

a. For shorter wavelengths, or higher frequencies, because the cross section for the interaction of the electrons in the cathode of the photomultiplier with photons is very small, very few electrons are emitted. In fact, when the wavelength is smaller than a certain value λ_1, the number of electrons emitted is practically zero, i.e., the photomultiplier has zero sensitivity.

b. For longer wavelengths, or lower frequencies, the window of the photomultiplier strongly absorbs the incident photons. Also, if the wavelength is greater than a certain value λ_2, the energy $h\nu$ of the photon is too small to knock off electrons from the cathode, again there is no sensitivity.

c. A possible circuit diagram is shown in Fig. 3.11, where R_LC is in the range τ to 5τ with τ=20 ns.

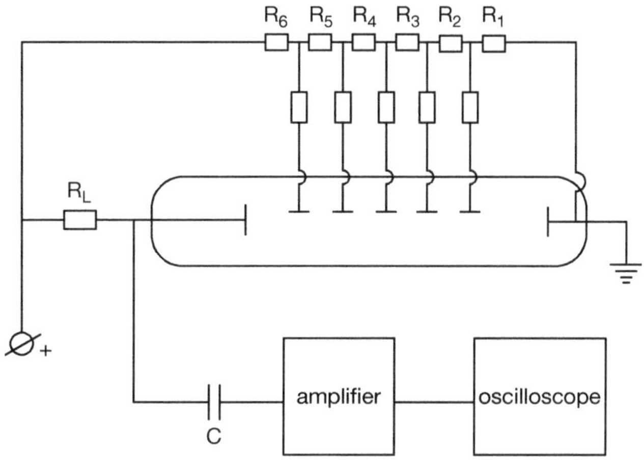

Fig. 3.11

3048

Consider a vertical film of soapy water across a vertical wire loop. The gravitational acceleration is g, the fluid density is ρ, the surface tension is σ. Find equations for the fluid pressure and film width as functions of height in the film.

Sol: It is easier to consider the three-dimensional case of a vertical rectangular wire loop with sides very long along the y-direction but short along

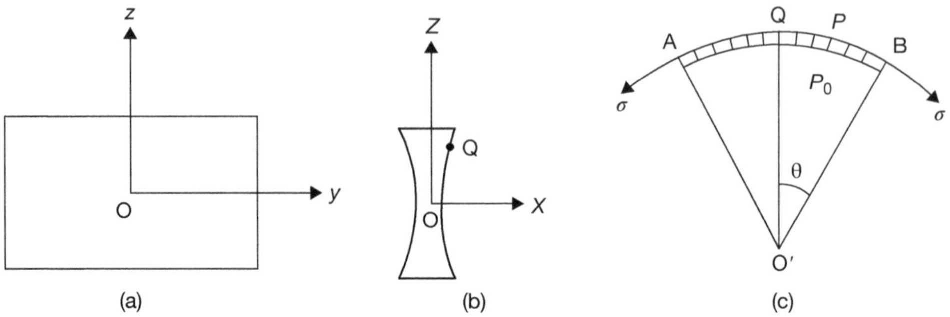

Fig. 3.12

the z-direction as shown in Fig. 3.12(a). The shape of the film for any xz cross-section, as shown in Fig. 3.12(b), can then be regarded as the same for all values of y.

Suppose the atmospheric pressure is p_0 and the fluid pressure near any point Q on the film surface is p. The boundary condition along x-direction is then

$$p = p_0 - \sigma/R,$$

R being the radius of curvature of the film.

This can be briefly proved as follows. Consider a section of the film parallel to the xz plane of unit thickness (along the y direction). Its cross section in the xz plane is bounded by two straight lines parallel to the x-axis and two arcs as shown in Fig. 3.12(b). Consider an element of arc AB centered at Q. As shown in Fig. 3.12(c), the resultant surface tension force is along QO′ with magnitude $2\sigma \sin\theta$, and is balanced by a force due to the excess pressure on the element of arc. Thus

$$2R\theta(p - p_0) + 2\sigma \sin\theta = 0.$$

Letting $\theta \to 0$, we have

$$p - p_0 = -\frac{\sigma}{R},$$

as stated above.

Let the thickness of the film be $t = 2f(z)$. Then the curvature of the arc is

$$R^{-1} = f'' / \left(1 + f'^2\right)^{3/2},$$

where $f' = \frac{df}{dz}$, and thus

$$p = p_0 - \sigma f'' / \left(1 + f'^2\right)^{3/2}.$$

But as the vertical film is in static equilibrium, the pressure satisfies the equation

$$p(z) = p(z = 0) - \rho g z.$$

Hence

$$p(z=0) - \rho gz = p_0 - \sigma f'' / (1 + f'^2)^{3/2}.$$

Note that p_0 and $p(z=0)$ are constants and

$$\Delta p = p_0 - p(z=0)$$

is also a constant, in terms of which

$$(1 + f'^2)^{-\frac{3}{2}} \frac{df'}{dz} = \frac{\Delta p + \rho gz}{\sigma}.$$

Integration yields

$$f' / (1 + f'^2)^{1/2} = \frac{1}{\sigma}\left(\Delta pz + \frac{1}{2}\rho gz^2\right) + C_1,$$

where C_1 is a constant. At $z=0$ the tangent to the arc is parallel to the z-axis and so $f' = 0$, giving $C_1 = 0$. Hence

$$f' / (1 + f'^2)^{1/2} = \frac{1}{\sigma}\left(\Delta pz + \frac{1}{2}\rho gz^2\right) \approx f',$$

as $f' \ll 1$ on account of the film being very thin. Integrating again gives

$$f(z) = \frac{1}{\sigma}\left(\frac{\Delta p}{2}z^2 + \frac{1}{6}\rho gz^3\right) + C_2.$$

If the film thickness at $z=0$ is $2f_0$, $C_2 = fo$. Hence

$$p(z) = p(z=0) - \rho gz = p_0 - \Delta p - \rho gz,$$

and the thickness is

$$t = 2f(z) = \frac{2}{\sigma}\left(\frac{\Delta p}{2} + \frac{1}{6}\rho gz\right)z^2 + 2f_0.$$

3.3

MATHEMATICAL TECHNIQUES (3049–3056)

3049

Consider the inhomogeneous differential equation

$$f''(x) + 2zf'(x) + k^2 f(x) = \delta(x - x_0)$$

where k and $z > 0$ are real constants and $\delta(x)$ is the Dirac δ-function. Find the general solution for $k^2 > z^2$.

Sol: We first find the general solution for the corresponding homogeneous differential equation

$$f''(x) + 2zf'(x) + k^2 f(x) = 0.$$

Let $f(x) = e^{\lambda x}$. Substitution gives

$$\lambda^2 + 2z\lambda + k^2 = 0,$$

which has solutions

$$\lambda = -z \pm i\sqrt{k^2 - z^2}\,.$$

Thus the general solution is

$$f(x) = A_1 e^{(-z+i\sqrt{k^2-z^2})x} + A_2 e^{(-z-i\sqrt{k^2-z^2})x},$$

where A_1 and A_2 are constants.

We next find a special solution of the original equation. Expressed as Fourier transforms,

$$f(x) = \frac{1}{2\pi}\int_{-\infty}^{\infty} g(p)\, e^{ip(x-x_0)}\, dp,$$

$$\delta(x - x_0) = \frac{1}{2\pi}\int_{-\infty}^{\infty} e^{ip(x-x_0)}\, dp.$$

Substitution in the inhomogeneous differential equation gives

$$g(p) = \frac{1}{k^2 + 2izp - p^2}.$$

The special solution is therefore

$$f(x) = -\frac{1}{2\pi}\int_{-\infty}^{\infty} \frac{e^{ip(x-x_0)}\,dp}{p^2 - 2izp - k^2}.$$

As $p^2 - 2izp - k^2 = (p - p_1)(p - p_2)$, where

$$p_1 = iz + \sqrt{k^2 - z^2}, \quad p_2 = iz - \sqrt{k^2 - z^2},$$

with $z > 0$, $k^2 > z^2$, there are two singularities p_1, p_2 in the upper half of the complex plane.

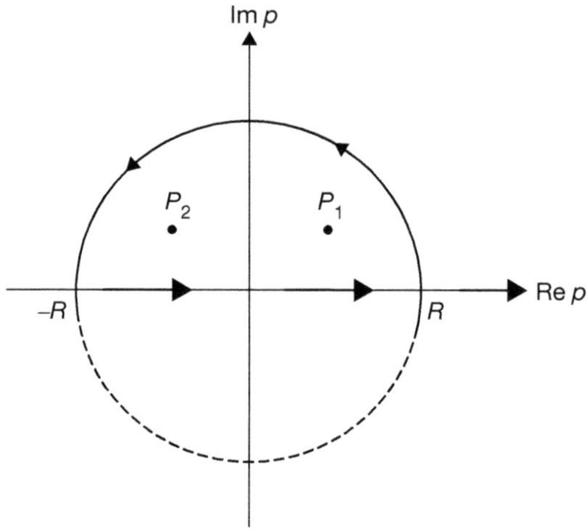

Fig. 3.13

For $x > x_0$, take for the contour of integration C the upper half-circle in the complex p-plane shown in Fig. 3.13, for which we find (cf. **Problem 3050** (c))

$$\begin{aligned} f(x) &= \lim_{R\to\infty}\left[-\frac{1}{2\pi}\oint_C \frac{e^{ip(x-x_0)}\,dp}{p^2 - 2izp - k^2}\right] \\ &= -\frac{1}{2\pi}\cdot 2\pi i\left[\frac{e^{i(x-x_0)\left(iz+\sqrt{k^2-z^2}\right)}}{2\sqrt{k^2-z^2}} - \frac{e^{i(x-x_0)\left(iz-\sqrt{k^2-z^2}\right)}}{2\sqrt{k^2-z^2}}\right] \\ &= \frac{\sin\left[(x-x_0)\sqrt{k^2-z^2}\right]}{\sqrt{k^2-z^2}}\,e^{-(x-x_0)z}. \end{aligned}$$

For $x < x_0$, we have to employ for the contour of integration the lower half-curve shown in Fig. 3.13, as use of the upper half-circle would give rise to a factor

$$e^{-i(\mathrm{Re}\,p+i\mathrm{Im}\,p)|x-x_0|} = e^{\mathrm{Im}\,p|x-x_0|}e^{-i\,\mathrm{Re}\,p|x-x_0|}$$

which goes to infinity as $R \to \infty$. However, since the lower half-circle encloses no singularity, $f(x) = 0$ for $x < x_0$.

Hence the general solution is

$$f(x) = A_1 e^{\left(-z+i\sqrt{k^2-z^2}\right)x} + A_2 e^{\left(-z-i\sqrt{k^2-z^2}\right)x} + \Theta(x-x_0)\frac{\sin\left[(x-x_0)\sqrt{k^2-z^2}\right]}{\sqrt{k^2-z^2}} e^{-(x-x_0)z},$$

where

$$\Theta(x-x_0) = \begin{cases} 1 & \text{if } x > x_0, \\ 0 & \text{if } x \le x_0. \end{cases}$$

3050

a. Sum the series $y = 1 + 2x + 3x^2 + 4x^3 + \ldots (|x| < 1)$

b. If $f(x) = xe^{-x/\lambda}$ over the interval $0 < x < \infty$, find the mean and most probable values of x.

c. Evaluate

$$I = \int_0^\infty \frac{dx}{4 + x^4}.$$

d. Find the eigenvalues and normalized eigenvectors of the matrix

$$\begin{pmatrix} 1 & 2 & 4 \\ 2 & 3 & 0 \\ 5 & 0 & 3 \end{pmatrix}.$$

Are the eigenvectors orthogonal? Comment on this.

Sol:

a. As

$$y = 1 + 2x + 3x^2 + 4x^3 + \ldots + nx^{n-1} + \ldots,$$

we have

$$xy = x + 2x^2 + 3x^3 + 4x^4 + \ldots + nx^n + \ldots$$

and thus

$$y - xy = 1 + x + x^2 + x^3 + \ldots + x^{n-1} + \ldots = \frac{1}{1-x}$$

since $|x| < 1$. Hence

$$y = \frac{1}{(1-x)^2}.$$

b. The mean value of x is

$$\bar{x} = \frac{\int_0^\infty xf(x)dx}{\int_0^\infty f(x)dx} = \frac{\int_0^\infty x^2 e^{-x/\lambda}dx}{\int_0^\infty xe^{-x/\lambda}dx} = \frac{2\lambda\int_0^\infty xe^{-x/\lambda}dx}{\int_0^\infty xe^{-x/\lambda}dx} = 2\lambda,$$

where integration by parts has been used for the numerator and $\lambda > 0$ has been assumed. The most probable value of x, x_{m}, is given by

$$\left(\frac{dt}{dx}\right)_{x_m} = 0,$$

whence $x_{\mathrm{m}} = \lambda$.

Note that if $\lambda < 0$, $f(x)$ is infinite at ∞ and the mean and most probable values are also infinite.

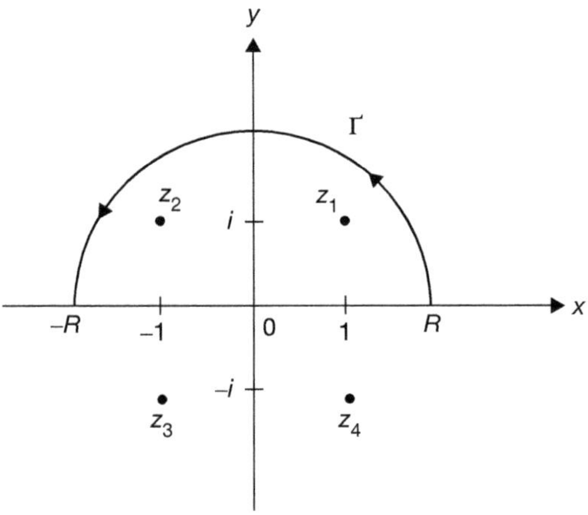

Fig. 3.14

c. Write

$$I = \frac{1}{2}\int_0^\infty \frac{dx}{x^4+4} + \frac{1}{2}\int_0^\infty \frac{dx}{x^4+4}$$
$$= \frac{1}{2}\int_{-\infty}^0 \frac{dx}{x^4+4} + \frac{1}{2}\int_0^\infty \frac{dx}{x^4+4} = \frac{1}{2}\int_\infty^\infty \frac{dx}{x^4+4}$$

and consider

$$\oint_C \frac{dz}{z^4+4},$$

where C is the closed contour of Fig. 3.14 consisting of the line from $-R$ to R and the semicircle Γ, traversed in anticlockwise sense, in the complex plane. As

$$\frac{1}{z^4+4} = \frac{1}{(z-z_1)(z-z_2)(z-z_3)(z-z_4)},$$

where

$$z_1 = 1+i, \qquad z_2 = -1+i, \qquad z_3 = -1-i, \qquad z_4 = 1-i,$$

the residue theorem gives

$$\oint_C \frac{dz}{z^4+4}$$
$$= 2\pi i\left[\frac{1}{(z_1-z_2)(z_1-z_3)(z_1-z_4)} + \frac{1}{(z_2-z_1)(z_2-z_3)(z_2-z_4)}\right]$$
$$= 2\pi i\left[\frac{1}{2(2+2i)\cdot 2i} - \frac{1}{2\cdot 2i(-2+2i)}\right] = \frac{\pi}{4}.$$

But

$$\oint_C \frac{dz}{z^4+4} = \int_{-R}^{R} \frac{dx}{x^4+4} + \int_\Gamma \frac{dz}{z^4+4} \to \int_{-\infty}^{\infty} \frac{dx}{x^4+4}$$

as $R \to \infty$. Hence

$$I = \frac{1}{2}\cdot\frac{\pi}{4} = \frac{\pi}{8}.$$

d. The characteristic equation of the matrix

$$\begin{vmatrix} 1-\lambda & 2 & 4 \\ 2 & 3-\lambda & 0 \\ 5 & 0 & 3-\lambda \end{vmatrix} = 0,$$

or $(\lambda-3)(\lambda+3)(\lambda-7)=0$, has roots $\lambda_1 = 3$, $\lambda_2 = -3$ and $\lambda_3 = 7$. These are the eigenvalues of the matrix. For $\lambda_1 = 3$, the matrix equation

$$\begin{pmatrix} -2 & 2 & 4 \\ 2 & 0 & 0 \\ 5 & 0 & 0 \end{pmatrix}\begin{pmatrix} x_1 \\ x_2 \\ x_2 \end{pmatrix} = 0$$

has solution

$$\begin{cases} x_1 = 0, \\ x_2 = -2x_3. \end{cases}$$

Hence the corresponding eigenvector is

$$\begin{pmatrix} 0 \\ -2x_3 \\ x_3 \end{pmatrix},$$

and the normalized eigenvector is

$$\frac{1}{\sqrt{5}}\begin{pmatrix} 0 \\ -2 \\ 1 \end{pmatrix}.$$

For $\lambda_2 = -3$, the matrix equation

$$\begin{pmatrix} 4 & 2 & 4 \\ 2 & 6 & 0 \\ 5 & 0 & 6 \end{pmatrix} \begin{pmatrix} x_1 \\ x_2 \\ x_3 \end{pmatrix} = 0$$

has solution

$$\begin{cases} x_2 = -\frac{1}{3}x_1, \\ x_3 = -\frac{5}{6}x_1, \end{cases}$$

which gives the normalized eigenvector

$$\frac{6}{\sqrt{65}} \begin{pmatrix} 1 \\ -\frac{1}{3} \\ -\frac{5}{6} \end{pmatrix}.$$

For $\lambda_3 = 7$, the matrix equation

$$\begin{pmatrix} -6 & 2 & 4 \\ 2 & -4 & 0 \\ 5 & 0 & -4 \end{pmatrix} \begin{pmatrix} x_1 \\ x_2 \\ x_3 \end{pmatrix} = 0$$

has solution

$$\begin{cases} x_2 = \frac{1}{2}x_1, \\ x_3 = \frac{5}{4}x_1, \end{cases}$$

which gives the normalized eigenvector

$$\frac{4}{3\sqrt{5}} \begin{pmatrix} 1 \\ \frac{1}{2} \\ \frac{5}{4} \end{pmatrix}.$$

These three eigenvectors are not orthogonal, since the matrix

$$\begin{pmatrix} 1 & 2 & 4 \\ 2 & 3 & 0 \\ 5 & 0 & 3 \end{pmatrix}$$

is not Hermitian, which is obvious as $a_{ij} \neq a_{ji}$ generally.

3051

Evaluate

$$\int_0^{2\pi} \frac{dx}{1 - 2p\cos x + p^2} \quad \text{for } 0 < p < 1.$$

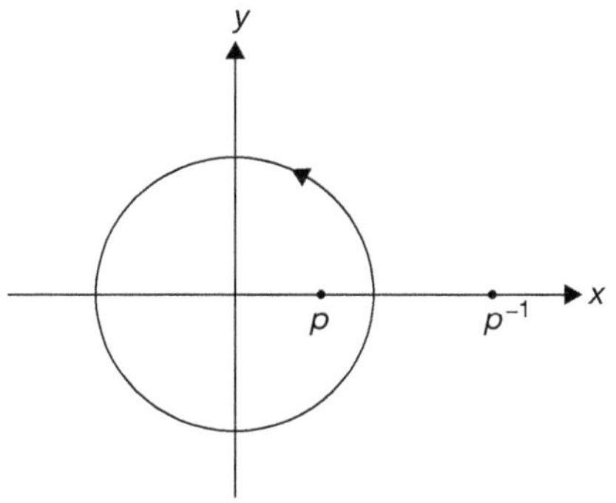

Fig. 3.15

Sol: Let $z = e^{ix}$. Then $dx = dz / iz$, $\cos x = \frac{1}{2}(z + z^{-1})$, and

$$\begin{aligned} I &= \int_0^{2\pi} \frac{dx}{1 - 2p\cos x + p^2} \\ &= \oint_C \frac{dz}{iz[1 - p(z + z^{-1}) + p^2]} = -\frac{1}{ip}\oint_C \frac{dz}{(z - p^{-1})(z - p)}, \end{aligned}$$

where C is the circle with unit radius shown in Fig. 3.15. There is only one pole, p, lying inside the circle. The residue at this pole is

$$\frac{1}{p - p^{-1}}.$$

Hence by the residue theorem we have

$$I = -\frac{1}{ip} \cdot \frac{2\pi i}{p - p^{-1}} = \frac{2\pi}{1 - p^2}.$$

3052

Evaluate the infinite sum $S(x) = 1 + 2x^2 + 3x^3 + \ldots$ for $|x| < 1$.

Sol: Rewriting the sum, we have

$$\begin{aligned} S(x) &= (x + 2x^2 + 3x^3 + \ldots) + (1 - x) \\ &= x(1 + 2x + 3x^2 + \ldots) + (1 - x) \\ &= x\frac{d}{dx}(x + x^2 + x^3 + \ldots) + (1 - x) \\ &= x\frac{d}{dx}\left(\frac{x}{1 - x}\right) + (1 - x) \\ &= \frac{x}{1 - x} + \frac{x^2}{(1 - x)^2} + (1 - x) \\ &= \frac{x}{(1 - x)^2} + (1 - x). \end{aligned}$$

3053

A circular polyisoprene elastomer membrane given by the equation $x_1^2 + x_2^2 = 1$, undergoes deformation represented by the matrix $A = \begin{bmatrix} 4 & \sqrt{3} \\ \sqrt{3} & 2 \end{bmatrix}$ such that point $P\,(x_1 x_2)$ in the circular membrane gets transformed into point $Q\,(y_1 y_2)$ after deformation when stress is applied.

a. Determine the factors of extension or contraction and the principal axis of deformation.

b. Find out the shape of the membrane after deformation.

Sol:

a. To find the principal directions or axes, we have to determine a position vector **x** for the point P that is parallel or antiparallel to the position vector **y** of Q, i.e., $\mathbf{y} = \lambda \mathbf{x}$.

We know that the deformation is given by $\mathbf{y} = A\mathbf{x}$, which implies $\mathbf{Ax} = \lambda \mathbf{x}$. This now becomes a simple eigenvalue problem where the eigenvalues give the factors of extension or contraction and the eigenvectors give the principal axis of deformation.

Let's find the eigenvalues from the characteristic equation, det $(A - \lambda I) = 0$

$$\det\left\{\begin{bmatrix} 4 & \sqrt{3} \\ \sqrt{3} & 2 \end{bmatrix} - \lambda\begin{bmatrix} 1 & 0 \\ 0 & 1 \end{bmatrix}\right\} = \begin{vmatrix} 4-\lambda & \sqrt{3} \\ \sqrt{3} & 2-\lambda \end{vmatrix} = 0$$

$$(4-\lambda)(2-\lambda) - 3 = 0$$

$$\lambda^2 - 6\lambda + 5 = 0$$

$$\lambda = 1, 5$$

To find the eigenvector, we have to solve $(A - \lambda I)\,x = 0$

$$\begin{bmatrix} 4 & \sqrt{3} \\ \sqrt{3} & 2 \end{bmatrix} - \lambda\begin{bmatrix} 1 & 0 \\ 0 & 1 \end{bmatrix} = \begin{bmatrix} 4-\lambda & \sqrt{3} \\ \sqrt{3} & 2-\lambda \end{bmatrix} = 0$$

Let's substitute $\lambda = 1$ in $(A - \lambda I)\,x = 0$ to determine the first eigenvector

$$\begin{bmatrix} 3 & \sqrt{3} \\ \sqrt{3} & 1 \end{bmatrix}\begin{bmatrix} x_1 \\ x_2 \end{bmatrix} = 0$$

$$3x_1 + \sqrt{3}\,x_2 = 0 \text{ and } \sqrt{3}\,x_1 + x_2 = 0$$

i.e., $x_2 = -\sqrt{3}\,x_1$

First eigenvector is $\begin{bmatrix} 1 \\ -\sqrt{3} \end{bmatrix}$

The angle between this vector and the positive x_1 axis is $\theta = \tan^{-1}\left(\frac{-\sqrt{3}}{1}\right) = -60°$

Let's substitute λ = 5 in $(A - \lambda\, I)\, x = 0$ to determine the second eigenvector

$$\begin{bmatrix} -1 & \sqrt{3} \\ \sqrt{3} & -3 \end{bmatrix} \begin{bmatrix} x_1 \\ x_2 \end{bmatrix} = 0$$

$-x_1 + \sqrt{3}\, x_2 = 0$ and $\sqrt{3}\, x_1 - 3\, x_2 = 0$

i.e., $x_1 = \sqrt{3}\, x_2$

Second eigenvector is $\begin{bmatrix} \sqrt{3} \\ 1 \end{bmatrix}$

The angle between this vector and the positive x_1 axis is $\theta = \tan^{-1}\left(\frac{1}{\sqrt{3}}\right) = 30°$

These two vectors give the principal directions of deformation and the elastomer is stretched along $\begin{bmatrix} \sqrt{3} \\ 1 \end{bmatrix}$ directions by a factor of 5 and neither expand nor contract along $\begin{bmatrix} 1 \\ -\sqrt{3} \end{bmatrix}$

b. Before deformation, $x_1^2 + x_2^2 = 1$, i.e., $(\cos^2\phi + \sin^2\phi) = 1$,

where $r\cos\phi = x_1$ and $r\sin\phi = x_2$ with $r = 1$

After deformation $x \rightarrow y$, $\cos\phi = y_1$ and $5\sin\phi = y_2$

$$\frac{y_1^2}{1^2} + \frac{y_2^2}{5^2} = 1$$

This is an equation of the ellipse. So after deformation, the circle becomes an ellipse as shown in the following figure.

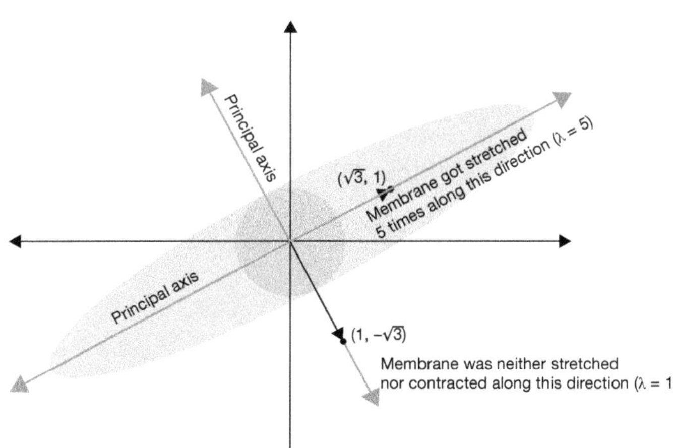

3054

The series $\sum_{n=0}^{\infty}(-1)^n n! z^n$ is divergent. Nevertheless, it can be "summed" in the sense that one can find a function $F(z)$ which has the property

$$\lim_{|z|\to 0} \frac{1}{n!}\frac{d^n F(z)}{dz^n} = (-1)^n n! \qquad (1)$$

for $|\arg z| \le \pi - \varepsilon$, $\varepsilon > 0$, and any given $n = 0,1,2,\ldots$.

a. Use the formula $\Gamma(n+1) = n! = \int_0^\infty d\lambda\, e^{-\lambda}\lambda^n$ in order to obtain $F(z)$ in the form of an integral representation.

b. Write $F(z)$ as a Hilbert representation

$$F(z) = \int dx \frac{\rho(x)}{x - z}.$$

c. Consider functions of the form $f(z) = c\exp\left(\frac{-b}{z^a}\right)$, with b real and positive. Under certain conditions, $f(z)$ may be added to $F(z)$ without affecting the limits (1) for any given finite value of $n = 0,1,2,\ldots$ and any $\varepsilon > 0$. What are these condition? (Assume a is real)

Sol:

a. For a given n, suppose for $|z| \to 0$

$$\frac{d^n F_n}{dz^n} = (-1)^n (n!^2).$$

Integrating once gives

$$\frac{d^{n-1} F_n}{dz^{n-1}} = (-1)^n (n!)^2 z + A,$$

or

$$(-1)^{n-1}[(n-1)!]^2 = (-1)^n (n!)^2 z + A,$$

where the constant of integration A is obtained by putting $z = 0$. Continuing the process we have

$$F_n(z) = \sum_0^n (-1)^r r! z^r.$$

In particular, for $n \to \infty$ we have

$$\begin{aligned} F(z) &= \sum_0^\infty (-1)^n n! z^n \\ &= \int_0^\infty d_\lambda e^{-\lambda} \sum_{n=0}^\infty (-1)^n \lambda^n z^n. \end{aligned}$$

With $|\, z \,| \to 0$ we can always take $|\lambda z| < 1$, for which

$$\sum_{n=0}^\infty (-1)^n \lambda^n z^n = \sum_{n=0}^\infty (-\lambda z)^n = \frac{1}{1+\lambda z}.$$

Hence

$$F(z) = \int_0^\infty \frac{e^{-\lambda}\, d\lambda}{1+\lambda z}.$$

b. Let $\lambda = -\frac{1}{x}$. The above becomes

$$F(z) = \int_{-\infty}^0 \frac{\left(\frac{1}{x} e^{\frac{1}{x}}\right) dx}{x - z},$$

which is a Hilbert representation.

c. The requirement is that, as $|z| \to 0$, $|f(z)|$ goes to 0 faster than $|F(z)|$, and hence faster than any power of z. As

$$\begin{aligned} f(z) &= c \exp\left[-b\left(|z|\, e^{i\theta}\right)^{-a}\right] \\ &= c \exp\left\{-b\, |z|^{-a} [\cos(a\theta) - i \sin(a\theta)]\right\}, \\ |f(z)| &= |c| \exp[-b\, |z|^{-a} \cos(a\theta)]. \end{aligned}$$

Then $|f(z)| \to 0$ as $|z| \to 0$ if

$$a \geq \delta \quad \text{with} \quad \delta > 0$$

and

$$\cos(a\theta) > 0, \quad \text{or} \quad a|\theta| \leq \frac{\pi}{2} - \delta' \quad \text{with} \quad \delta' > 0.$$

We are given that

$$|\theta| \leq \pi - \varepsilon \quad \text{with } \varepsilon > 0,$$

where $\theta = \arg z$, or

$$a|\theta| \leq a\pi - a\varepsilon.$$

All the above conditions can be satisfied if

$$0 < \delta \leq a \leq \frac{1}{2}, \quad \delta' = \varepsilon.$$

3055

a. Solve the differential equation

$$\left(\frac{d}{dt}+2\right)\left(\frac{d}{dt}+1\right)y=1,$$

with initial conditions $\frac{dy}{dt}|_{t=0}=y|_{t=0}=0.$

b. Calculate the definite integral

$$\int_{-\infty}^{\infty} dx \frac{1}{(a^2+x^2)^2}$$

with "a" real.

c. What are the eigenvalues of the matrix

$$M=\begin{pmatrix} 1 & 0 & -i \\ 0 & 2 & 0 \\ i & 0 & -1 \end{pmatrix}?$$

Sol:

a. The equation is usually written as

$$\frac{d^2y}{dt^2}+3\frac{dy}{dt}+2y=1,$$

whose eigenequation

$$\lambda^2+3\lambda+2=0$$

has solutions

$$\lambda_1=-2, \quad \lambda_2=-1.$$

Hence the general solution of the homogeneous equation is

$$y(t)=Ae^{-2t}+Be^{-t}.$$

A special solution is $y=\frac{1}{2}$, giving the general solution of the differential equation as

$$y=Ae^{-2t}+Be^{-t}+\frac{1}{2}.$$

The initial conditions then give

$$A+B+\frac{1}{2}=0, \qquad -2A-B=0,$$

which have solution

$$A = \frac{1}{2}, \quad B = -1.$$

Therefore

$$y(t) = \frac{1}{2}e^{-2t} - e^{-t} + \frac{1}{2}.$$

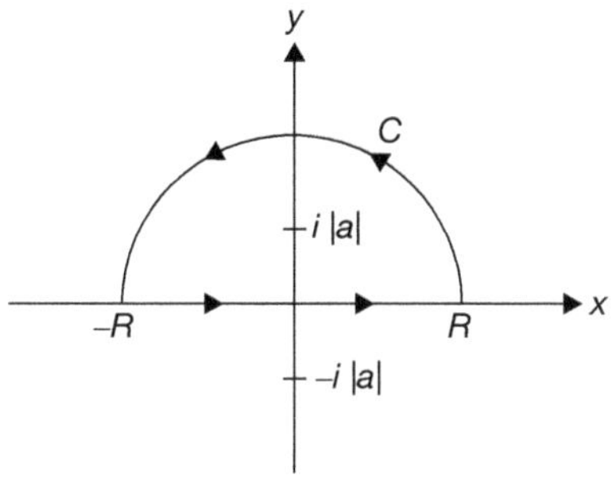

Fig. 3.16

b. Consider

$$\oint_C \frac{dz}{(z^2 + a^2)^2},$$

where C is the semicircle shown in Fig. 3.16.

As

$$(z^2 + a^2)^{-2} = (z - i|a|)^{-2}(z + i|a|)^{-2},$$

the integrand has a pole of order 2 enclosed by C at $z_1 = i|a|$. The residue at z_1 is

$$\lim_{z \to i|a|} \frac{1}{(2-1)!} \frac{d^{2-1}}{dz^{2-1}} \left[\frac{(z - i|a|)^2}{(z^2 + a^2)^2} \right] = \frac{-2}{(z + i|a|)^3} \Big|_{z=i|a|} = \frac{1}{4i\,|a|^3}.$$

Hence

$$\oint_C \frac{dz}{(z^2 + a^2)^2} = \frac{2\pi i}{4i\,|a|^3} = \frac{\pi}{2\,|a|^3}.$$

On the other hand,

$$\oint_C = \int_{-R}^{R} \frac{dx}{(x^2 + a^2)^2} + \int_\Gamma \frac{dz}{(z^2 + a^2)^2} \to \int_{-\infty}^{\infty} \frac{dx}{(x^2 + a^2)^2} + 0$$

as $R \to \infty$. Therefore

$$\int_{-\infty}^{\infty} \frac{dx}{(x^2 + a^2)^2} = \frac{\pi}{2\,|a|^3}.$$

c. The eigenequation

$$\begin{vmatrix} 1-\lambda & 0 & -i \\ 0 & 2-\lambda & 0 \\ i & 0 & -1-\lambda \end{vmatrix} = -(\lambda-2)(\lambda^2-2) = 0$$

has solutions

$$\lambda_1 = 2, \quad \lambda_2 = \sqrt{2}, \quad \lambda_3 = -\sqrt{2}\,.$$

These are the eigenvalues of the matrix M.

3056

Consider a linear system in one dimension whose displacement $y(t)$ is governed by the equation

$$c_1\ddot{y} + c_2\dot{y} + c_3 y = F(t),$$

where c_1, c_2, c_3 are time independent but unknown. At $t = 0$, the system is at rest at $y = 0$.

You are given the response of the system for all times to a constant driving force of magnitude F_0 and lasting between $t = 0$ and $t = 10T : y_0(t)$.

Find, in terms of what is given,

a. the displacement at $t = 50T$ to a driving force of magnitude $F_0/5$ acting between $t = 0$ and $t = 50T$,

b. the displacement at $t = 6T$ to a driving force of magnitude F_0 acting between $t = 0$ and $t = T$.

c. Consider now an arbitrary driving force $F(t)$ acting between $t = 0$ and $t = T$, where $F(0) = F(T) = 0$. Show that the displacement at $t = 6T$ can in fact be found and that it can be written in the form

$$y(6T) = A\int_0^T y_0(6T - t)\dot{F}(t)dt,$$

where A is a constant.

Sol:

a. We are given

$$c_1\ddot{y}_0(t) + c_2\dot{y}_0(t) + c_3 y_0(t) = F(0)\ ,$$

or

$$c_1\frac{\ddot{y}_0(t)}{5}+c_2\frac{\dot{y}_0(t)}{5}+c_3\frac{y_0(t)}{5}=\frac{F_0}{5}\ \text{ for } 0\le t\le 10T$$

and

$$c_1\ddot{y}_0(t)+c_2\dot{y}_0(t)+c_3y_0(t)=0 \quad \text{ for }\ 10T<t,$$

and are required to find $y(t)$ which satisfies

1. $$\begin{cases} c_1\ddot{y}(t)+c_2\dot{y}(t)+c_3y(t)=\dfrac{F_0}{5}, & 0\le t\le 50T,\\ c_1\ddot{y}(t)+c_2\dot{y}(t)+c_3y(t)=0, & 50T<t.\end{cases}$$

Since the system is linear, its response $y(t)$ to a constant driving force $\frac{F_0}{5}$ which acts for the interval 0–50T can be considered as a linear superposition of the responses to five constant forces each of magnitude $\frac{F_0}{5}$ acting in turn for the intervals 0–10T, 10T–20T, 20T–30T, 30T–40T, 40T–50T. Thus

2. $$\begin{cases} c_1\dfrac{\ddot{y}_0(t)}{5}+c_2\dfrac{\dot{y}_0(t)}{5}+c_3\dfrac{y_0(t)}{5}=\dfrac{F_0}{5} & 0\le t\le 10T,\\ c_1\dfrac{\ddot{y}_0(t)}{5}+c_2\dfrac{\dot{y}_0(t)}{5}+c_3\dfrac{y_0(t)}{5}=0, & t>10T\,;\end{cases}$$

3. $$\begin{cases} c_1\dfrac{\ddot{y}_0(t-10T)}{5}+c_2\dfrac{\dot{y}_0(t-10T)}{5}+c_3\dfrac{y_0(t-10T)}{5}=\dfrac{F_0}{5}, & 10T\le t\le 20T,\\ c_1\dfrac{\ddot{y}_0(t-10T)}{5}+c_2\dfrac{\dot{y}_0(t-10T)}{5}+c_3\dfrac{y_0(t-10T)}{5}=0, & t>20T\,;\end{cases}$$

4. $$\begin{cases} c_1\dfrac{\ddot{y}_0(t-20T)}{5}+c_2\dfrac{\dot{y}_0(t-20T)}{5}+c_3\dfrac{y_0(t-20T)}{5}=\dfrac{F_0}{5}, & 20T\le t\le 30T,\\ c_1\dfrac{\ddot{y}_0(t-20T)}{5}+c_2\dfrac{\dot{y}_0(t-20T)}{5}+c_3\dfrac{y_0(t-20T)}{5}=0, & t>30T\,;\end{cases}$$

5. $$\begin{cases} c_1\dfrac{\ddot{y}_0(t-30T)}{5}+c_2\dfrac{\dot{y}_0(t-30T)}{5}+c_3\dfrac{y_0(t-30T)}{5}=\dfrac{F_0}{5}, & 30T\le t\le 40T,\\ c_1\dfrac{\ddot{y}_0(t-30T)}{5}+c_2\dfrac{\dot{y}_0(t-30T)}{5}+c_3\dfrac{y_0(t-30T)}{5}=0, & t>40T\,;\end{cases}$$

6. $$\begin{cases} c_1\dfrac{\ddot{y}_0(t-40T)}{5}+c_2\dfrac{\dot{y}_0(t-40T)}{5}+c_3\dfrac{y_0(t-40T)}{5}=\dfrac{F_0}{5}, & 40T\le t\le 50T,\\ c_1\dfrac{\ddot{y}_0(t-40T)}{5}+c_2\dfrac{\dot{y}_0(t-40T)}{5}+c_3\dfrac{y_0(t-40T)}{5}=0, & t>50T.\end{cases}$$

Combining (2)–(6) we have (1) with

$$y(t) = \frac{1}{5}\big[y_0(t) + y_0(t - 10T)\theta(t - 10T) + y_0(t - 20T)\theta(t - 20T) \\ + y_0(t - 30T)\theta(t - 30T) + y_0(t - 40T)\theta(t - 40T)\big], \quad t \geq 0,$$

Where

$$\theta(t - t_0) = \begin{cases} 1, & t \geq t_0, \\ 0 & t < t_0. \end{cases}$$

The required displacement is $y(50T)$ as given by the above formula with $t = 50T$.

b. Considering the effect of F_0 acting in the interval $0 \leq t \leq 10T$ and the effect of F_0 acting in the interval $T \leq t \leq 11T$, we have

$$\begin{cases} c_1\ddot{y}_0(t) + c_2\dot{y}_0(t) + c_3 y_0(t) = F_0, & 0 \leq t \leq 10T, \\ c_1\ddot{y}_0(t) + c_2\dot{y}_0(t) + c_3 y_0(t) = 0, & t > 10T. \end{cases}$$

and

$$\begin{cases} c_1\ddot{y}_0(t - T) + c_2\dot{y}_0(t - T) + c_3 y_0(t - T) = F_0, & T \leq t \leq 11T, \\ c_1\ddot{y}_0(t - T) + c_2\dot{y}_0(t - T) + c_3 y_0(t - T) = 0, & t > 11T. \end{cases}$$

Subtracting we obtain

$$c_1\big[\ddot{y}_0(t) - \ddot{y}_0(t - T)\theta(t - T)\big] + c_2\big[\dot{y}_0(t) - \dot{y}_0(t - T)\theta(t - T)\big] \\ + c_3\big[y_0(t) - y_0(t - T)\theta(t - T)\big] = \begin{cases} F_0, & 0 \leq t \leq T, \\ 0, & T < t \leq 10T, \\ -F_0, & 10T < t \leq 11T, \\ 0, & t > 11T. \end{cases}$$

Since an earlier motion is not affected by conditions applied at a later time, the first two of the last set of equations are valid for $0 \leq t \leq 10T$. Hence

$$y(t) = y_0(t) - y_0(t - T), \quad 0 \leq t \leq 10T,$$

and thus

$$y(6T) = y_0(6T) - y_0(5T).$$

c. In a similar way, consider the effects of F_0 acting in $0 \leq t \leq 10T$ and in $T \leq t \leq (10 + \tau)T$, where $\tau > 0$, and obtain

1. $c_1\ddot{y}_0(t) + c_2\dot{y}_0(t) + c_3 y_0(t) = F_0, \qquad 0 \leq t \leq 10T,$

2. $c_1\ddot{y}_0(t - \tau) + c_2\dot{y}_0(t - \tau) + c_3 y_0(t - \tau) = F_0, \qquad \tau \leq t \leq (10 + \tau)T.$

(1) – (2) gives

$$c_1\frac{d^2}{dt^2}\big[y_0(t) - y_0(t - \tau)\theta(t - \tau)\big] + c_2\frac{d}{dt}\big[y_0(t) - y_0(t - \tau)\theta(t - \tau)\big] \\ + c_3\big[y_0(t) - y_0(t - \tau)\theta(t - \tau)\big] = \begin{cases} F_0, & 0 \leq t \leq \tau, \\ 0, & \tau < t \leq 10T. \end{cases}$$

Using the definition of time derivative

$$y_0(t) - y_0(t-\tau)\theta(t-\tau) = \begin{cases} \dot{y}_0(t)t, & 0 \le t \le \tau, \\ \dot{y}_0(t)\tau, & \tau < t. \end{cases}$$
$$\approx \dot{y}_0(t)\tau \quad \text{for } \tau \to 0,$$

we can write the above as

$$c_1 \frac{d^2}{dt^2}\dot{y}_0(t) + c_2 \frac{d}{dt}\dot{y}_0(t) + c_3\dot{y}_0(t) = \begin{cases} \frac{F}{\tau}, & 0 \le t \le \tau \\ 0, & t > \tau. \end{cases}$$

With $\tau \to 0$, this becomes

$$c_1 \frac{d^2}{dt^2}\dot{y}_0(t) + c_2 \frac{d}{dt}\dot{y}_0(t) + c_3\dot{y}_0(t) = B\delta(t), \qquad 0 \le t \le 10T,$$

where B is a constant. Replacing t by $t-\tau$, multiplying both sides by $F(\tau)d\tau$ and integrating over τ from 0 to T, we obtain

$$c_1 \frac{d^2}{dt^2}\left[\int_0^T \dot{y}_0(t-\tau)F(\tau)d\tau\right] + c\frac{d}{dt}\left[\int_0^T \dot{y}_0(t-\tau)F(\tau)d\tau\right]$$
$$+ c_3 \int_0^T \dot{y}_0(t-\tau)F(\tau)d\tau = B\int_0^T \delta(t-\tau)F(\tau)d\tau.$$

We are required to find $y(t)$ given that

$$c_1\ddot{y}(t) + c_2\dot{y}(t) + c_3 y(t) = \begin{cases} F(t), & 0 \le t \le T, \\ 0, & t > T, \end{cases}$$
$$= \int_0^T \delta(t-\tau)F(\tau)d\tau.$$

Comparing the last two differential equations, we have for $t \le 10T$

$$y(t) = B^{-1}\int_0^T \dot{y}_0(t-\tau)F(\tau)d\tau$$
$$= -B^{-1}F(\tau)y_0(t-\tau)|_0^T + B^{-1}\int_0^T y_0(t-\tau)\dot{F}(\tau)d\tau$$
$$= A\int_0^T y_0(t-\tau)\dot{F}(\tau)d\tau$$

as $F(0) = F(T) = 0$. Note that the constant B^{-1} has been replaced by A. Hence

$$y(6T) = A\int_0^T y_0(6T-\tau)\dot{F}(\tau)d\tau.$$

INDEX TO PROBLEMS

www.ingramcontent.com/pod-product-compliance
Ingram Content Group UK Ltd.
Pitfield, Milton Keynes, MK11 3LW, UK
UKHW020359150926
8624IPUK00004B/245

9 789811 291494